Fuzhong Weng

Passive Microwave Remote Sensing of the Earth

Wiley Series in Atmospheric Physics and Remote Sensing

Series Editor: Alexander Kokhanovsky

Wendisch, M. / Brenguier, J.-L. (eds.)

Airborne Measurements for Environmental Research

Methods and Instruments

2013

Coakley Jr., J. A. / Yang, P.

Atmospheric Radiation

A Primer with Illustrative Solutions

2014

Stamnes, K. / Stamnes, J. J.

Radiative Transfer in Coupled Environmental Systems

An Introduction to Forward and Inverse Modeling

2015

Tomasi, C. / Fuzzi, S. / Kokhanovsky, A.

Atmospheric Aerosols

Life Cycles and Effects on Air Quality and Climate

2016

Weng, F.

Passive Microwave Remote Sensing of the Earth

for Meteorological Applications

2017

Forthcoming:

Kokhanovsky, A. / Natraj, V.

Analytical Methods in Atmospheric Radiative Transfer

Huang, X. / Yang, P.

Radiative Transfer Processes in Weather and Climate Models

North, G. R. / Kim, K.-Y.

Energy Balance Climate Models

Davis, A. B. / Marshak, A.

Multi-dimensional Radiative Transfer

Theory, Observation, and Computation

Minnis, P. *et al.*

Satellite Remote Sensing of Clouds

Zhang, Z. *et al.*

Polarimetric Remote Sensing

Aerosols and Clouds

Fuzhong Weng

Passive Microwave Remote Sensing of the Earth

for Meteorological Applications

Author

Dr. Fuzhong Weng
3025 Carlee Run Court
Maryland
United States

A book of the Wiley Series in Atmospheric Physics and Remote Sensing

The Series Editor

Dr. Alexander Kokhanovsky
EUMETSAT
Remote sensing Products
EUMETSAT-Allee 1
64295 Darmstadt
Germany

and

Moscow Engineering Physics Institute (MEPhI)
National Research Nuclear University
Kashirskoe Str. 31
115 409 Moscow
Russia

Cover
Suomi NPP satellite–NASA/NOAA

Library of Congress Card No.: applied for

British Library Cataloguing-in-Publication Data
A catalogue record for this book is available from the British Library.

Bibliographic information published by the Deutsche Nationalbibliothek
The Deutsche Nationalbibliothek lists this publication in the Deutsche Nationalbibliografie; detailed bibliographic data are available on the Internet at <http://dnb.d-nb.de>.

Print ISBN: 978-3-527-33627-2
ePDF ISBN: 978-3-527-33629-6
ePub ISBN: 978-3-527-33630-2
Mobi ISBN: 978-3-527-33631-9
oBook ISBN: 978-3-527-33628-9

Cover Design Schulz Grafik-Design, Fußgönheim, Germany
Typesetting SPi Global, Chennai, India
Printing and Binding
CPI Group (UK) Ltd, Croydon, CR0 4YY

Printed on acid-free paper

To my wife and children.

Contents

Preface

Essential to the understanding of global climate change and improved weather forecasts is the gathering of basic data of variables such as temperature, pressure, wind, and the distribution of water vapor, clouds, and other active constituents. Microwave, which we consider as a generic term to include the centimeter, millimeter, and submillimeter regions of the spectrum, plays a special role in the remote sensing of these variables. Microwave signal penetrates through clouds and provides a measurement capability under all weather conditions. It also provides a direct means for the determination of cloud water content and precipitation. In the past years, I had been working on a variety of microwave instruments for retrieving the atmospheric and surface parameter. I began with the Special Sensor Microwave Imager (SSM/I) and developed several algorithms for operational uses. Then, I have also developed a suite of microwave products from Advanced Microwave Sounding Unit (AMSU), which started in 1998 with continuing efforts for further improvements of these satellite retrievals. Today, we have a flood of new satellite microwave data that are available from Microwave Humidity Sounder (MWHS), WindSat, SSMIS, AMSR-E, AMSR2, MHS, and ATMS, to name a few. These instruments provide unprecedented observations of Earth's environments and offer many unique opportunities to further improve our understanding of weather and climate changes and to significantly benefit to the numerical weather prediction. The WindSat measures the four Stokes channel components at 10, 18, and 37 GHz and allows simultaneous retrievals of wind speed and direction over oceans. The Special Sensor Microwave Imager and Sounder (SSMIS) observes, for the first time, the atmospheric temperature and water vapor profiles from a conical scanning mode. SSMIS and MHS have some channels in millimeter wavelengths ranging from 89 to 191 GHz, which are vital for the improvement in precipitation analysis and estimation. The observations from AMSR-E, AMSR2, and Windsat at 6 and 10 GHz over land are used to retrieve soil moisture content and other land surface parameters. My early work from the 1990s to 1999s was mostly centered on developing the operational products from various operational and research satellite microwave sensors. It is emphasized that the formulation of the remote sensing algorithms could be sensor specific. In the early 2000s, I began simplifying the radiative transfer process so that the inversion process can be more physically based and the retrieval algorithms can be easily adapted for new instruments under most environmental conditions. Recently, a general variational approach has also been developed for deriving the atmospheric and surface parameters from microwave sounding instruments.

While writing this book, I was faced with the usual problems encountered in compiling a work on a rapidly advancing field of microwave remote sensing and applications. In response, I first lay the fundamentals in the microwave remote sensing and then develop problem-specific algorithms and applications. This book condenses a large amount of information into a limited space, including microwave spectroscopy, particle scattering, radiative transfer, instrument calibration, and data applications in weather and climate research. It is intended for the broad class of scientists, researchers, and students who are interested in the environment and wish to be acquainted with the impact that microwave technology has on remote sensing. I have attempted to make this book as self-contained and thus have the first three chapters on microwave radiometer systems, atmospheric absorption and scattering, and radiative transfer modeling in the atmosphere. Today, the radiative transfer model is becoming a powerful tool in microwave instrument calibration and sensor data validation. Thus, Chapter 4 is fully developed for simulations of microwave observations using the Community Radiative Transfer Model (CRTM). Microwave instrument calibration is also introduced in Chapter 5. In Chapter 6, the algorithms for detecting the radio-frequency interference at microwave frequencies are illustrated. Since the scope of microwave remote sensing is very broad, I choose some typical parameters to illustrate the principle of algorithms. Chapter 7 discusses remote sensing of surface parameters including ocean wind vector and sea surface temperature, land emissivity. Chapters 8 and 9 are devoted to the microwave algorithms for retrieving clouds and atmospheric temperature and water vapor profiles. For applications of microwave data, I illustrated the impacts of assimilation in numerical weather prediction models in Chapter 10 and the uses of microwave data in deriving atmospheric temperature and water vapor trend in chapter 11.

I am indebted to many researchers in the field who have contributed materials and offered encouragement in preparing this book. I would also like to offer special thanks to Professor Xiaolei Zou for many of her elegant research work and artistic figures and to Dr Lin Lin, who was kind enough to review the whole manuscript. Also, I am happy to acknowledge here the moral support of my wife, which was essential to carry me through the years of work on the manuscript.

Maryland, United States
December 10, 2016

Dr. Fuzhong Weng

1 Introduction

Microwave remote sensing infers the physical parameters from satellite radiometers that operate at wavelengths ranging from millimeters to centimeters (1.0 mm to 20.0 cm). The microwave radiometers are generally of two types: imagers that have channels in the window regions of the spectrum to monitor surface, cloud, and precipitation; and sounders that have channels in opaque spectral regions to profile atmospheric temperature and water vapor under all weather conditions. All the current and planned instruments are flown aboard low Earth-orbiting satellites. The vantage point of the geosynchronous orbit would be valuable for obtaining the synoptic and continuous views of the atmosphere provided by optical and infrared sensors.

This chapter introduces some fundamental parameters used in microwave remote sensing.

1.1 A Microwave Radiometer System

A typical microwave radiometer uses the so-called heterodyne principle applied at radio frequencies. A heterodyne receiver is one in which the received signal, called the radio frequency (RF) signal, is translated to a different and usually lower frequency (the intermediate-frequency (IF)) signal before it is detected. The simplest heterodyne radiometer is shown in Figure 1.1. It is an example of a total-power radiometer and illustrates the features common to most microwave radiometers. As a signal at some frequency is incident upon the antenna of the radiometer, it couples the RF signal into a transmission line (a waveguide, for example), the function of which is to carry the RF signal to and from the various elements of the circuit. In the example, this signal is introduced directly into a mixer, which is a nonlinear circuit element in which the RF signal is combined with a constant-frequency signal produced at the output of this element because of its nonlinearity. These products include a signal whose frequency is the difference between the RF and local oscillator (LO) frequencies. This signal has the important property that its power is proportional to the power in the RF signal under the condition that the

Passive Microwave Remote Sensing of the Earth: For Meteorological Applications, First Edition. Fuzhong Weng.

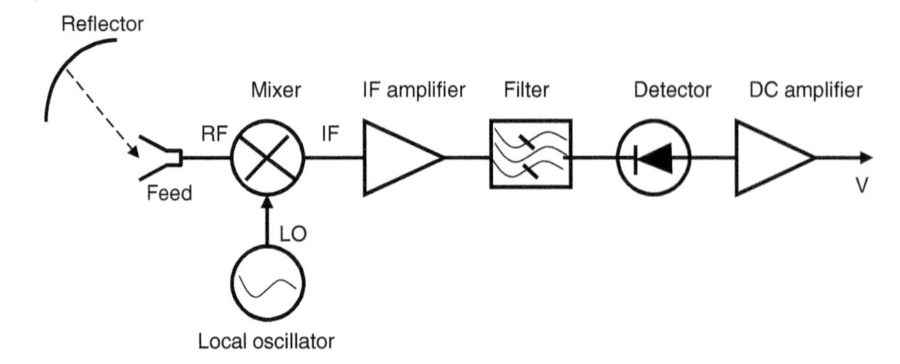

Figure 1.1 Schematic of a microwave radiometer with a circuit that produces an output voltage proportional to the received signal power.

latter is much weaker than the LO signal. It is then filtered to exclude the unwanted products of the mixing and amplified to produce the IF output signal.

The total-power radiometer produces an output voltage V, which is a polynomial function of the input current I:

$$V = a_1 I + a_2 I^2 + a_3 I^3 + a_4 I^4. \tag{1.1}$$

For a perfect square-law detector, the last two terms vanish. After integration over time and considering the current as a sinusoidal function of time, the average voltage is a function of the current squared:

$$\overline{V} = (a_2 + 3a_4 I^2) I^2. \tag{1.2}$$

Using the Nyquist theorem, this current squared is related to the total power input to the IF system, which is the sum of the thermal radiation $R(T_A)$ from the measurement target and the noise $R(T_n)$:

$$I^2 = KBG[R(T_A) + R(T_N)], \tag{1.3}$$

where T_A, G, B, and T_N are the temperature of the measurement target, amplifier gain, bandwidth, and noise temperature, respectively, and K is the Boltzmann constant.

It is seen from Eq. (1.3) that for the total-power radiometer, the amplifier gain and noise affect the mean voltage. In order to reduce the effect of the internal amplifier noise on the output stability, Dicke [1] introduced a radiometer circuit that can eliminate the noise term through differentiating the signals from the measuring target and an internal load with a known temperature during one integration cycle. The Dicke radiometer was a great invention and was used to measure the low power levels associated with thermal microwave radiation. The use of an internal noise diode injecting noise at a known temperature into the receiver can also reduce the effects of the gain instability and internal noise on the output of the total-power radiometer.

Combining Eqs. (1.2) and (1.3) results in

$$V = b_0 + b_1 R(T_A)[1 + \mu R(T_A)], \tag{1.4a}$$

where μ is the nonlinear parameter, and b_0 and b_1 are the parameters that can be directly determined from two-point calibration. They are mathematically expressed as

$$b_0 = [a_2 + 3a_4 KBR(T)]KBGR(T), \tag{1.4b}$$

$$b_1 = [a_2 + 6a_4 KBR(T)]BG, \tag{1.4c}$$

$$\mu = 3a_4 \frac{KBG}{a_2}. \tag{1.4d}$$

1.2 Blackbody Emission

A blackbody is an object that absorbs light at a certain wavelength and also emits radiation at the same wavelength. The total amount of energy radiated by a blackbody can be described through Planck's law in a special function. The function is valid for electromagnetic radiation pervading any medium, regardless of its constitution, that is in thermodynamic equilibrium at a definite temperature. If the medium is homogeneous and isotropic, then the radiation is homogeneous, isotropic, unpolarized, and incoherent. The law is named after Max Planck, who originally proposed it in 1900. It is a pioneer result of modern physics and quantum theory. For a wave number υ, Planckian radiation or spectral radiance (in unit: $\mathrm{W/m^2/sr/cm}$) is expressed as

$$B_\upsilon(T) = \frac{2hc^2\upsilon^3}{\exp\left(\frac{hc\upsilon}{kT}\right) - 1} \equiv \frac{C_1\upsilon^3}{\exp\left(\frac{C_2\upsilon}{T}\right) - 1}, \tag{1.5}$$

where the Planck constant $h = 6.626 \times 10^{-34}$ J s; the Boltzmann constant $k = 1.381 \times 10^{-23}$ J/K; c is the speed of light, $C_1 = 2hc^2 = 1.1909 \times 10^{-8}\ \mathrm{W/m^2/sr/cm^3}$, and $C_2 = \frac{hc}{k} = 1.438786$ cm/K.

It should be pointed out that the Planck function can be expressed in terms of wavelength or frequency, but the resultant unit is different. When the energy is integrated within a wavelength, wave number, or frequency domain, the unit for the radiance should be all the same ($\mathrm{W/m^2/sr}$). For example, in a frequency domain f,

$$B_f(T) = \frac{2hf^3}{c^2} \frac{1}{\exp\left(\frac{hf}{kT}\right) - 1}, \tag{1.6}$$

which represents the energy in W/m^2/sr/Hz. Alternatively, it can be written in terms of the wavelength λ, as

$$B_\lambda(T) = \frac{2hc^2}{\lambda^5} \frac{1}{\exp\left(\frac{hc}{\lambda kT}\right) - 1}, \tag{1.7}$$

which represents the energy in W/m^2/sr/cm.

An integration for radiance within a spectrum can be derived using any function from Eqs. (1.5)–(1.7) with a relationship to the frequency, wavelength, and wave number for changing the limits in the integration. For instance, we can use Eq. (1.6) with $f = c\upsilon$ and $f = \frac{c}{\lambda}$

$$I = \int_{f_1}^{f_2} B_f(T)\mathrm{d}f = \int_{\upsilon_1}^{\upsilon_2} B_\upsilon(T)\mathrm{d}\upsilon = \int_{\lambda_1}^{\lambda_2} B_\lambda(T)\mathrm{d}\lambda. \tag{1.8}$$

Thus, we can also understand the equivalence of the Planck function expressed as Eqs. (1.5)–(1.7), which are interchangeable through

$$B_\upsilon(T) \equiv B_f(T)c, \tag{1.9}$$

$$B_\lambda(T) \equiv B_f(T)\frac{c^2}{\lambda^2}, \tag{1.10}$$

$$B_\upsilon(T) \equiv B_\lambda(T)\lambda^2. \tag{1.11}$$

1.3 Linearized Planck Function

Assuming $\frac{C_2 v}{T} \ll 1$, the exponential term in the Planck function can be expressed as a Taylor series:

$$\exp\left(\frac{C_2\upsilon}{T}\right) = 1 + \frac{C_2\upsilon}{T} + \frac{1}{2}\left(\frac{C_2\upsilon}{T}\right)^2 + \cdots + \frac{1}{n!}\left(\frac{C_2\upsilon}{T}\right)^n + \cdots. \tag{1.12}$$

Substituting the first-order approximation of the given Taylor expansion into Eq. (1.5), one obtains the following linear relationship between the blackbody temperature (T) and radiance $B_\upsilon(T)$, which is also referred to as the Rayleigh–Jeans (RJ) approximation:

$$B_\upsilon^{RJ}(T) = \frac{C_1\upsilon^2}{C_2}T. \tag{1.13}$$

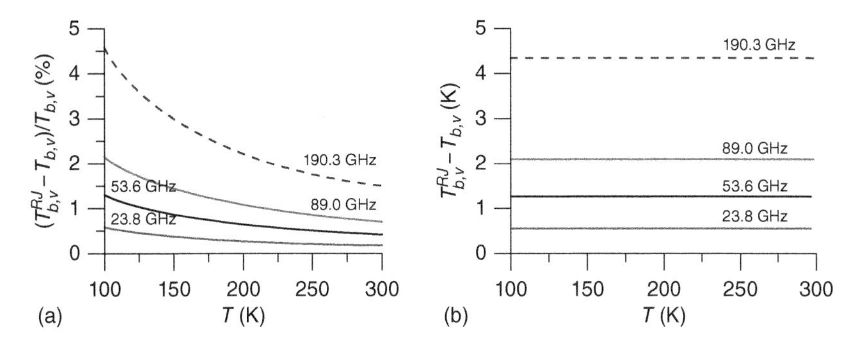

Figure 1.2 (a) Relative and (b) absolute variations of the brightness temperature with blackbody temperature varying from 100 to 300 K at frequencies 23.8, 53.6, 89.0, and 190.3 GHz. (Weng and Zou 2013 [2]. Reproduced with permission of Optical Society of America.)

The accuracy of the radiance calculated from Eq. (1.5) and the linear approximation Eq. (1.13) varies with the frequency and temperature. Figure 1.2 shows the relative accuracy of the first-order approximation of the Planck function $(B_\nu^{RJ} - B_\nu)/B_\nu$ with respect to temperature at four arbitrarily selected frequencies of 23.8, 53.6, 89.0, and 190.3 GHz. At a fixed temperature, the higher the frequency, the larger the error. Alternatively, at a fixed frequency, the lower the temperature, the larger the error. At a high frequency near 190.3 GHz, there is a 4.5% error in radiance. The error decreases rapidly with an increase in temperature. More analyses of the RJ approximation can be found in Weng and Zou [2].

1.4 Stokes Vector and Its Transformation

When an electromagnetic wave propagates in space, both its electric and magnetic fields are expressed as vectors, and they travel through space by exciting the field of each other. As a result, the radiation field is a Stokes vector with four elements which are related to the amplitudes of the electric field in the form

$$\mathbf{I} = (I, Q, U, V) \tag{1.14a}$$

or

$$\mathbf{I} = (I_\parallel, I_\perp, U, V) \tag{1.14b}$$

where

$$I = \langle E_\parallel E_\parallel^* \rangle + \langle E_\perp E_\perp^* \rangle, \tag{1.15a}$$

$$Q = \langle E_\parallel E_\parallel^* \rangle - \langle E_\perp E_\perp^* \rangle, \tag{1.15b}$$

$$I_\parallel = \langle E_\parallel E_\parallel^* \rangle, \tag{1.15c}$$

$$I_\perp = \langle E_\perp E_\perp^* \rangle, \tag{1.15d}$$

$$U = \langle E_\parallel E_\perp^* \rangle + \langle E_\perp E_\parallel^* \rangle, \tag{1.15e}$$

$$V = i\langle E_\parallel E_\perp^* \rangle - \langle E_\perp E_\parallel^* \rangle, \tag{1.15f}$$

where $E_\parallel$ and $E_\perp$ are the horizontal and vertical components of the electric field, respectively, the star ($*$) denotes the conjugate of a complex value, and the angular brackets indicate the time average over an interval longer than the oscillation period of the electric field.

The component expressed by Eq. (1.15a) also represents the total energy from the electromagnetic field and thus is also referred to as the radiation intensity. In the microwave remote sensing field, the subscripts $\parallel$ and $\perp$ are often replaced with h and v in the first two Stokes components. Thus, in this textbook, we use the following notations for the Stokes brightness temperature components:

$$\mathbf{T}_b = (T_b^h, T_b^v, T_b^3, T_b^4), \tag{1.16}$$

where the superscripts (3, 4) are used in the brightness temperature components in Eq. (1.16) to replace the third and fourth Stokes components in Eqs. (1.15c) and (1.15d) for avoiding the repetition of the superscripts used in the first two components. The four brightness temperature components are related to the Stokes parameters in Eq. (1.15) through the Planck function.

The law of transformation of the Stokes parameters for a rotation of the axes through an angle θ is derived in two forms. For the Stokes parameters expressed in Eq. (1.14a), the transformation matrix [3] is given as

$$\mathbf{L}_i = \begin{pmatrix} 1 & 0 & 0 & 0 \\ 0 & \cos 2i & -\sin 2i & 0 \\ 0 & \sin 2i & \cos 2i & 0 \\ 0 & 0 & 0 & 1 \end{pmatrix}. \tag{1.17a}$$

Alternatively, it can be written as

$$\mathbf{L}_i = \begin{pmatrix} \cos^2 i & \sin^2 i & \frac{1}{2}\sin 2i & 0 \\ \sin^2 i & \cos^2 i & -\frac{1}{2}\sin 2i & 0 \\ -\sin 2i & \sin 2i & \cos 2i & 0 \\ 0 & 0 & 0 & 1 \end{pmatrix}, \tag{1.17b}$$

when the Stokes parameters in Eq. (1.14b) are used [4].

Equation (1.17b) is often used in microwave remote-sensing applications. For a Stokes vector having zero third and fourth components in one coordinate system, its rotation to other coordinate systems may result in nonzero third and fourth components. The third and fourth Stokes components can also contribute to the first and second components through a coordinate transformation. For a microwave scanning radiometer, Eq. (1.17b) is applied for converting the brightness temperature components at the local coordinate to those at a scan angle θ with respect to the nadir position such that

$$T_b^{Qh} = T_b^h \cos^2\theta + T_b^v \sin^2\theta - T_b^3 \frac{1}{2} \sin 2\theta, \tag{1.18a}$$

and

$$T_b^{Qv} = T_b^h \sin^2\theta + T_b^v \cos^2\theta + T_b^3 \frac{1}{2} \sin 2\theta, \tag{1.18b}$$

respectively, where the subscripts on the left-hand side of Eq. (1.18) denote the first two components of the Stokes vector after the coordinate transformation. The third terms on the right-hand side are normally neglected in microwave calibration [5]. However, for the window channels over the polarized surfaces, the third term could be significant, and further studies are needed to quantify the magnitude in microwave instrument calibration.

1.5 Microwave Spectrum

In microwave remote sensing, the spectrum ranges from centimeter wavelength to millimeter wavelength. The microwave spectrum is usually defined as electromagnetic energy ranging from ~1 to 200 GHz in frequency, but older usage includes lower frequencies. Most common applications are within the 1–60 GHz range. Microwave frequency bands, as defined by the Radio Society of Great Britain (RSGB), are shown in Table 1.1. L-band technology is used for remote sensing of soil moisture. C- and X-bands are more utilized for remote sensing of ocean properties such as sea-surface temperature and wind vector. K-band is more sensitive to atmospheric clouds, water vapor, and precipitation. V-band is explored for probing the atmospheric temperature profile. W- and G-bands are used for remote sensing of ice cloud and of atmospheric moisture profile. F-band is also considered as an alternative for atmospheric temperature sounding at a high spatial resolution.

The microwave bands used for meteorological applications are always protected from other emission sources operating at the same frequencies for commercial and military applications. The growing competition may result in strong interferences of the weather microwave instruments with active sources. It is important to develop some techniques to detect the RF interferences on microwave instruments [6, 7].

Table 1.1 Microwave frequency bands used for Earth environmental and atmospheric remote sensing.

Band	Frequency range (GHz)	Major remote sensing applications	RFI sources
L	1–2	Soil moisture content, water salinity	Radar, cell phone, garage door
S	2–4	No applications in passive microwave remote sensing	Radar
C	4–8	Sea surface temperature, soil moisture	Radar, speed monitor
X	8–12	Sea surface wind, sea surface temperature	Direct TV
Ku	12–18	No applications in passive microwave remote sensing	
K	18–26.5	Precipitation, atmospheric water vapor, cloud water, sea surface wind	Direct TV
Ka	26.5–40	Precipitation, water vapor, cloud water, sea surface wind	
Q	30–50	No applications in passive microwave remote sensing	
U	40–60	No applications in passive microwave remote sensing	
V	50–75	Atmospheric temperature profile	
E	60–90	No applications in passive microwave remote sensing	
W	75–110	Precipitation over land, cloud water, and cloud ice	
F	90–140	Atmospheric temperature and moisture profile	
D	110–170	No applications in passive microwave remote sensing	
G	150–190	Atmospheric moisture profile	

1.6 Spectral Response Function

For a typical instrument, the radiance received at a central wavelength corresponds to the radiation within a spectral band and can be expressed as

$$I_f = \int_{f-\Delta f}^{f+\Delta f} B_x(x)S(x)\mathrm{d}x, \tag{1.19}$$

where f is the instrument central frequency, and $2\Delta f$ is the instrument bandwidth, which is determined by the 3-dB window in $S(x)$.

In the past, the instrument spectral response function was not well characterized for many microwave sensors. Thus, it is often assumed as a boxcar function or a Gaussian distribution function. As shown in Figure 1.3, the boxcar, $S(x)$, for the Advanced Technology Microwave Sounder (ATMS) at some channels could be significantly different from the one measured from the laboratory data.

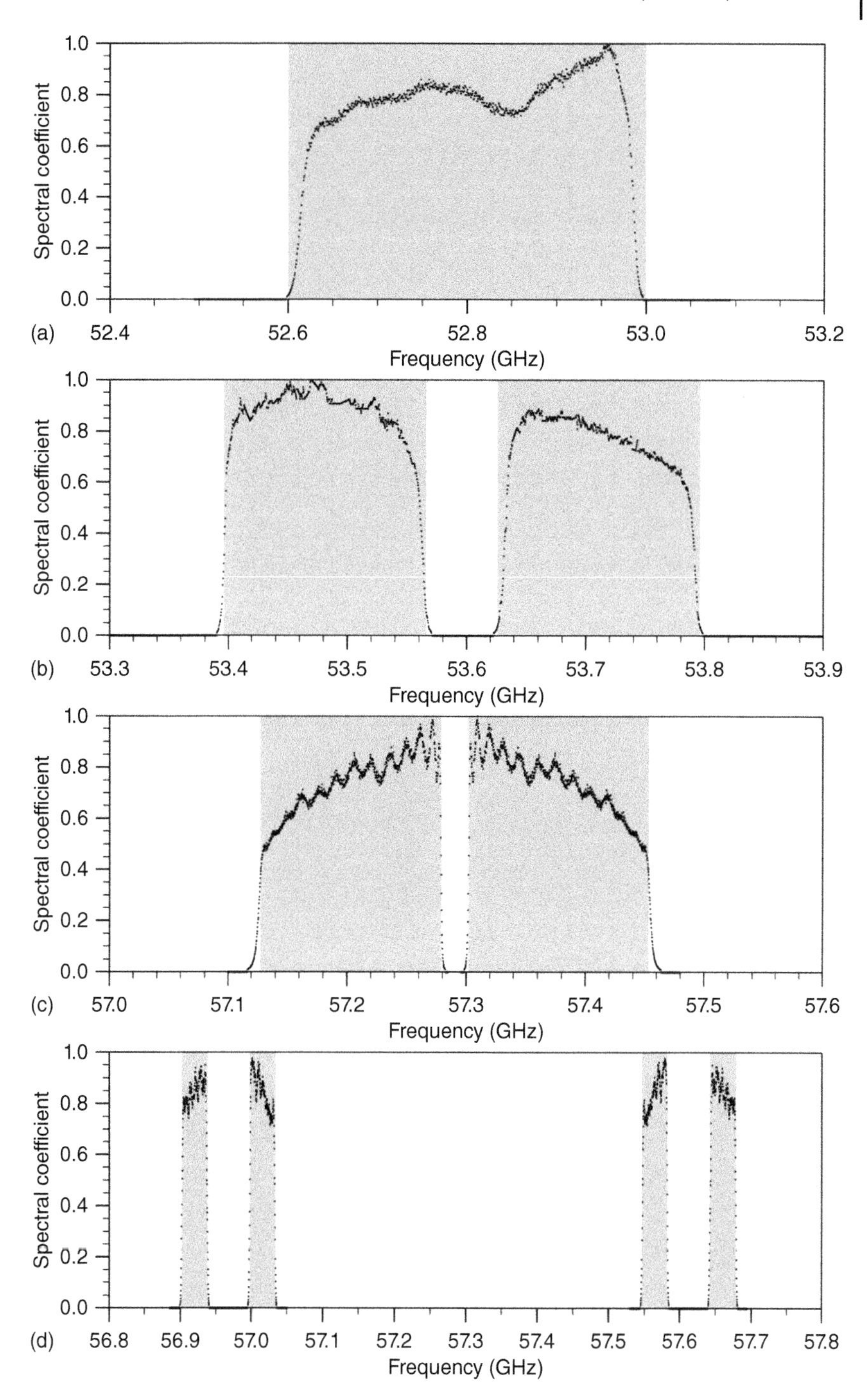

Figure 1.3 Laboratory-measured SRF used in line-by-line model after −20 dB truncation (black line) and the boxcar SRF used in CRTM (shading) for ATMS channel (a) 5, (b) 6, (c) 10, and (d) 12. (Replotted from Zou *et al.* 2014 [8].)

The radiance differences between a boxcar spectral response function (SRF) and the laboratory SRF were recently studied and could be significant for some instruments [8].

1.7 Microwave Antenna Gain and Distribution Function

A microwave radiometer system requires an antenna to collect the radiative energy from the Earth and other targets. The antenna gain or efficiency is determined by its power distribution function, as shown in Figure 1.4. An antenna having a higher gain tends to have a main lobe or beam with a larger power value. The secondary and third peaks are called the side lobes. For an antenna subsystem, the magnitude of the side lobes depends on both the frequency and the antenna size. In general, the higher the frequency and the larger the antenna size, the narrower the main beam and the smaller the side-lobe effect. The antenna pattern is often shown at a decibel scale through normalizing the measured intensity in each direction by the sum of the co- and cross-polarized radiations.

The antenna's main beam width θ is defined through the half-power points in the antenna power function. Based on the normalized antenna pattern, the main beam efficiency is integrated with the antenna angles within 2.5 times the beam width, as follows:

$$\eta_m^{pp} = \int_{\Omega_m} G^{pp} d\Omega / \int_{\Omega} (G^{pp} + G^{qp}) d\Omega. \tag{1.20}$$

The antenna side-lobe efficiency is similarly derived as

$$\eta_s^{pp} = \int_{\Omega_s} G^{pp} d\Omega / \int_{\Omega} (G^{pp} + G^{qp}) d\Omega, \tag{1.21}$$

where the superscripts pp and qp stand for either v or h polarization state; G^{pp} and G^{qp} are the normalized antenna gains in the far field (i.e., Earth and cold space)

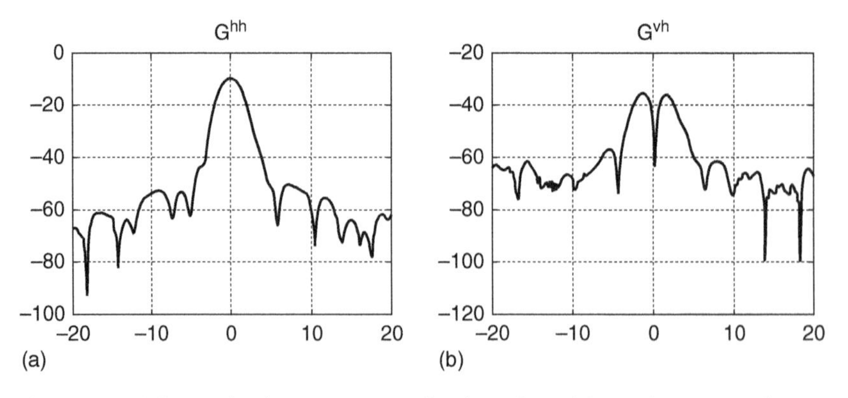

Figure 1.4 ATMS normalized antenna pattern for channel 1 at (a) co-polarization and (b) cross-polarization. (Replotted from Weng *et al.* 2013 [9].)

for co- and cross-polarization, respectively; and Ω_m and Ω_s are the solid angles corresponding to the antenna's main beam and side lobes, respectively. Normally, the side-lobe efficiency is derived according to the portions that intercept the energy from the unwanted targets [9].

1.8 Microwave Instrument Scan Geometry

Similar to a radar system, a satellite microwave instrument also requires an antenna subsystem to scan and collect the energy within its field of view (FOV). Currently, two scanning modes (stepwise or continuous) are deployed in space for cross-track scanning instruments, with the two specific examples illustrated here. The most important parameters are listed in Table 1.2.

The Advanced Microwave Sounding Unit (AMSU-A) is a stepwise scanning instrument. It has a scan angle of ±48.33° with respect to the nadir direction. The instantaneous field of view (IFOV) of each channel is ~57.6 mrad (3.3°),

Table 1.2 Parameters from stepwise and continuous scanning microwave radiometer systems.

Parameter	Symbol	Definition	AMSU (1–15)	ATMS (3–16)
Instantaneous field of view	IFOV	Angular range corresponding to 3-dB points in antenna gain distribution function	3.3°	2.1°
Effective field of view	EFOV	Angular range during which the antenna collects the signal to produce a mean signal	3.3°	2.3°
Angular swath width	θ_m	The maximum value of the instrument scan angle	±48.33°	±52.75°
Number of field of views	N_s	The number of scan-angle positions across each scan	30	96
Scan cycle	T	Total time for antenna viewing Earth, calibration target, then returning to its original position	8 s	8/3 s
Earth view time	T_c	The time of antenna viewing the Earth scene	6 s	1.73 s
Sample time	ΔT	The time for the two consecutive samples made within a scan line	200 ms	18 ms
Sampling angle	θ_s	The time for the two consecutive samples made within a scan line	3.3°	1.1°
Integration time	Δt	The time during which the antenna collects the signal to produce a mean signal	158 ms	17.6 ms
Earth view scan rate	R	The scan rate with which the antenna views the earth scene. These parameters are often related with each other	n/a	61.6 °/s

leading to a circular IFVO size close to 47.63 km at nadir and a swath width of ±1026.31 km for a nominal satellite altitude of 833 km. Its sampling time is 200.0 ms, with 160 ms for integration and 40 ms for stepping to the next scan position. The sampling angular interval is close to 58.18 mrad (3.3333°). The distance between two consecutive scans is ~52.69 km. On each scan line, there are 30 Earth views. Each scan takes 8.0 s to complete.

Microwave instruments can continuously scan across the track. For ATMS, channels 3–16 have a beam width of 2.2°, which is smaller than that of the AMSU-A channels 1–15. However, the beam width for ATMS surface channels 1 and 2 is 5.2°, which is much larger than that of the corresponding AMSU channels 1 and 2. Six of the seven ATMS channels above 60 GHz, channels 17–22, have a beam width of 1.1°, which is the same as that of the water vapor sounding channels of AMSU-A and the microwave humidity sounder (MHS).

The aforementioned differences of the beam width between ATMS and AMSU channels, along with the difference in the satellite altitudes between Suomi NPP (National Polar-orbiting Partnership) (824 km) and its predecessors such as NOAA-19 (870 km), result in significant differences in FOV sizes between ATMS and AMSU. At ATMS channels 1 and 2, its single FOV is ~1.6 times the AMSU FOV in diameter, which is mostly determined by the beam width differences between the two instruments. There is no overlap between the neighboring FOVs and between the neighboring scan lines of AMSU, but significant overlaps occur for ATMS FOVs and scan lines of channels 1 and 2 (Figure 1.5). For example, the FOV48 has overlaps with the neighboring four FOVs and four scan lines.

A single AMSU FOV for channels 3–15 is about 1.5 times larger than that of ATMS channels 3–16. At these channels, a single ATMS FOV overlaps with its surrounding four FOVs. The differences in FOVs for water vapor channels between ATMS and AMSU are rather small. There is a small difference in the integration time between ATMS (18 ms) and MHS (19 ms).

The oversampling features of ATMS allow the estimation of brightness temperatures at resolutions higher or lower than the raw ATMS data resolution. However,

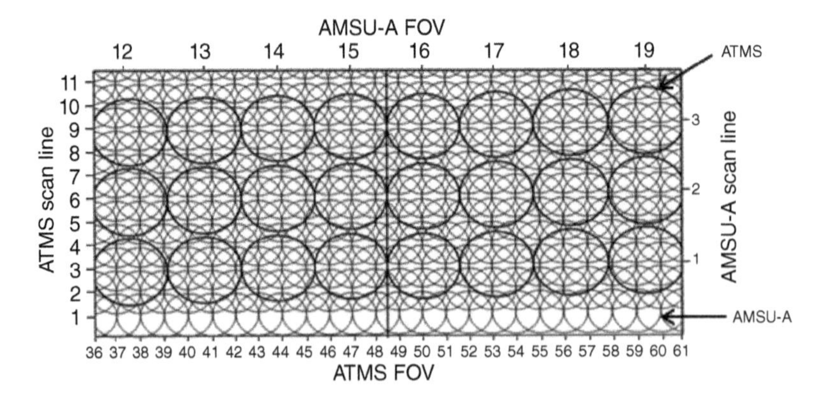

Figure 1.5 Comparison of FOVs from stepwise and continuous scanning microwave radiometers (AMSU-A vs ATMS).

an optimal balance between the desirable resolution and the resulting data noise must be taken into consideration when developing such an estimate for investigating specific weather systems.

1.9
Microwave Data Records and Their Terminology

A microwave radiometer measures the energy averaged within its FOV. Its antenna brightness temperature is defined as

$$T_a^p = \int_{\Omega_{me}} (G^{pp} T_b^p + G^{qp} T_b^q) \mathrm{d}\Omega + \int_{\Omega_{se}} (G^{pp} T_b^p + G^{qp} T_b^q) \mathrm{d}\Omega + \int_{\Omega_{sc}} (G^{pp} T_b^p + G^{qp} T_b^q) \mathrm{d}\Omega + \int_{\Omega_{ss}} (G^{pp} T_b^p + G^{qp} T_b^q) \mathrm{d}\Omega, \quad (1.22)$$

where T_b^p is the brightness temperature at the polarization state of p. Note that T_b^p has a spatial distribution covering the Earth, cold space, and the spacecraft. Ω_{se} and Ω_{ss} are the solid angles of the side lobes facing the cold space and the satellite platform, respectively. The solid angle Ω_{me} corresponds to 2.5 times the main beam width. The first term in Eq. (1.22) is the Earth radiation entering into the receiver system through the main beam; the second term is the Earth radiation through the side lobes that are out of the main beam but within the Earth view sector, the third term is the cold space radiation through the side lobes, and the fourth term is the radiation scattered and emitted from the satellite platform to the receiver. Equation (1.22) can also be expressed in terms of the antenna efficiency associated with each term as

$$T_a^p = \eta_{me}^{pp} T_b^p + \eta_{me}^{qp} T_b^q + \eta_{se}^{pp} T_{b,se}^q + \eta_{se}^{qp} T_{b,se}^q + \eta_{sc}^{pp} T_{b,sc}^q + \eta_{sc}^{qp} T_{b,sc}^q + \eta_{ss}^{pp} T_{b,ss}^p + \eta_{ss}^{qp} T_{b,ss}^q. \quad (1.23)$$

The sensor brightness temperature is derived from Eq. (1.23) by correcting the effects of the cross-polarization and side-lobe contributions from the Earth, cold space, and spacecraft on the antenna brightness temperature:

$$T_b^p = \frac{1}{\eta_{me}^{pp}} (T_a^p - \eta_{me}^{qp} T_b^q - \eta_{se}^{pp} T_{b,se}^q - \eta_{se}^{qp} T_{b,se}^q - \eta_{sc}^{pp} T_{b,sc}^p - \eta_{sc}^{qp} T_{b,sc}^q - \eta_{ss}^{pp} T_{b,ss}^p - \eta_{ss}^{qp} T_{b,ss}^q). \quad (1.24)$$

Determination of the contributions from cross-polarization and side lobes requires both accurate antenna gain efficiencies and brightness temperatures within various sectors viewed by the antenna.

2 Atmospheric Absorption and Scattering

2.1 Introduction

Atmospheric absorption and scattering models are required for radiative transfer modeling. For microwave remote sensing applications, a number of spectroscopy databases have been developed as part of the radiative transfer model. For example, HITRAN (an acronym for high-resolution transmission) is a compilation of spectroscopic parameters that a variety of computer codes use to predict and simulate the transmittance and emission of light in the atmosphere [10]. The line-by-line radiative transfer model (LBLRTM) is an accurate, efficient, and highly flexible model for calculating spectral transmittance and radiance [11]. In addition, GEISA (Gestion et Etude des Informations Spectroscopiques Atmosphériques) is a publically accessible spectroscopic database designed for accurate forward radiative transfer calculations using a line-by-line (LBL) and layer-by-layer approach. However, the LBL computations of atmospheric transmittance are very costly and must be further parameterized for fast but accurate radiative transfer simulations required in satellite data assimilation systems [12–14]. In this chapter, we provide a comprehensive overview of the absorption lines in microwave to millimeter wave regions and also present our unique efforts in the parameterization of the absorption lines for fast and efficient computations. In addition, the techniques developed for fast computations of scattering from aerosol, cloud, and precipitation particles are discussed [15, 16]. The fast atmospheric transmittance models and the lookup table for particle scatterings are the key components in the community radiative transfer model (CRTM). The concept of CRTM was conceived when the United States Joint Center for Satellite Data Assimilation (JCSDA) was established in 2002. The CRTM is a software library for computing the satellite instrument radiances and their gradients with respect to various atmospheric and surface state variables required in the data assimilation systems. The CRTM is based on a design framework that emphasizes modularity and code reuse. Some modules are based on those components that already existed in the NCEP global forecast system

Passive Microwave Remote Sensing of the Earth: For Meteorological Applications, First Edition. Fuzhong Weng.

(GFS), while others were developed to handle more sophisticated atmospheric and surface radiative transfer problems. From its initial conception, the JCSDA CRTM library was designed to be both platform independent and assimilation system independent for ease of implementation by the JCSDA partners. The CRTM has progressed into a system for use primarily at infrared and microwave frequencies, with development opportunities for visible wavelengths.

2.2 Microwave Gaseous Absorption

For a clear atmospheric condition, transmittance, Υ, at altitude z with respect to the top of atmosphere (TOA) is related to the optical depth, τ, as follows:

$$\Upsilon_{\nu}(z) = \exp\left(-\frac{\tau_{\nu}}{\mu}\right) = \exp\left(-\int_{z}^{\infty}\frac{\kappa(\nu, z)}{\mu}\mathrm{d}z\right), \tag{2.1}$$

where $\kappa(\nu, z)$ is the volumetric absorption coefficient expressed in the unit of cm^{-1}, $\mu = \cos(\theta)$, and θ is the observing angle. The absorption coefficient is usually derived from laboratory measurements of transmittance through layers of gas at fixed temperatures. All the physics related to the interaction of radiation with matter (gases and particles) is contained within the absorption coefficient. The optical depth in Eq. (2.1) is related to the integration of the volumetric absorption coefficient, given in units of inverse distance, over the thickness of the layer. In many applications, the gaseous mass absorption coefficients are used and are given in the unit of $\mathrm{cm}^2\mathrm{g}^{-1}$. A conversion from the mass absorption coefficient to the volumetric coefficient is performed through the gaseous density, $\rho(\mathrm{z})$. For the atmospheric sounding, the integration in Eq. (2.1) is often derived through a summation of all the absorptions within each thin layer. Usually, the absorption characteristics of the gaseous constituents are a function of the pressure and temperature or a function of the partial pressure of other gases. The quantity of gas within a planet's atmosphere increases exponentially with depth (due to hydrodynamic equilibrium) and may also be a function of depth due to the chemical and photolysis processes.

2.2.1 Absorption Line and Shape

In the troposphere, the absorption coefficient is dominated by pressure broadening due to collisions with the molecules of all species in the atmosphere. Each absorption line can be represented by a Lorentz line shape with a half-width [17]:

$$\kappa(\nu - \nu_0) = \frac{\gamma/\pi}{(\nu - \nu_i)^2 + (\gamma)^2}, \tag{2.2}$$

where $\gamma = 1/2\pi t_c$ and t_c is the mean time a molecule spends in the perturbed state. The absorption coefficient at a given wave number requires a summation over all lines in a given spectral interval. Note that the far wings of a large number of lines can contribute to absorption even in window regions as such

$$\kappa(\nu, p, T) = \sum_{i=1}^{N_i} \frac{n_i S_i}{\pi} \frac{\gamma_i}{(\nu - \nu_i)^2 + (\gamma_i)^2}, \tag{2.3}$$

where S_i is the absorption line intensity, n_i is the number of absorbers, ν_i is the central wave number of absorption line, and γ_i is the pressure-broadened absorption line half-width, which is

$$\gamma_i \cong \gamma_i^0 \frac{p}{p_0} \sqrt{\frac{T}{T_0}}, \tag{2.4}$$

For low pressures ($p \leq 1$ hPa), the following corrections must be included in computing the absorption. Thermal broadening due to Doppler shifting of lines is important at high temperatures and low pressures. If T is the kinetic temperature, the probability of finding a molecule with a velocity V in the range of $(V, V + \delta V)$ is given by a Maxwellian distribution

$$W(V)\mathrm{d}V = \frac{1}{\sqrt{\pi} V_0} \exp\left(-\frac{V^2}{V_0^2}\right) \mathrm{d}V, \tag{2.5}$$

where $V_0 = \sqrt{2k_B T/m}$, k_B is the Boltzmann constant, and m is the mass of the molecule. When observed at wave number ν, the absorbing wave number is given by $\nu[1 - (V/c)]$. The Doppler frequency shift from the distribution of molecular velocities causes absorption line broadening and results in Gaussian line shape as

$$\kappa_d(\nu - \nu_0) = \frac{1}{\gamma_D \sqrt{\pi}} \exp\left[-\frac{(\nu - \nu_0)^2}{\gamma_D^2}\right]. \tag{2.6}$$

When the pressure-broadened Lorenz half-width becomes comparable to the Doppler width, the broadened absorption coefficient, κ_d, is given by a convolution integral of the unbroadened absorption coefficient, κ, Doppler-shifted by the velocity distribution

$$\kappa_d(\nu) = \int_{-\infty}^{\infty} \kappa(\nu - \nu V/c) W(V) \mathrm{d}V. \tag{2.7}$$

Thus, Doppler broadening results in a Voigt line profile, which does not have any particular analytic form of line shape [18].

2.2.2
Oxygen Absorption

The primary gaseous absorbers in the troposphere and lower stratosphere at frequencies below 300 GHz are diatomic oxygen (O_2) and water vapor (H_2O). Carbon dioxide (CO_2), carbon monoxide (CO), nitrous oxide (N_2O), and ozone (O_3) exhibit resonant absorptions, and nitrogen (N_2) displays a weak nonresonant absorption at microwave frequencies. In general, these trace species can be neglected at altitudes and frequencies of concern in microwave remote sensing.

The absorption of the linear molecule oxygen, O_2, in the electronic ground state arises from a fine-structure transition caused by the interaction of the molecule's permanent magnetic dipole with the magnetic field produced by the rotor's orbital angular momentum [19, 20]. Approximately 33 transitions of significant strength in the atmosphere are located between 50 and 70 GHz, and a single isolated transition is located at 118.75 GHz. Zeeman splitting of microwave O_2 lines by the Earth's magnetic field is important at altitudes starting in the upper stratosphere [21, 22]. Absorption by the isotopic species $^{16}O^{18}O$ is normally negligible for the purpose of passive tropospheric and lower stratospheric remote sensing.

The microwave absorption spectrum of H_2O is due to rotational transitions induced by the interaction of external fields with the molecule's permanent electric-dipole moment. Water vapor resonances at microwave frequencies are modeled by the van Vleck–Weisskopf (VVW) line shape [23]. An additional absorption contribution by water vapor takes the form of a continuum that varies slowly with frequency. Empirical characterization of H_2O continuum and resonant absorption was developed by Liebe [24–26].

Currently, the absorption coefficients for microwave spectrum were considered primarily from Rosenkranz [27] and Liebe *et al.* [28]. For molecular oxygen, the absorption arises from complicated magnetic-dipole transitions in which the two unpaired electron spins of the electronic ground state change alignment with respect to the rotational angular momentum, which is given by the quantum number, N. The allowed transitions are from $J=N$ to $J=N\pm 1$, where J is the resultant angular momentum quantum number. The absorption coefficients are complicated functions of the frequency (f in GHz), pressure (p in mb), temperature (T in K), and amount of O_2 derived from water vapor density, and can be written in terms of the line parameters

$$\kappa^{O_2}(f,p,T,\rho_v) = \kappa_f^{O2-NR} + 10^{-10} n_{O_2} \cdot \sum_{j\geq 0} S_j(T) F(f, f_{0,j}), \tag{2.8}$$

where κ_f^{O2-NR} is the absorption from nonresonant absorption, and n_{O_2} is the number density of O_2

$$n_{O_2} = 0.503384 \times 10^{22} p_{dry} \cdot \theta, \tag{2.9}$$

where $\theta = 300/T$, $p_{dry} = p - p_{H_2O}$, and $p_{H_2O} = \rho_w T/217$.

The nonresonant absorption arises from the relaxation spectrum of oxygen's magnetic dipole moment, which is related to the imaginary component of the refractivity ($N = N' + iN''$)

$$\kappa_f^{O_2-NR} = 0.04191 \cdot f \cdot N''$$

$$\equiv 2.57327 \cdot 10^{-6} \cdot p \cdot \theta^2 \cdot \frac{f^2 \gamma_{NR}}{f^2 + \gamma_{NR}^2}, \text{ nepers/km} \tag{2.10}$$

The line half-width parameter for nonresonant absorption, w_{NR}, has a value of 0.56 GHz/bar (bar is a non-Si unit of pressure, 1 bar = 100 000, Pa = 1000 mb) [28], and with pressure given in mb, the pressure-broadened half-width is given by

$$\gamma_{NR}(j) = (w_{NR}/1000) \cdot (p_{dry} \cdot \theta^{0.8} + 1.1 p_{H_2O} \cdot \theta) \tag{2.11}$$

For pressures greater than 0.1 mb, the pressure-broadened line shape is given as follows [27]:

$$F(f, f_{0,j}) = \frac{1}{\pi} \left(\frac{f}{f_{0,j}} \right)^2 \left[\frac{\gamma_c + (f - f_{0,j}) \cdot Y(j)}{(f - f_{0,j})^2 + \gamma_c^2} + \frac{\gamma_c - (f + f_{0,j}) \cdot Y(j)}{(f + f_{0,j})^2 + \gamma_c^2} \right] \tag{2.12}$$

where the overlap correction is given by

$$Y(j) = \frac{p}{1000} \cdot \theta^{0.8} \cdot [y(j) + (\theta - 1) \cdot y(j)], \tag{2.13}$$

and the line strength is given by

$$Sj(T) = S' \cdot Q_{elec} \cdot Q_{vib} \cdot Q_{rot} \cdot \exp[b(j) \cdot (1 - \theta)], \tag{2.14}$$

where $y(j), fb(j)$, and other parameters in Eqs. (2.12)–(2.14) are shown in [27].

Figure 2.1 displays an absorption spectrum for frequencies ranging from 50 to 70 GHz, corresponding to a pressure of 50 mb and a temperature of 211 K. There are a total of 34 rotational absorption lines in this region. Each is named with the O_2 rotational energy transition quantum, as summarized in Table 2.1. Those lines at frequencies less than 58 GHz are widely used for atmospheric temperature sounding; for example, the Advanced Technology Microwave Sounder (ATMS) onboard Suomi National Polar-orbiting Partnership (SNPP) satellite with its channels 3–10 located in the region. Normally, the channels are selected at the valleys of two resonant absorption lines where the instrument noise can be designed to be lower through a large band pass. In addition, the channel having a higher absorption coefficient is sensitive to the upper atmospheric temperature.

However, other ATMS channels from 11 and 15 are located on the slopes of two absorption lines at 56.968 and 57.612 GHz, respectively, as shown in Figure 2.2. In addition, there are two to four subbands for each channel, as indicated by the identical channel numbers. Use of more subbands allows for noise reduction.

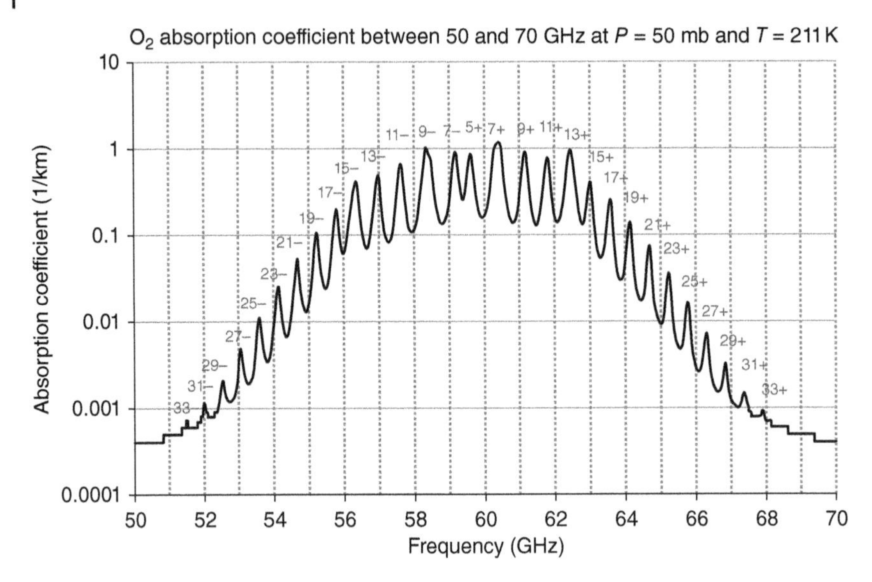

Figure 2.1 Oxygen absorption coefficients between 50 and 70 GHz at pressure of 50 mb and temperature of 211 K. Labeled in numeric are the resonant frequency locations where the magnetic-dipole transitions occur with +sign for quantum number J from N to $N+1$ and −sign from N to $N-1$.

Table 2.1 Frequency location with the magnetic dipole quantum number.

$N-$	Frequency	$N+$	Frequency
1−	118.7503	1+	56.2648
3−	62.4863	3+	58.4466
5−	60.3061	5+	59.5910
7−	59.1642	7+	60.4348
9−	58.3239	9+	61.1506
11−	57.6125	11+	61.8002
13−	56.9682	13+	62.4112
15−	56.3634	15+	62.9980
17−	55.7838	17+	63.5685
19−	55.2214	19+	64.1278
21−	54.6712	21+	64.6789
23−	54.1300	23+	65.2241
25−	53.5958	25+	65.7648
27−	53.0670	27+	66.3021
29−	52.5424	29+	66.8368
31−	52.0215	31+	67.3695
33−	51.5034	33+	67.9008

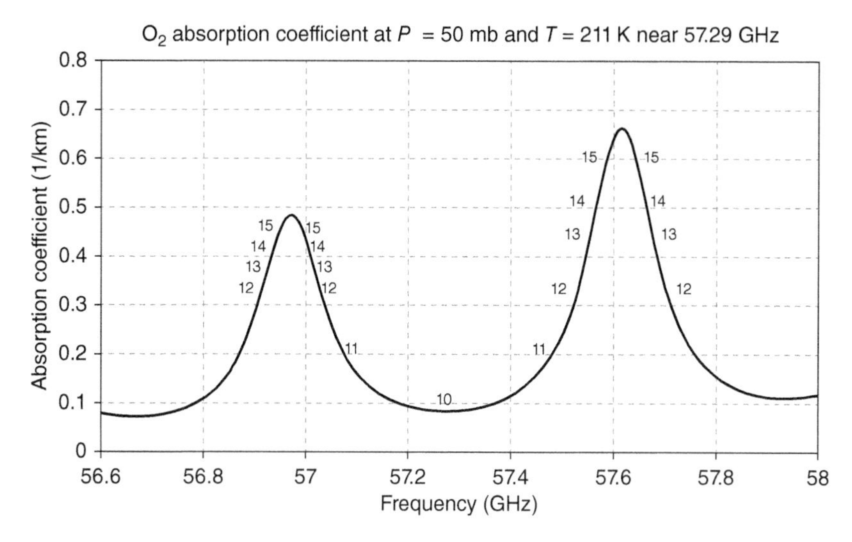

Figure 2.2 Oxygen absorption coefficient near the central frequency of 57.29 GHz at pressure of 50 mb and temperature of 211 K. Labeled in numeric are the frequency locations at ATMS channels 10–15.

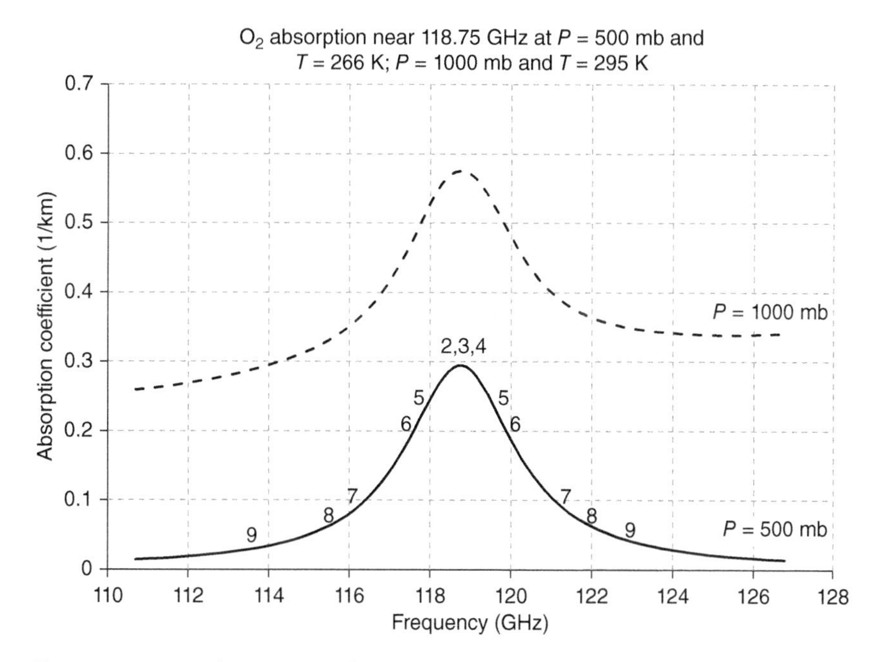

Figure 2.3 Oxygen absorption coefficient near 118 GHz at pressure of 50 mb and temperature of 211 K. The channels of Microwave Humidity Sounder (MWHS) on-board Fengyun-3C satellite are also indicated.

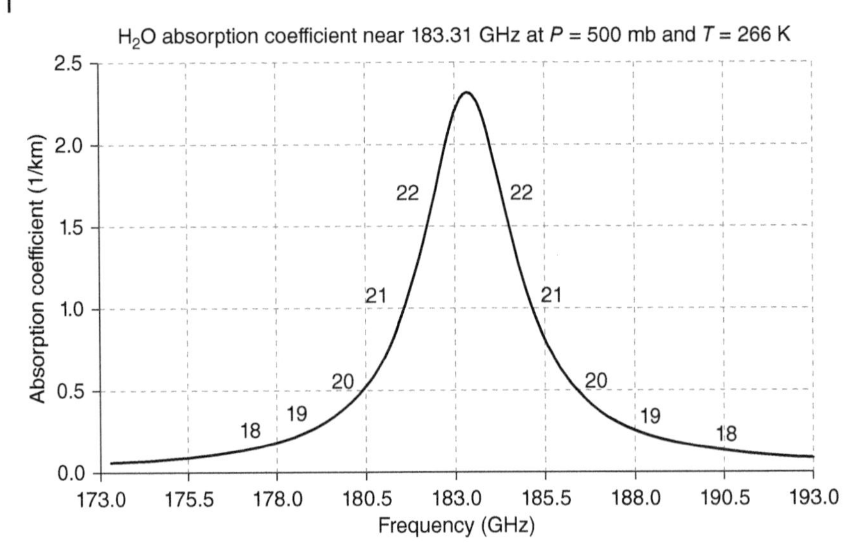

Figure 2.4 Volumetric oxygen absorption coefficient near 118 GHz at pressure of 50 mb and temperature of 211 K. The channel locations for MWHS are also indicated.

However, the unbalanced subband receiver gains can introduce other issues such as bias, if not properly characterized. It is shown that for channels 13–15, the absorption coefficients at the subbands near 56.968 GHz are different from those at 57.612 GHz. Thus, the radiative components received at the four subbands arise from different atmospheric levels. A fast radiative transfer model parameterized at the central frequency for accurate calculation must carefully reflect the subtle effects from the instrument subband design.

There is growing interest in using 118.75 GHz oxygen absorption lines for space remote sensing (Figures 2.3 and 2.4). However, the absorption line is often complicated by the continuum water absorption, as discussed in Section 2.2.3.

2.2.3 Water Vapor Absorption

The absorption coefficient for water vapor is a function of frequency (f in GHz), pressure (p in mb), temperature (T in K), and amount of H_2O derived from water vapor density (ρ_v in g/M^3). It is given by Rosenkranz [27] as follows:

$$\kappa^{H_2O}(f, p, T, \rho_v) = \kappa_f^{H_2O,con} + 10^{-4} \cdot n_f \sum_{j\geq 0} S_j(T) \cdot F(f, f_{0,j}). \tag{2.15}$$

The constant 10^{-4} is a conversion from 10^{-9} GHz/Hz and 10^5cm/km. The water number density, n_w, is given in units of molecules cm^{-3} by

$$n_w = 3.343 \times 10^{16} \rho_v. \tag{2.16}$$

The continuum absorption coefficient, $\kappa_f^{H_2O,con}$, is also developed by Rosenkranz [27] as

$$\kappa_f^{H_2O,con} = p_{H_2O}(4.74 \times 10^{-10} p_{dry}\theta^3 + 1.50 \times 10^{-8} p_{H_2O} \times \theta^{10.5}) f^2. \tag{2.17}$$

The strength term, $S_j(T)$, is given by Eq. (2.14) with $Q_{rot} = \theta^{2.5}$, and $f_0(j)$, $S'(j)$, and $b(j)$ are given by Rosenkranz [22]. For $p \geq 0.1$ mb, the pressure-broadened VVW line shape is

$$F(f, f_{0,j}) = \frac{1}{\pi}\left(\frac{f}{f_{0,j}}\right)^2 \left[\frac{\gamma_c}{(f - f_{0,j})^2 + \gamma_c^2} + \frac{\gamma_c}{(f + f_{0,j})^2 + \gamma_c^2}\right] \tag{2.18}$$

and

$$\gamma_c(j) = \frac{w(j)}{1000}(p_{dry}\theta^{x(j)} + 4.8 \times p_{H_2O} \times \theta^{0.8}) \tag{2.19}$$

2.2.4
Nitrogen and Ozone Absorption

The nitrogen absorption coefficient in nepers/km (note: dB/nepers = 8.68588, nepers = ln(a/b), db = 20 log(a/b)) is given as a function of frequency (f in GHz), pressure (p in mb), and temperature (T in K) as

$$\kappa^{N_2}(f, p, T) = 6.4 \times 10^{-14} \times p^2 \times f^2 \times (300/T)^{3.55}, \tag{2.20}$$

and the absorption coefficient for ozone is a function of frequency (f in GHz), pressure (p in mb), temperature (T in K), and amount of O_3 derived from ozone density, ρ_o

$$\kappa^{O_3}(f, p, T, \rho_o) = 10^{-4} \times n_{O_3} \times \sum_{j \geq 0} S_j(T) \times F(f, f_{0,j}), \tag{2.21}$$

where the ozone number density, $n_{O_3} = 1.255 \times 10^{16} \rho_{O_3}$, and $S_j(T)$ and $F(f, f_{0,j})$ are the line intensity and line shape, respectively, for O_3 absorption [27]. The ozone lines dominate in the stratosphere and are quite narrow ($\approx$ 60 MHz). The net effect of the ozone lines on a broadband instrument is negligible in a typical microwave remote sensing problem.

2.2.5
Line-by-Line Radiative Transfer Model (LBLRTM)

LBLRTM was developed for computing the transmittance and radiances across all the spectra [11]. Figure 2.5 shows an atmospheric transmittance spectrum

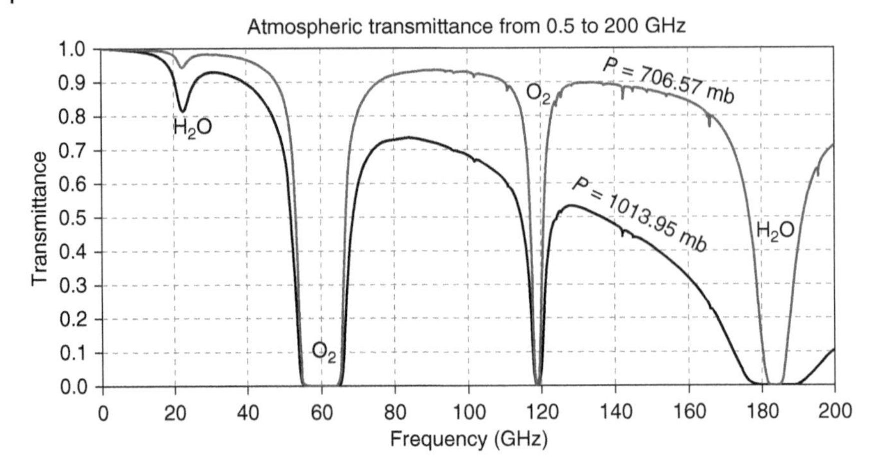

Figure 2.5 Atmospheric transmittance spectrum computed from the U.S. standard atmosphere using line-by-line radiative transfer model (LBLRTM) [11]. The HITRAN spectroscopy data base is used in LBLRTM calculation.

from 0 to 200 GHz at the U.S. standard atmospheric pressure levels of 1013.95 and 706.57 mb, respectively. The frequency allocation for most of the satellite microwave instruments is based on the transmittance spectrum as indicated. For an instrument located at the higher transmittance region, it allows for sensing the surface or lower atmospheric features. To obtain atmospheric temperature and water vapor profiles, we need to select a spectral range within which the transmittance strongly varies with frequency. In the frequency domain, there are several water vapor absorption regions near 22.23, 183, 325, and 380 GHz, which are associated with variable transmittance values. Two oxygen absorption bands are located near 60 and 118 GHz, respectively. At 100 mb, the smaller transmittance values at frequencies above 100 GHz are due to ozone absorption.

2.2.6 Zeeman Splitting Absorption

If magnetic fields are present, the atomic energy levels are split into a larger number of levels, and the spectral lines are also split. This splitting is called the Zeeman effect and can be characterized by the orbital angular momentum quantum number, J, of the atomic level. J can take nonnegative integer values, and the number of split levels in the magnetic field is $2J + 1$.

Atomic physicists use the abbreviation "S" for a level with $J = 0$, "p" for $J = 1$, and "d" for $J = 2$, and so on. It is also common to precede this designation with the integer principle quantum number, n. Thus, the designation "2p" means a level that has $n = 2$ and $J = 1$. For an atom having its lowest level of an "S" level, it has $J = 0$ and $2J + 1 = 0$, so it is not split in the magnetic field, while the first excited state has $J = 1$ ("p" level), so it is split into $2J + 1 = 3$ levels by the magnetic field. Thus, a single transition is split into three transitions by the magnetic field, according to

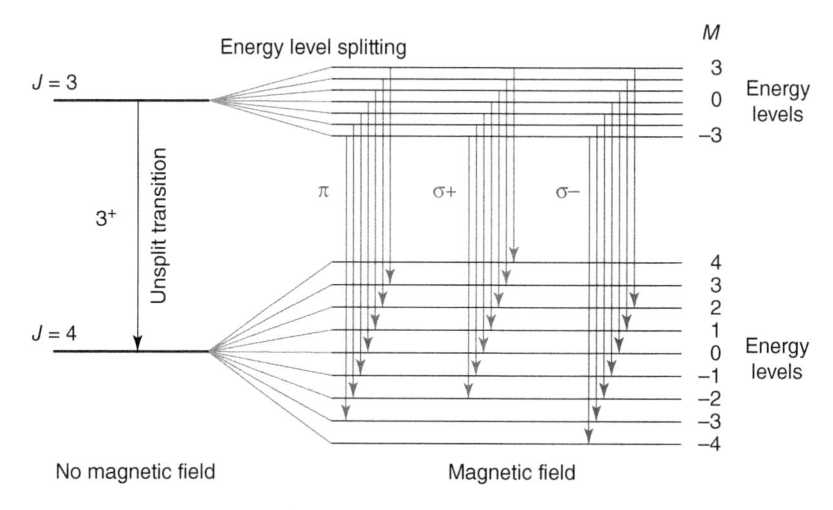

Figure 2.6 Energy transition formulating three major categories of lines.

the azimuthal quantum number $M = -J, \dots, 0, \dots, J$. The selection rules permit transitions with $\Delta J = \pm 1$ and $\Delta M = 0, \pm 1$, or

$$\Delta M = \begin{cases} 0 \to \pi \\ +1 \to \sigma^+ . \\ -1 \to \sigma^- \end{cases} \tag{2.22}$$

Figure 2.6 displays the energy level transition, and the resulting absorption lines are shown in Figure 2.7 [29]. The line center frequency of a single Zeeman component (in MHz) is

$$v_{z,N^\pm} = v_{N^\pm} + 2.8026 B_e \alpha(N^\pm, M, \Delta M), \tag{2.23}$$

where v_0 is the center frequency of the unsplit resonance line corresponding to $N^\pm$, B_e is the Earth's magnetic field strength in Gauss (~0.22–0.65 near the Earth's surface), and α is the coefficient that depends on N, M, and ΔM [21] as follows:

$$\alpha(N^+, M, +1) = \frac{1}{(N+1)}\left(1 + M\frac{N-1}{N}\right), \tag{2.24a}$$

$$\alpha(N^-, M, +1) = -\frac{1}{N}\left(1 + M\frac{N+2}{N}\right), \tag{2.24b}$$

$$\alpha(N^+, M, 0) = \frac{M}{(N+1)}\frac{N-1}{N}, \tag{2.24c}$$

$$\alpha(N^-, M, 0) = -\frac{M}{N}\frac{N+2}{N+1}, \tag{2.24d}$$

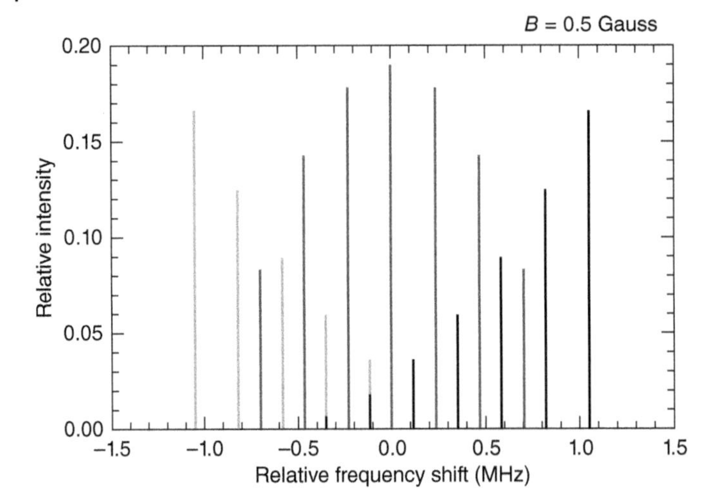

Figure 2.7 Absorption line intensity from Zeeman splitting effect for 3+ lines under a magnetic field of $B = 0.5$ Gauss. The moderate black lines correspond to the azimuthal quantum number $M = 0$ (π), light gray lines to $M = 1$ (σ^+), and dark black lines to $M = -1$ (σ^-).

$$\alpha(N^+, M, -1) = \frac{1}{(N+1)}\left(-1 + M\frac{N-1}{N}\right), \tag{2.24e}$$

$$\alpha(N^-, M, -1) = -\frac{1}{N}\left(-1 + M\frac{N+2}{N+1}\right). \tag{2.24f}$$

The line intensities of radiation emitted by the π and $\sigma\pm$, components are given by

$$I_{\sigma\pm} \sim |\mu(N, M, \Delta J, \pm 1)|^2 \rho_{\sigma\pm}, \tag{2.25a}$$

and

$$I_{\pi} \sim |\mu(N, M, \Delta J, 0)|^2 \rho_{\pi}, \tag{2.25b}$$

respectively, where θ is the angle between the magnetic field direction and the direction of propagation, and

$$\rho_{\sigma\pm} = \begin{bmatrix} 1 & \mp i\cos\theta \\ \pm i\cos\theta & \cos^2\theta \end{bmatrix}, \tag{2.26a}$$

$$\rho_{\pi} = \begin{bmatrix} 0 & 0 \\ 0 & \sin^2\theta \end{bmatrix}, \tag{2.26b}$$

$$|\mu(N, M, +1, +1)|^2 = \frac{3N(N+M+1)(N+M+2)}{4(N+1)^2(2N+1)}, \tag{2.26c}$$

$$|\mu(N,M,-1,+1)|^2 = \frac{3(N+1)(N-M)(N-M-1)}{4N^2(2N+1)}, \tag{2.26d}$$

$$|\mu(N,M,+1,0)|^2 = \frac{3N[(N+1)-M^2]}{(N+1)^2(2N+1)}, \tag{2.26e}$$

$$|\mu(N,M,-1,0)|^2 = \frac{3(N+1)(N^2-M^2)}{4N^2(2N+1)}, \tag{2.26f}$$

$$|\mu(N,M,+1,-1)|^2 = \frac{3N(N-M+1)(N-M+2)}{4(N+1)^2(2N+1)}, \tag{2.26g}$$

$$|\mu(N,M,-1,-1)|^2 = \frac{3(N+1)(N+M)(N+M-1)}{4N^2(2N+1)}. \tag{2.26h}$$

The effect of the Zeeman splitting on the 3+ resonance line is shown in Figure 2.7.

Since the Zeeman splitting occurs in the upper atmosphere, the split line spectrum is also broadened by Doppler shifting and pressure-induced collision. To take these two processes into account, Lenior [21] obtained the combined absorption profile by convolving the collision-broadened line-shape function of van Vleck–Weisskopf [23] with the Doppler line-shape function and then derived the total attenuation for σ and π components of given N^+ and N^- transition lines as

$$A_{N\pm,\sigma\pm}(\nu) = C\frac{p_{mb}\nu^2}{T^3}e^{-E_N/T}\rho_{\sigma\pm} \times \sum_{M=-N}^{N} |\mu(N,M,\pm1,\pm1)|^2$$
$$\times F(\nu,\nu_0,\Delta\nu_c,\Delta\nu_D), \tag{2.27a}$$

$$A_{N\pm,\pi}(\nu) = C\frac{p_{mb}\nu^2}{T^3}e^{-E_N/T}\rho_{\pi} \times \sum_{M=-N}^{N} |\mu(N,M,\pm1,0)|^2$$
$$\times F(\nu,\nu_0,\Delta\nu_c,\Delta\nu_D), \tag{2.27b}$$

where p_{mb} is the pressure in millibars; ν is the frequency in GHz; T is the temperature in K; E_n is the energy of the nth rotational state and equals 2.0685N $(N+1)$ in K; F is the Voigt function with the frequency being the shifted resonance frequency, as described in Eq. (2.23); and $C = 0.229$. The total attenuation matrix (in the unit of km^{-1}) for a given transition line is given by Lenior [21]:

$$A_{N^+}(\nu) = A_{N^+,\sigma^+}(\nu) + A_{N^+,\sigma^-}(\nu) + A_{N^+,\pi}(\nu). \tag{2.28}$$

Overall, the Zeeman-induced line spectrum overlays on the major resonant absorption line, as shown in Figure 2.8, at 61.15056 GHz. The transmittance spectrum without the magnetic field is a typical resonant shape through Doppler broadening. Under the magnetic field, the resonant shape flattens with small local resonant shapes. In addition, the angle between the wave propagation direction

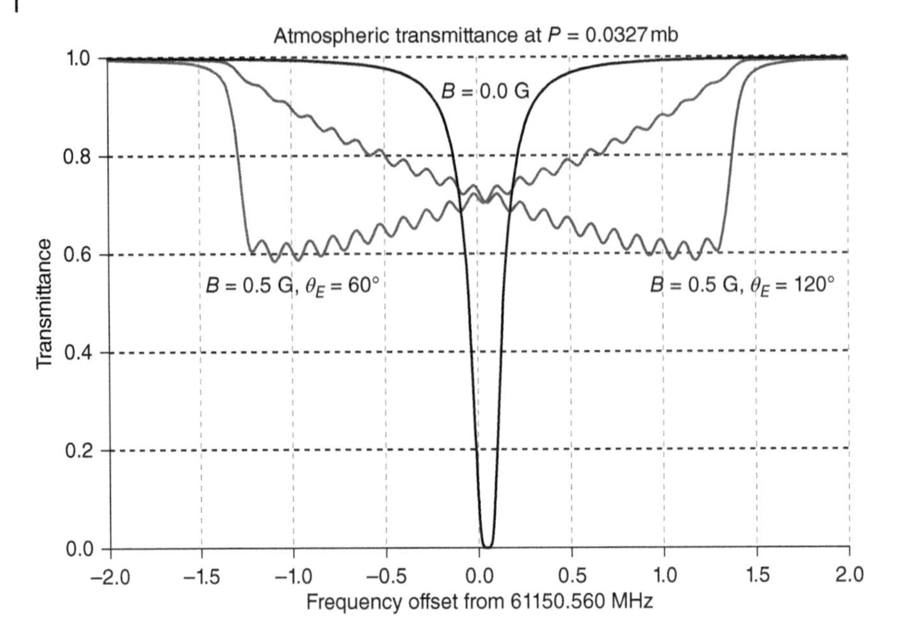

Figure 2.8 Oxygen absorption at a pressure of 0.0327 mb near 61.1506 GHz for unsplit (black) and Zeeman split lines of right-hand circularly polarized radiation for an angle of 60° and 120° (gray curves) between the propagation direction and magnetic field vector. The magnetic intensity is 0.5 Gauss. For a left-hand circularly polarized radiation, the shape is symmetric to the left-hand circularly polarized pattern.

and the magnetic vector further affects the Zeeman splitting spectrum. The double sides of the frequencies near each resonant line can be used to reduce the impact of the Zeeman splitting on microwave sounding.

The Zeeman splitting effects are important applications in upper atmospheric remote sensing. AMSU-A channel 14 and ATMS channel 15 are located near O_2 rotational lines 11− and 13−. SSMIS channel 19 has two narrow passbands centered on O_2 rotational lines 15+ and 17+, and SSMIS channel 20 has two passbands on lines 7+ and 9+. SSMIS channels 21 and 24 have four passbands each, paired and situated symmetrically on the opposite sides of the transition lines 7+ and 9+ with increased bandwidths and band frequency offsets in the line centers as the channel number increases. These channels are mostly affected by the Zeeman splitting absorption, and the simulations of the atmospheric transmittance at these channels must take the Zeeman splitting absorption profiles into account, as shown in Figure 2.8 [29].

2.2.7 Parameterized Transmittance Model

In satellite data assimilation system and remote sensing applications, atmospheric gaseous absorption must be calculated fast and accurately. A common approach is to derive a parametric relationship between the transmittance and pressure, temperature, and water vapor mixing ratio.

The polychromatic gas absorption model begins with the channel layer-to-space transmittance, Υ_{ch}, which is defined as the convolution of the monochromatic transmittance, $\Upsilon(\nu)$, with the channel spectral response function (SRF), $\phi(\nu)$:

$$\Upsilon_{ch} = \int_{\Delta\nu} \Upsilon(\nu)\phi(\nu)\mathrm{d}\nu. \tag{2.29}$$

Let $\Upsilon_w(\nu)$ and $\Upsilon_o(\nu)$ be the monochromatic transmittances of water vapor and ozone, respectively, and $\Upsilon_d(\nu)$ the transmittance of the dry gas, which is a collective component including all the absorbing gases except water vapor and ozone. Then, the total monochromatic transmittance can be expressed as the product of these three components:

$$\Upsilon(\nu) = \Upsilon_w(\nu)\Upsilon_o(\nu)\Upsilon_d(\nu). \tag{2.30}$$

The total channel transmittance Υ_{ch} defined in (2.29) can also be expressed in a similar form through an effective transmittance:

$$\Upsilon_{ch} = \Upsilon_{ch,w}\Upsilon^*_{ch,o}\Upsilon^*_{ch,d}, \tag{2.31}$$

where $\Upsilon_{ch,w}$ is the channel transmittance of water vapor and is defined as

$$\Upsilon_{ch,w} = \int \Upsilon_w(\nu)\phi(\nu)\mathrm{d}\nu, \tag{2.32}$$

and $\Upsilon^*_{ch,d}$ and $\Upsilon^*_{ch,o}$ are the effective channel transmittances of the dry gas and ozone, respectively. The effective dry gas $\Upsilon^*_{ch,d}$ is defined as

$$\Upsilon^*_{ch,d} = \Upsilon_{ch,d+w}/\Upsilon_{ch,w}, \tag{2.33}$$

where $\Upsilon_{ch,d+w}$ is the channel transmittance of the combined dry gas and water vapor,

$$\Upsilon_{ch,d+w} = \int \Upsilon_d(\nu)\Upsilon_w(\nu)\phi(\nu)\mathrm{d}\nu, \tag{2.34}$$

and the effective ozone transmittance $\Upsilon^*_{ch,o}$ is defined as

$$\Upsilon^*_{ch,o} = \Upsilon_{ch}/\Upsilon_{ch,d+w}. \tag{2.35}$$

Equation (2.31) is used to derive the channel transmittance. The three transmittance components, $\Upsilon_{ch,w}$, $\Upsilon^*_{ch,d}$, and $\Upsilon^*_{ch,o}$, are estimated using the regression technique. For simplicity, let the index i represent water vapor, ozone, or dry gas and $\Upsilon_{ch,i}(A_i)$ represent one of the three transmittance components, $\Upsilon_{ch,w}$, $\Upsilon^*_{ch,d}$, and$\Upsilon^*_{ch,o}$, at the level with the integrated absorber amount A_i (from space to the pressure level p), which is computed as

$$A_i = \int_0^p \frac{r_i}{g\cos(\theta)} \mathrm{d}p', \tag{2.36}$$

where r_i is the gas specific amount, θ the zenith angle, and g the gravitation constant. With the symbols defined, the transmittance is calculated as

$$\Upsilon_{ch,i}(A_i) = e^{-\int_0^{A_i} k_{ch,i}(A_i')\mathrm{d}A'}, \tag{2.37}$$

where

$$\ln(k_{ch,i}(A_i)) = c_{i,0}(A_i) + \sum_{j=1}^{6} c_{i,j}(A_i)x_{i,j}(A_i). \tag{2.38}$$

In Eqs. (2.37) and (2.38), $k_{ch,i}(A_i)$ is the absorption coefficient and ln() is the natural logarithm. The predictors $x_{ij}(A_i)(j = 1, 6)$ are the functions of atmospheric state variables, and the coefficients $c_{i,0}(A_i)$ and $c_{i,j}(A_i)$ are polynomial functions of A_i in the form

$$c_{i,j}(A_i) = \sum_{n=0}^{N} a_{i,j,n} \ln{(A_i)^n}, \tag{2.39}$$

where $a_{i,j,n}$ are the regression coefficients (also referred to as transmittance coefficients). The set of six predictors varies among different channels and is selected from a 29-predictor pool, as listed in Table 2.2. The predictor pool includes 11 standard predictors, which are not specific to any of the three transmittance components, and 18 integrated predictors, which are evenly divided into three subsets, each belonging to a particular transmittance component. Let u represent the atmosphere pressure P or temperature T; the integrated predictors for the component i may be expressed as

$$u_i^*(A_i) = \frac{\int_0^{A_i} u(A_i')\mathrm{d}A_i'}{\int_0^{A_i} \mathrm{d}A_i'}, \tag{2.40a}$$

$$u_i^{**}(A_i) = \frac{\int_0^{A_i} u(A_i')A_i'\mathrm{d}A_i'}{\int_0^{A_i} A_i'\mathrm{d}A_i'}, \tag{2.40b}$$

$$u_i^{***}(A_i) = \frac{\int_0^{A_i} u(A_i')A_i'^2\mathrm{d}A_i'}{\int_0^{A_i} A_i'^2\mathrm{d}A_i'}. \tag{2.40c}$$

Table 2.2 Predictors for the optical thickness at SSMIS Zeeman splitting channels.

Channel	Predictors
19,20	$\theta, \theta B^{-1}, \cos^2(\theta_B), \theta\cos^2(\theta_B), B^{-1}, B^{-2}, \cos^2(\theta_B)B^2$
21,22	$\theta, \cos^2(\theta_B), B, B^3, \cos^2(\theta_B)B, \cos^2(\theta_B)B^2$
23	$\theta, \theta^2, \cos^2(\theta_B), B$
24	θ, θ^2

$\theta = T/300$, T is temperature in Kelvin; B, the Earth's magnetic field strength; θ_B, angle between magnetic field and propagation direction.

The transmittance coefficients, $a_{i,j,n}$, in Eq. (2.39) are obtained through a training process with a statistical data ensemble, in which the predictands and predictors are calculated from a set of diversified atmospheric profiles. For the dry gas component, the mixing ratio profile does not change in different atmospheric states. Because of this, the dry gas is also called fixed gas. An exhausting search is performed for each gas component and channel to select the best set of predictors and order $N(\leq 10)$ of the polynomial function, which minimize the fitting residual. Low order is considered if the fitting accuracy is not degraded significantly for better computational stability. In addition, an automated procedure is adopted to make sure that the set of predictors with strong correlations between the selected predictors is not selected, which may cause the transmittance calculation to become unstable (Table 2.3).

The fitting errors for AMSU channels on NOAA 16 are shown in Figure 2.9. The fitting errors are measured with the brightness temperature calculated with the radiative transfer under a clear-sky condition. On average, the errors are less than 0.1 K.

Using Rosenkranz's models [22], Han *et al.* [29] parameterized an averaged Zeeman optical thickness within an SSMIS frequency passband and predicted

Table 2.3 Standard and integrated predictors.

Standard predictors		Integrated predictors			
1	T	1	T_w^*	12	P_o^{***}
2	P	2	T_w^{**}	13	T_d^*
3	T^2	3	T_w^{***}	14	T_d^{**}
4	P^2	4	P_w^*	15	T_d^{***}
5	$T\,P$	5	P_w^{**}	16	P_d^*
6	$T^2\,P$	6	P_w^{***}	17	P_d^{**}
7	$T\,P^2$	7	T_o^*	18	P_d^{***}
8	$T^2\,P^2$	8	T_o^{**}		
9	$\sqrt[4]{P}$	9	T_o^{***}		
10	Q	10	P_o^*		
11	$Q/\sqrt{T}$	11	P_o^{**}		

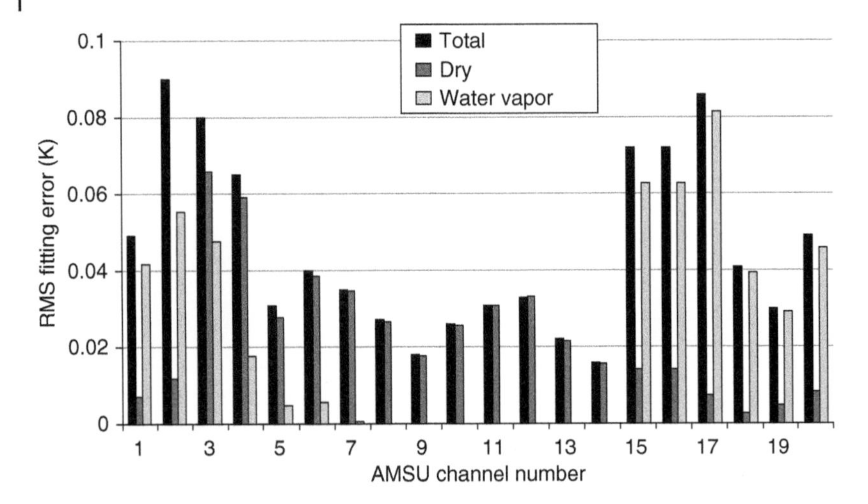

Figure 2.9 RMS fitting errors for AMSU on NOAA-16.

with predictors of atmospheric temperature, geomagnetic field strength, and angle between the geomagnetic field vector and the electromagnetic wave propagation direction. The coefficients of each predictor are trained with an LBL model [11] that accurately computes the absorption at each Zeeman splitting frequency using 48 atmospheric profiles. It is shown that

$$\tau_i = c_{i,0} + \sum_{j=1}^{m} c_{i,j} X_{i,j}, \tag{2.41}$$

where m is the number of predictors. For SSMIS affected by Zeeman splitting, Han *et al.* [29] used the predictors as listed in Table 2.2.

The fast model produced very accurate computations for brightness temperature simulations at SSMIS upper-air sounding channels, compared with LBL computations (see Figure 2.10).

2.3
Cloud Absorption and Scattering

2.3.1
Scattering Parameters

It is worth mentioning about the general particle scattering from Mie theory [30]. For a sphere having a radius of r, the scattering efficiency (dimensionless) can be written as

$$Q_s = \frac{2}{x^2} \sum_{n=1}^{\infty} (2n+1)(|a_n|^2 + |b_n|^2), \tag{2.42}$$

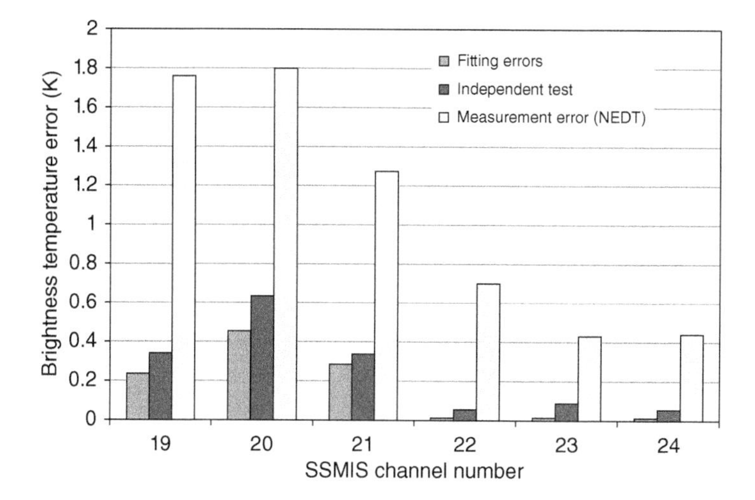

Figure 2.10 The performance of fast Zeeman splitting models.

and the extinction efficiency is

$$Q_e = \frac{2}{x^2}\sum_{n=1}^{\infty}(2n+1)\mathrm{Re}(a_n + b_n) \tag{2.43}$$

where $x = 2\pi r/\lambda$ is the size parameter and λ is the wavelength of incident radiation. For an arbitrary shape of particle, the particle scattering phase matrix **S** has 16 components and can be generalized as follows:

$$\mathbf{S} = k\begin{bmatrix} S_{11} & S_{12} & S_{13} & S_{14} \\ S_{21} & S_{22} & S_{23} & S_{24} \\ S_{31} & S_{32} & S_{33} & S_{34} \\ S_{41} & S_{42} & S_{43} & S_{44} \end{bmatrix}, \tag{2.44}$$

where k is the normalization factor. In general, the elements in **S** are of nonzero values for nonspherical particle scattering. For a spherical particle, the scattering matrix is expressed as

$$S = \frac{\lambda^2}{\pi\sigma_s}\begin{bmatrix} S_{11} & S_{12} & 0 & 0 \\ S_{12} & S_{11} & 0 & 0 \\ 0 & 0 & S_{33} & S_{34} \\ 0 & 0 & -S_{34} & S_{33} \end{bmatrix}, \tag{2.45}$$

where $\sigma_s = \pi r^2 Q_s$ is the particle scattering cross section. Each element, S_{ij}, in Eq. (2.45) depends on the Mie scattering functions, which is related by

$$S_{11}(\Theta) = S_1(\Theta)S_1^*(\Theta) + S_2(\Theta)S_2^*(\Theta), \tag{2.46a}$$

$$S_{12}(\Theta) = S_2(\Theta)S_2^*(\Theta) - S_1(\Theta)S_1^*(\Theta), \tag{2.46b}$$

$$S_{33}(\Theta) = \mathrm{Re}[S_2(\Theta)S_1^*(\Theta)], \tag{2.46c}$$

$$S_{34}(\Theta) = \mathrm{Im}[S_2(\Theta)S_1^*(\Theta)], \tag{2.46d}$$

where Θ is the scattering angle, and Re and Im are used to derive the real and imaginary parts of a given complex value. The scattering functions $S_1(\Theta)$ and $S_2(\Theta)$ are derived from the Mie scattering theory.

The aforementioned equations are only applied for the scattering phase matrix of a single spherical particle. When the radiation transfers in a medium containing spherical particles with different radii, the radiation interference among the particles can be neglected if separation distances between the particles are assumed to be sufficiently large in comparison to the electromagnetic wavelength. The elements of the scattering matrix for polydispersed spherical particles with a distribution density, $n(r)$, are derived by adding individual elements of the single particle at every scattering angle, that is,

$$S_{ij}(\Theta) = \int_0^{\infty} s_{ij}(\Theta, r)n(r)dr, \quad ij = 11,22,33,34, \tag{2.47}$$

where r is the particle diameter. Note that the phase matrix of polydispersed particles is similar to that in Eq. (2.45) but should be scaled with $\frac{\lambda^2}{\pi\beta_s}$, where β_s is the scattering coefficient with a unit of inverse length. In addition, the scattering and extinction coefficients, (β_e), and single-scattering albedo (ω) are derived from

$$\beta_s = \int_0^{\infty} \sigma_s(r)n(r)\mathrm{d}r, \tag{2.48}$$

$$\beta_e = \int_0^{\infty} \sigma_e(r)n(r)\mathrm{d}r, \tag{2.49}$$

$$\omega = \frac{\beta_s}{\beta_e}, \tag{2.50}$$

where σ_s and σ_e are the scattering and extinction cross sections, respectively. For a single spherical particle, the extinction and scattering cross sections are given according to Mie's theory [30]. For nonspherical particles having their maximum dimension, which are randomly oriented in space, the phase matrix only has six independent nonzero elements and takes the same form as Eq. (2.45) [23].

2.3.2
Particle Size Distribution

In computing the scattering and absorption parameters, the size spectrum of hydrometeors in Eqs. (2.47)–(2.49) can be modeled by a modified gamma distribution

$$n(D) = N_0(\Lambda D)^P \exp(-(\Lambda D)^Q), \tag{2.51}$$

where $n(D)$ ($m^{-3}mm^{-1}$) is the number of particles with the equivalent spherical diameters between D and $D + dD$ (mm), and N_0, Λ (mm^{-1}), P, and Q are free parameters. The mean diameter D, diameter variance σ_D, mode diameter D_M, total number density of particles n_0, fractional volume f, and water (liquid or ice) content M follow from Eq. (2.51):

$$D = \frac{1}{\Lambda}\frac{\Gamma\left(\frac{P+2}{Q}\right)}{\Gamma\left(\frac{P+1}{Q}\right)}\ (\text{mm}), \tag{2.52a}$$

$$\sigma_D = \frac{1}{\Lambda}\sqrt{\frac{\Gamma(P+3/Q)}{\Gamma(P+1/Q)} - \frac{\Gamma^2(P+2/Q)}{\Gamma^2(P+1/Q)}}\ (\text{mm}), \tag{2.52b}$$

$$D_M = \frac{1}{\Lambda}\frac{P}{Q}^{1/Q},\ (\text{mm}), \tag{2.52c}$$

$$n_0 = \frac{N_0}{\Lambda Q}\Gamma\left(\frac{P+1}{Q}\right)\ (\text{m}^{-3}), \tag{2.52d}$$

$$f = \pi \times 10^{-9}\frac{N_0}{6\Lambda^4 Q}\Gamma\left(\frac{P+4}{Q}\right), \tag{2.52e}$$

$$M = 10^6 f\ (\text{g/mm}^3) \tag{2.52f}$$

where $\Gamma(n+1) = n!$ is the gamma function. The parameter P describes the rate of increase of particles at small diameters, and the parameter Q describes the falloff in particle concentration at large diameters. When computing these parameters using numerical quadrature, care must be taken so that the Mie resonances are adequately sampled within the region of support of the particle size spectrum and that concentrations up to several mean particle diameters are included.

The natural size distribution of raindrops was defined by Marshall and Palmer (MP) as a simple exponential law over a wide range of meteorological conditions and precipitation rates, that is,

$$n(D) = N_0 \exp(-\Lambda D), \tag{2.53}$$

where

$$N_0 = 8 \times 10^3\ (\text{m}^{-3}\text{mm}^{-1}), \tag{2.54}$$

and

$$\Lambda = 4.1R^{-0.21}(\text{mm}^{-1}), \tag{2.55}$$

where R is the rain rate in mm/h. Similar exponential distributions were measured by Laws and Pawson and by Joss *et al.*, who classified the size spectra according to the type of precipitation: drizzle, widespread, or thunderstorm. Sekhon and Srivastava (SS) derived an exponential relationship for the equivalent liquid-sphere size distribution of snowflakes near the ground with the different parameters

$$N_0 = 2.50 \times 10^3 R^{-0.94}(\text{m}^{-3}\text{mm}^{-1}), \tag{2.56}$$

and

$$\Lambda = 2.29R^{-0.45}(\text{mm}^{-1}), \tag{2.57}$$

where R is the equivalent liquid-water precipitation rate in mm/h. Theoretical calculations of spherical hydrometeor absorption and scattering coefficients using the MP size distribution for liquid and the SS size distribution for ice are shown, for various precipitation rates at 0 °C, in Figure 2.11. The liquid calculations compare favorably with the independent calculation by Salvage, as indicated by specific points in Figure 2.11.

In microwave remote sensing of ice clouds, a full set of single-scattering properties of nonspherical particles are used in radiative transfer calculations [31–34].

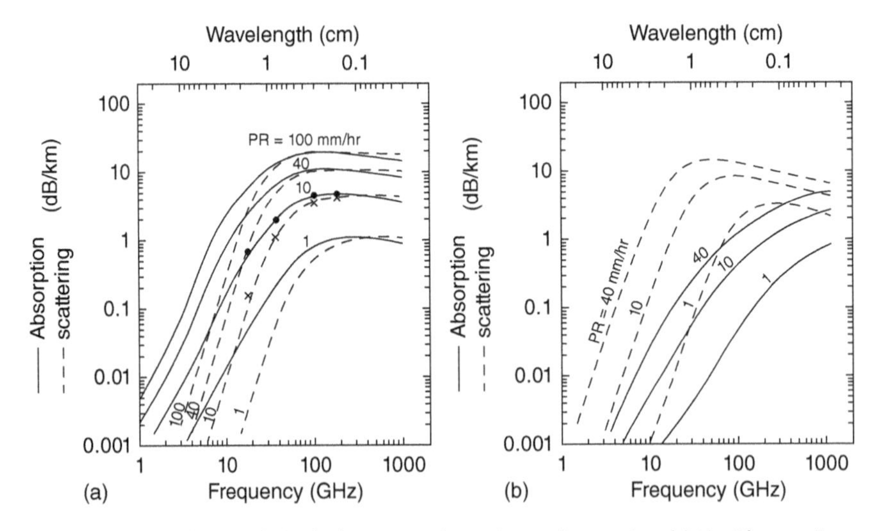

Figure 2.11 Polydispersed Mie hydrometeor absorption and scattering. (a) Liquid, assuming a Marshall–Palmer drop-size distribution. Computations from Salvage for absorption (•) and scattering (×) are plotted for comparison. (b) Ice, assuming a Sekhon and Srivastava distribution. Calculations are shown for precipitation rates of 1, 10, and 40 mm/h for both phases and 100 mm/h for liquid.

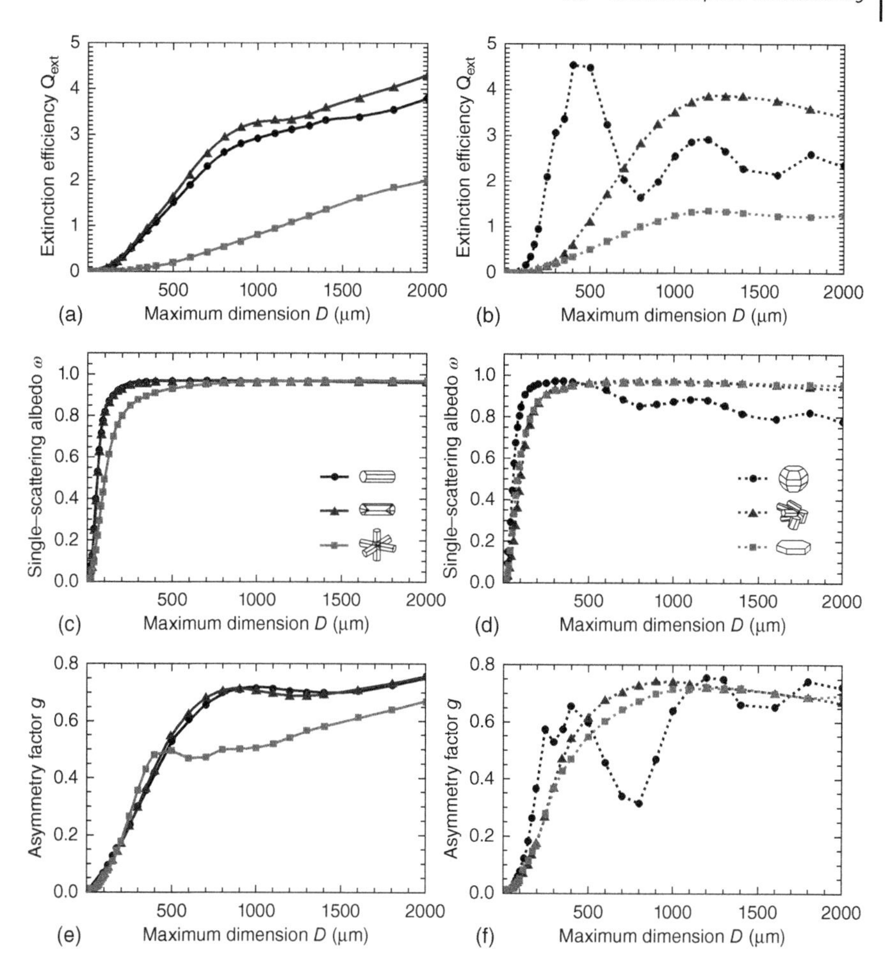

Figure 2.12 Extinction efficiency Q_{ext}, single-scattering albedo ω, and asymmetry factor g as a function of maximum dimension D at 640 GHz for the six nonspherical ice habits. (Hong *et al.* 2009 [34]. Reproduced with permission of Wiley.)

For six nonspherical ice habits (e.g., hexagonal solid and hollow columns, hexagonal plates, 3D six-branch solid bullet rosettes, aggregates, and droxtals) are computed from the discrete dipole approximation (DDA) [34]. Figure 2.12a–f shows the extinction efficiency, single-scattering albedo, and asymmetry factor as a function of maximum dimension size. In general, the single-scattering properties differ between the various habits except for the hexagonal solid and hollow columns, which have similar features. In the particle size range considered in this study, pronounced oscillations of scattering parameters tend to occur for droxtals at high frequencies. The distribution of phase matrix elements with scattering angles is also derived and can be found in the studies [33, 34].

To obtain the bulk-scattering properties that may be more representative of ice clouds, the single-scattering properties of the individual particles are integrated over the particle size distribution (PSD) for an assumed habit mixture. The effective particle size D_e (e.g., [35–37]) is computed as

$$D_e = \frac{3}{2}\frac{\int_{D_{\min}}^{D_{\max}} \left[\sum_{i=1}^{M} f_i(D)\,V_i(D)\right] N(D)\mathrm{d}D}{\int_{D_{\min}}^{D_{\max}} \left[\sum_{i=1}^{M} f_i(D)\,S_i(D)\right] N(D)\mathrm{d}D}, \tag{2.58}$$

where $N(D)$ is the number density of ice particles with D, $\sum_{i=1}^{M} f_i(D) = 1$, $f_i(D)$ is the ice particle habit percentage for habit i at D for up to M habits, $V_i(D)$ and $S_i(D)$ are the volume and project area, respectively, of the habit i for a given D, and $D_{\min}$ and $D_{\max}$ are the minimum and maximum sizes of D in the given $N(D)$, respectively.

Bulk ice-cloud scattering properties were first developed by assuming that the cloud is composed of a single habit (i.e., $M = 1$ in Eq. (2.53)) [34]. Figure 2.13 shows the mean optical parameters of ice clouds composed of each of six habits at $D_e = 100$ μm as a function of frequency. Both the mean extinction efficiency and asymmetry factor increase monotonically with frequency, for all habits. The mean single-scattering albedo also increases with frequency until it reaches its asymptotic value.

2.3.3 Rayleigh Approximation

For most of the cloud droplets in liquid phase, the radius (~10 μm) is much smaller than the wavelength at microwave (~1 cm). Consequently, scattering becomes negligible and the extinction reduces to absorption. At the small particle limit, the absorption cross section,

$$\sigma_s = \frac{8}{3}x^4\left|\frac{m^2-1}{m^2+2}\right|^2, \tag{2.59}$$

and

$$\sigma_e = 4x\,\mathrm{Im}\left\{\frac{m^2-1}{m^2+2}\left[1+\frac{x^2}{15}\times\frac{m^2-1}{m^2+2}\times\frac{m^4+27m^2+38}{2m^2+3}\right]\right\} + \frac{8}{3}x^4\mathrm{Re}\left\{\left(\frac{m^2-1}{m^2+2}\right)^2\right\}, \tag{2.60}$$

and absorption coefficient after neglecting the smaller terms in Eq. (2.60) is

$$\sigma_a = 4x\,\mathrm{Im}\left\{\frac{m^2-1}{m^2+2}\right\} \tag{2.61}$$

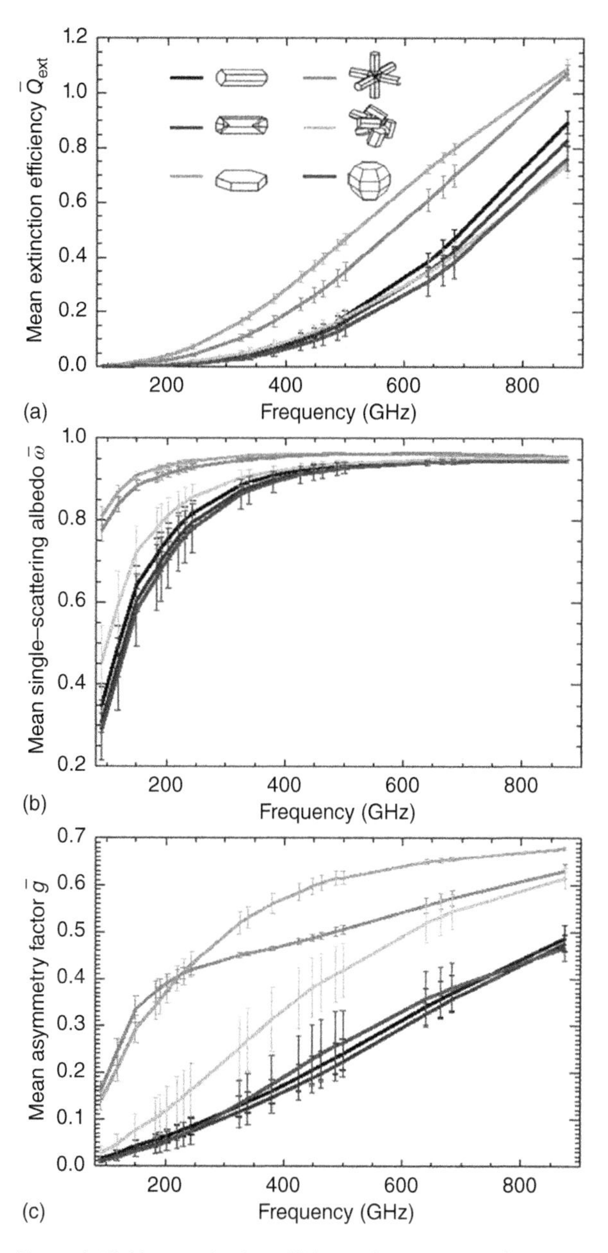

Figure 2.13 Mean extinction efficiency Q_{ext}, mean single-scattering albedo ω, and mean asymmetry factor g as a function of frequency for the six nonspherical ice habits for $D_e = 100\,\mu m$. The error bars indicate the standard deviations. (Hong *et al.* 2009 [34]. Reproduced with permission of Wiley.)

Thus, for polydispersed cloud size distribution, we can derive the total extinction from Rayleigh scattering as

$$
\begin{aligned}
\beta_e &= \int_0^{\infty} \pi r^2 Q_a(r) n(r) \mathrm{d}r \\
&= \frac{6\pi}{\lambda} \mathrm{Im}\left\{\frac{m^2-1}{m^2+2}\right\} \int_0^{\infty} \frac{4}{3}\pi r^3 n(r) Er \\
&= \frac{6\pi}{\lambda \rho_w} \mathrm{Im}\left\{\frac{m^2-1}{m^2+2}\right\} \mathrm{LWC}, \qquad (2.62)
\end{aligned}
$$

where $\mathrm{LWC} = \int_0^{\infty} \frac{4}{3}\pi r^3 \rho_w n(r) dr$ is the cloud liquid water content. Equation (2.62) has been extensively used for microwave remote sensing of cloud liquid water from space- and ground-based measurements. This is mainly because cloud optical thickness can be directly related to the vertically integrated water content as

$$
\begin{aligned}
\tau_L &= \int_{\Delta Z} \beta_e \mathrm{d}z \\
&= \int_{\Delta Z} \frac{6\pi}{\lambda \rho_w} \mathrm{Im}\left\{\frac{m^2-1}{m^2+2}\right\} \mathrm{LWC}\mathrm{d}z \\
&= \frac{6\pi}{\lambda \rho_w} \mathrm{Im}\left\{\frac{m^2-1}{m^2+2}\right\} \mathrm{LWP}, \qquad (2.63)
\end{aligned}
$$

where $\mathrm{LWP} = \int_{\Delta Z} \mathrm{LWC}(z)\mathrm{d}z$ and is also referred to as the liquid water path, that is, vertically integrated liquid water content. Here, we have assumed that the complex water dielectric constant is independent of height. In reality, the dielectric constant of water is also a function of height due to its temperature dependence. Thus, the mass absorption coefficient in Eq. (2.63) is also dependent on the temperature, as shown in Figure 2.14. Since the absorption coefficient decreases as the cloud temperature increases, a cloud for a given amount of liquid would be less opaque in tropics compared to middle latitudes. In addition, microwave radiation at lower frequencies is less affected by liquid-phase clouds.

Normally, the cloud-layer mean temperature is a prior parameter that can be estimated from other independent sources and is then used to compute the complex dielectric constant of clouds. From Eq. (2.63), we find that the optical depth of cloud droplets at microwave frequencies is proportional to the liquid water path and is not a function of droplet size. This physical mechanism provides a basic foundation for remote sensing of the cloud liquid water path. It is worthwhile mentioning that ice clouds nearly have minimal impacts on the radiation at low microwave frequencies. The imagery part of the refractive index m of ice is about 3 orders smaller than that of water at low frequencies. Consequently, the term $\mathrm{Im}\{\}$ for ice is close to 0. Therefore, the ice emission/absorption is generally negligible.

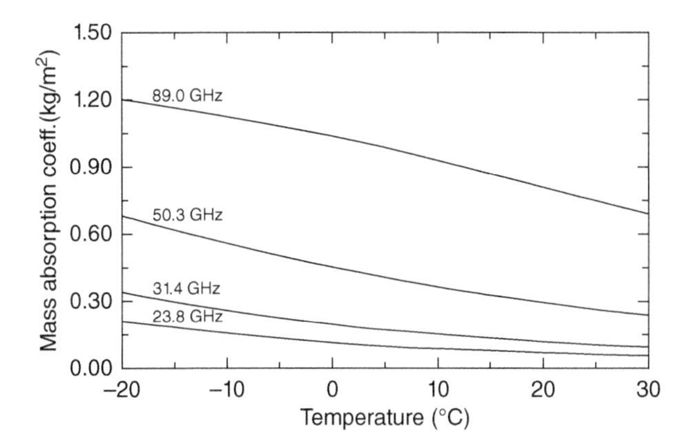

Figure 2.14 The mass absorption coefficient at Advanced Microwave Sounding Unit (AMSU) channels as a function of temperature.

This property allows liquid water alone to be inferred in a mixed-phase cloud or liquid water cloud covered by cirrus ice cloud. For rainfall remote sensing, the problem becomes more complicated compared to sensing of liquid water path of nonprecipitating clouds because of the following three factors. First, raindrops are comparable to microwave wavelength, and the scattering by raindrops cannot be ignored. Thus, the brightness temperature varies with not only liquid water path but also size distribution of raindrops. The τ_w − LWP relationship can no longer be expressed by an analytic equation, as is shown in Eq. (2.61). Second, the rainfall rate is a measure of water flux at the ground level, while the satellite-received radiation reflects the vertically integrated property of the column, resulting in that the brightness temperature is a function of the vertical profile of rain, not just the rainfall rate at the surface. The third complication arises from the fact that the rainfall field often has a very high horizontal variability, and the field of view (FOV) of a microwave sensor is large, typically several tens of kilometers, so that rainfall is not uniformly distributed within a satellite FOV. These problems are discussed in later chapters where we deal with cloud and precipitation retrievals.

The extinction of ice clouds at microwave frequencies is strongly dependent on the frequency. From Rayleigh approximation, the extinction from ice clouds is dominated by scattering since the imaginary part of the ice dielectric constant is very small and the absorption coefficient can be neglected (see Eq. (2.60)). With the derivation similar to Eq. (2.62), we have

$$\begin{aligned}\tau_{ice} &= \int_{\Delta Z} \mathrm{d}z \frac{128\pi^5}{3\lambda^4} \left[\frac{m^2-1}{m^2+2}\right]^2 \int_0^\infty r^6 n(r)\mathrm{d}r \\ &= \int_{\Delta Z} \mathrm{d}z \frac{24\pi^3}{3\lambda^4 \rho_i^2} \left[\frac{m^2-1}{m^2+2}\right]^2 \int_0^\infty n(r) M^2(r)\mathrm{d}r, \qquad (2.64)\end{aligned}$$

where M is the mass of an ice sphere with a radius of r. Unlike liquid water absorption, which is proportional to the mass of the liquid water content, the ice scattering cross section is proportional to the 6th power of r under Rayleigh approximation. Therefore, the ice optical depth, which is the vertical integration of ice cloud extinction, is not only related to ice water path (*IWP*) but also strongly depends on the PSD. Increasing either the amount or the size of particles will lead to an increase in ice optical depth and, in turn, a decrease in brightness temperature. To conduct ice water retrieval requires either a prior knowledge of PSD or performing simultaneous retrieval of both size distribution and IWP. In addition, the IWP retrieval problem is further complicated by the bulk density of ice particles, which may vary from less than 0.1 cm^{-3} to larger than 0.9 g/cm^3, depending on the ice particle type. The simultaneous retrieval of IWP and ice PSD may be achieved by observing the radiances at two high microwave frequencies [38, 39]. In addition, the large ice particles must be included in the retrievals through Mie calculations, which are discussed in the chapters on cloud property retrievals.

When the particle size is much smaller than the incident wavelength, the scattering amplitudes can also be approximated in the same manner as its scattering cross section as in Eqs. (2.59) and (2.61). As a result,

$$S_1 = \frac{3}{2}a_1, \tag{2.65}$$

$$S_2 = \frac{3}{2}a_1 cos(\Theta), \tag{2.66}$$

$$a_1 = -\frac{i2x^3}{3}\frac{m^2-1}{m^2+2}, \tag{2.67}$$

and the corresponding scattering matrix is

$$S(\Theta)_{ray} = \frac{3}{4}\cdot\begin{pmatrix} 1+\cos^2\Theta & 1-\cos^2\Theta & 0 & 0 \\ \cos^2\Theta-1 & 1+\cos^2\Theta & 0 & 0 \\ 0 & 0 & \cos\Theta & 0 \\ 0 & 0 & 0 & \cos\Theta \end{pmatrix}. \tag{2.68}$$

2.3.4
Henyey–Greenstein and Rayleigh Phase Matrix

A Henyey–Greenstein (HG) phase function is a parameterized phase function for radiation scattered by particles in the atmosphere and is often applied in visible and infrared ranges in which the asymmetry factor (i.e., the first moment) of the phase function represents the asymmetric forward and backward scattering. The formula for scattering through angle Θ is given by

$$S(\Theta, g) = \frac{1-g^2}{(1+g^2-2g\cos\Theta)^{3/2}}, \tag{2.69}$$

where g is the asymmetry factor.

For a small scatter, the Rayleigh phase function is widely applied in intensity and polarization calculations. It is normally valid for molecular scattering in visible and ultraviolet wavelength ranges and for nonprecipitation cloud scattering at low frequencies of the microwave range. In general, the actual phase function differs from the Rayleigh and HG phase functions. Here, a new analytical phase function called the HG and Rayleigh phase function is proposed for small asymmetry scattering. The HG – Rayleigh phase function is a normalized product of the Rayleigh phase function and the HG phase function with a modified asymmetry factor (G). The modified asymmetry factor is related to the original asymmetry factor (g) of the phase function through the following derivations. For intensity radiative transfer, the HG – Rayleigh phase function can be written as [40]

$$\begin{aligned} S(\Theta, g) &= CS_{ray}(\Theta) \times HG[\Theta, G(g)] \\ &= C\frac{3}{16}(1+\cos^2\Theta) \times \frac{1-G^2}{(1+G^2-2G\cos\Theta)^{3/2}}, \end{aligned} \tag{2.70}$$

where G is the modified asymmetry factor. Through normalizing the zeroth and the first moment of phase normalization function, we can obtain

$$\begin{aligned} C &= \frac{4}{2+G^2}, \\ G &= \frac{5}{9}g + \left\{\frac{1}{2}\left(\frac{10}{9}g + \frac{250}{729}g^3\right) + \sqrt{\Delta}\right\}^{1/3} \\ &\quad - \left\{\sqrt{\Delta} - \frac{1}{2}\left(\frac{10}{9}g + \frac{250}{729}g^3\right)\right\}^{1/3}, \end{aligned} \tag{2.71}$$

where

$$\Delta = \left[\frac{1}{2}\left(-\frac{10}{9}g - \frac{250}{729}g^3\right)\right]^2 + \left[\frac{1}{3}\left(4 - \frac{25}{27}g^2\right)\right]^3. \tag{2.72}$$

The modified asymmetry factor G can be calculated from the original asymmetry factor g in Eqs. (2.70) and (2.71) and is shown in Figure 2.15.

For polarimetric radiation, the HG – Rayleigh scattering matrix can be written as

$$\begin{aligned} S(\Theta) &= \frac{1}{2+G^2}\frac{1-G^2}{(1+G^2-2G\cos\Theta)^{3/2}} \\ &\quad \times \frac{3}{4} \times \begin{pmatrix} 1+\cos^2\Theta & 1-\cos^2\Theta & 0 & 0 \\ \cos^2\Theta - 1 & 1+\cos^2\Theta & 0 & 0 \\ 0 & 0 & \cos\Theta & 0 \\ 0 & 0 & 0 & \cos\Theta \end{pmatrix}, \end{aligned} \tag{2.73}$$

which is applied to the Stokes vector (I, Q, U, V). The phase matrix is then the product of the scattering matrix and the associated rotational matrices,

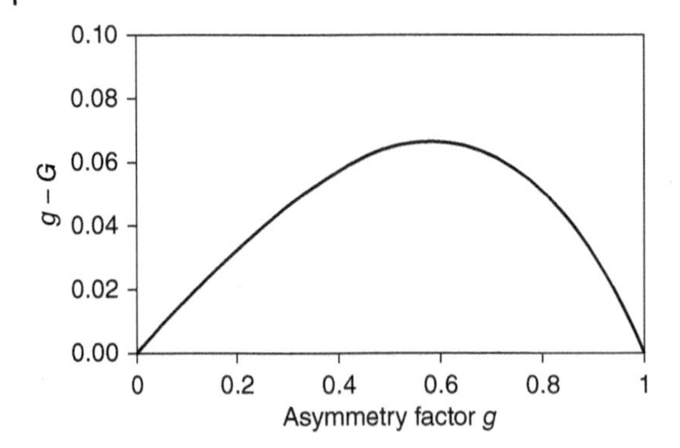

Figure 2.15 Modified asymmetry factor, G, in relation to g. (Liu and Weng 2006 [40]. Reproduced with permission of Optical Society of America.)

as discussed in Chapter 3. The main advantage of this combined function is that it allows the use of analytic forms of scattering matrix in radiative transfer calculation and maintains the properties of HG and Rayleigh scattering function.

2.4
Summary and Conclusions

Oxygen and water vapor are two main absorbers in microwave regions. Their resonant lines can be well characterized as Lorenz line shapes with the intensity and line width as a function of temperature and pressure. In addition, the continuum absorption in the microwave regions from water vapor is also parameterized. In atmospheric particle scattering, a general Mie theory can be used for spherical scatters, and a discrete dipole approximation is used for nonspherical scatters. The analytic phase matrices from Rayleigh and HG approximations are also discussed for general applications.

3
Radiative Transfer Modeling at Microwave Frequencies

3.1
Introduction

At microwave frequencies, effects of scattering and emission by atmospheric hydrometeors on upwelling radiance at the top of the atmosphere are simulated through various radiative transfer schemes. Since the first plane-parallel model was proposed by Wilheit *et al.* [41] to simulate microwave brightness temperature at 19.35 GHz, more advanced radiative transfer schemes were developed for simulations of microwave brightness temperatures at a wide range of microwave frequencies [42]. To account for the horizontal as well as the vertical variability over scale lengths that are often smaller than the radiometer footprint size, the outputs from a finite cloud model including precipitation-sized hydrometeors were used as inputs of a three-dimensional radiative transfer model [43]. A plane-parallel polarized radiative transfer model was solved using the doubling and adding scheme to compute the radiances exiting an atmosphere containing the randomly oriented particles [3]. This doubling and adding technique was improved for fast computations through an analytic formula for the layer source function [44].

The discrete-ordinate method was developed to solve the scalar radiative transfer equation [45], and later it was extended to a vector radiative transfer calculation [46–48]. A computationally efficient and accurate vector radiative transfer model is needed for radiance and Jacobian calculations of the polarized measurements used in numerical weather prediction (NWP) models. In this chapter, we introduce two radiative transfer schemes including the discrete-ordinate method and the doubling–adding technique. As part of the radiative transfer model, we also introduce microwave ocean and land emissivity models.

3.2
Radiative Transfer Equation

Radiative transfer theory has been discussed in many literatures and textbooks. Here, we emphasize the polarized or vector radiative transfer problem and pay

Passive Microwave Remote Sensing of the Earth: For Meteorological Applications, First Edition. Fuzhong Weng.

more attention to those areas that are specifically developed for microwave remote sensing of the environmental parameters. Since microwave instruments that have been deployed in space measure partial or full Stokes components, it is necessary to understand the complete radiative transfer process in scattering and emission atmosphere and surfaces.

Normally, when a particle intercepts radiation, the scattered radiation is related to the incident radiation through the relation

$$\mathbf{I}_s = \mathbf{P}\mathbf{I}_i, \tag{3.1}$$

where $\mathbf{I}_s$ and $\mathbf{I}_i$ are the scattering and incident radiative vectors, respectively, each of which is given by

$$\mathbf{I}_i = (I, Q, U, V)_i^T, \tag{3.2}$$

and

$$\mathbf{I}_s = (I, Q, U, V)_s^T, \tag{3.3}$$

where the superscript T executes a transpose to a row matrix that is used in the following sections. Four components in the radiative vectors are related to the amplitudes of the electric field, as shown in Eq. (1.15). For a scattering and absorbing atmosphere, the radiative vector $\mathbf{I}(\tau, \mu, \phi)$ emerging at the optical depth τ in the direction (μ, ϕ) is given by

$$\mu \frac{\mathrm{d}\mathbf{I}(\tau, \mu, \phi)}{d\tau} = -\mathbf{I}(\tau, \mu, \phi) + \frac{\omega(\tau)}{4\pi} \int_0^{2\pi} \int_{-1}^{1} \mathbf{M}(\tau, \mu, \phi, \mu', \phi') I(\tau, \mu', \phi') \mathrm{d}\mu' \mathrm{d}\phi' + \mathbf{S}(\tau, \mu, \phi, \mu_0, \phi_0), \tag{3.4}$$

where $\omega(\tau)$ is the single-scattering albedo, $\mathbf{M}(\tau, \mu, \phi, \mu', \phi')$ is the phase matrix, and the direct radiance is from the direction (μ_0, ϕ_0). Note that the wavelength dependencies of all variables are implicit. The source matrix includes single-scattering and thermal emission and is given by

$$\mathbf{S}(\tau, \mu, \phi, \mu_0, \phi_0) = (1 - \omega) B \begin{pmatrix} 1 \\ 0 \\ 0 \\ 0 \end{pmatrix} + \frac{\omega F_0}{4\pi} \exp(-\tau/\mu_0) \begin{pmatrix} M_{11}(\phi, \mu_0, \phi_0) \\ M_{12}(\phi, \mu_0, \phi_0) \\ M_{13}(\phi, \mu_0, \phi_0) \\ M_{14}(\phi, \mu_0, \phi_0) \end{pmatrix}, \tag{3.5}$$

where F_0 is the solar spectral constant; $B(T)$ the Planck function at a temperature T; F_0 the solar spectral constant; μ_0 the cosine of sun zenith angle; ω the single-scattering albedo; and τ the optical thickness. The phase matrix $\mathbf{M}(\tau, \mu, \phi, \mu', \phi')$ is derived from a linear transformation of the scattering matrix $\mathbf{S}(\Theta)$ in Eq. (2.4) according to spherical trigonometry [49] as follows:

$$\mathbf{M}(\tau, \mu, \phi, \mu', \phi') = \mathbf{L}(\pi - i_2)\mathbf{S}(\Theta)\mathbf{L}(i_1), \tag{3.6}$$

where $\mathbf{L}$ is expressed as Eq. (1.17a) and i_1 and i_2 are the angles at which the radiation vector rotates from the incident plan to the outgoing plan. For a spherical particle scattering, the phase matrix can be derived by rotating the Mie scattering matrix in Eq. (2.44) through Eq. (1.17a). Thus, Eq. (3.6) can be written as

$$\mathbf{M}(\tau, \mu, \phi, \mu', \phi') = \begin{bmatrix} S_{11} & S_{12}\cos 2i_1 & -S_{12}\sin 2i_1 & 0 \\ S_{12}\cos 2i_2 & S_{11}\cos 2i_1\cos 2i_2 - S_{33}\sin i_1 \sin 2i_2 & S_{11}\sin 2i_1\cos 2i_2 - S_{33}\cos i_1\sin 2i_2 & -S_{34}\sin 2i_2 \\ S_{12}\sin 2i_1 & S_{11}\cos 2i_1\sin 2i_2 + S_{33}\sin i_1\cos 2i_2 & -S_{11}\sin 2i_1\sin 2i_2 + S_{33}\cos i_1\cos 2i_2 & S_{34}\cos 2i_2 \\ 0 & -S_{34}\sin 2i_1 & -S_{34}\cos 2i_2 & S_{33} \end{bmatrix}.$$

Chandrasekhar [4] derived analytic expressions for all the 16 elements in Eq. (3.6) from the Rayleigh phase matrix shown in Eq. (2.70).

3.3 Vector Discrete-Ordinate Method

Equation (3.4) can be solved by some standard routines such as the multi-layer discrete-ordinate method [46, 48], the doubling–adding method [3], and the matrix operator method [50]. Dave [49] showed that the phase matrix $\mathbf{M}(\tau, \mu, \phi, \mu', \phi')$ can be expanded into the following Fourier cosine and sinusoidal series:

$$\mathbf{M}(\tau, \mu, \phi, \mu', \phi') = \sum_{m=0}^{2N-1} [\mathbf{M}_m^c(\tau, \mu, \mu')\cos m(\phi' - \phi) + \mathbf{M}_m^s(\tau, \mu, \mu')\sin m(\phi' - \phi)]. \tag{3.7}$$

where $\mathbf{M}_m^c$ and $\mathbf{M}_m^s$ are the mth cosine and sine mode Fourier expansion matrices, respectively.

The radiative vector may also be expanded in this manner to give

$$\mathbf{I}(\tau, \mu, \phi) = \sum_{m=0}^{2N-1} [\mathbf{I}_m^c(\tau, \mu)\cos m(\phi_0 - \phi) + \mathbf{I}_m^s(\tau, \mu)\sin m(\phi_0 - \phi)], \tag{3.8}$$

where $\mathbf{I}_m^c$ and $\mathbf{I}_m^s$ are the mth cosine and sine mode Fourier expansion vectors, respectively, and are related to the mth Fourier expansion coefficients of radiative components in the following expressions:

$$\mathbf{I}_m^c(\tau,\mu) = (I_m^c, Q_m^c, U_m^c, V_m^c)^T, \tag{3.9a}$$

and

$$\mathbf{I}_m^s(\tau,\mu) = (I_m^s, Q_m^s, U_m^s, V_m^s)^T. \tag{3.9b}$$

The source vector may be expanded to yield

$$\mathbf{S}(\tau,\mu,\phi) = \sum_{m=0}^{2N-1} [\mathbf{S}_m^c(\tau,\mu)\cos m(\phi_0-\phi) + \mathbf{S}_m^s(\tau,\mu)\sin m(\phi_0-\phi)], \tag{3.10}$$

Substituting Eqs. (3.7)–(3.10) into Eqs. (3.4) and (3.5), manipulating some integrations, and comparing the coefficients for the cosine and sine Fourier modes result in

$$\mu\frac{\mathrm{d}\mathbf{I}_m^c(\tau,\mu)}{\mathrm{d}\tau} = \mathbf{I}_m^c(\tau,\mu) - \frac{\omega(\tau)}{4}\int_{-1}^{1}[(1+\delta_{0m})\mathbf{M}_m^c\mathbf{I}_m^c - (1-\delta_{0m})\mathbf{M}_m^s\mathbf{I}_m^s]\mathrm{d}\mu' - \mathbf{S}_m^c(\tau,\mu), \tag{3.11a}$$

and

$$\mu\frac{\mathrm{d}\mathbf{I}_m^s(\tau,\mu)}{\mathrm{d}\tau} = \mathbf{I}_m^s(\tau,\mu) - \frac{\omega(\tau)}{4}\int_{-1}^{1}[(1-\delta_{0m})\mathbf{M}_m^c\mathbf{I}_m^s + (1-\delta_{0m})\mathbf{M}_m^s\mathbf{I}_m^c]\mathrm{d}\mu' - \mathbf{S}_m^s(\tau,\mu), \quad m=0,\ldots,(2N-1). \tag{3.11b}$$

A further discretization of integrals in Eq. (3.11) using Gaussian quadrature scheme is as follows:

$$\mu\frac{\mathrm{d}\mathbf{I}_m^c(\tau,\mu_i)}{\mathrm{d}\tau} = \mathbf{I}_m^c(\tau,\mu) - \frac{\omega(\tau)}{4}\sum_{j=-N}^{N} a_j[(1+\delta_{0m})\mathbf{M}_m^c(\tau,\mu_i,\mu_j)\mathbf{I}_m^c(\tau,\mu_j) - (1-\delta_{0m})\mathbf{M}_m^s(\tau,\mu_i,\mu_j)\mathbf{I}_m^s(\tau,\mu_j)] - \mathbf{S}_m^c(\tau,\mu_i), \tag{3.12a}$$

$$\mu\frac{\mathrm{d}\mathbf{I}_m^s(\tau,\mu_i)}{d\tau} = \mathbf{I}_m^s(\tau,\mu) - \frac{\omega(\tau)}{4}\sum_{j=-N}^{N} a_j[(1-\delta_{0m})\mathbf{M}_m^c(\tau,\mu_i,\mu_j)\mathbf{I}_m^s(\tau,\mu_j) - (1-\delta_{0m})\mathbf{M}_m^s(\tau,\mu_i,\mu_j)\mathbf{I}_m^c(\tau,\mu_j)] - \mathbf{S}_m^s(\tau,\mu_i), \tag{3.12b}$$

$$i = \pm 1, \ldots \pm N, m = 0, \ldots, (2N-1).$$

where a_j and μ_j are the Gauss quadrature expansion coefficients and points.

For a scattering sphere, Dave [49] proved that the elements M_{ij} of the phase matrix (ij = 11,12, 21,22, 33,34, 43,44) are even functions of $\phi' - \phi$, whereas those with ij of 13,14, 23,24, 31,32, 41,42 are odd functions of $\phi' - \phi$. This result means that

$$\mathbf{M}_m^c = \begin{pmatrix} M_{11m}^c & M_{12m}^c & 0 & 0 \\ M_{21m}^c & M_{22m}^c & 0 & 0 \\ 0 & 0 & M_{33m}^c & M_{34m}^c \\ 0 & 0 & M_{43m}^c & M_{44m}^c \end{pmatrix}, \tag{3.13a}$$

and

$$\mathbf{M}_m^s = \begin{pmatrix} 0 & 0 & M_{13m}^s & M_{14m}^s \\ 0 & 0 & M_{23m}^s & M_{24m}^s \\ M_{31m}^s & M_{32m}^s & 0 & 0 \\ M_{41m}^s & M_{42m}^s & 0 & 0 \end{pmatrix}. \tag{3.13b}$$

These matrix properties result in the radiative components $I_m^c(\tau, \mu)$ and $Q_m^c(\tau, \mu)$ of the cosine mode Fourier expansion vector $\mathbf{I}_m^c$ being coupled with the components $U_m^s(\tau, \mu)$ and $V_m^s(\tau, \mu)$ in the sine mode Fourier expansion vector $\mathbf{I}_m^s$. On the other hand, the radiative components $I_m^s(\tau, \mu)$ and $Q_m^s(\tau, \mu)$ of $\mathbf{I}_m^s$ are coupled with the components $U_m^c(\tau, \mu)$ and $V_m^c(\tau, \mu)$ of $\mathbf{I}_m^c$. Thus, we defined two $4N$-dimensional specific vectors for a combined cosine Stokes vector as

$$\mathbf{I}_m^c(\tau, -\mu_i) = \begin{pmatrix} I_m^c(\tau, -\mu_i) \\ Q_m^c(\tau, -\mu_i) \\ U_m^s(\tau, -\mu_i) \\ V_m^s(\tau, -\mu_i) \end{pmatrix}_{1\times 4N}, \quad \mathbf{I}_m^c(\tau, \mu_i) = \begin{pmatrix} I_m^c(\tau, +\mu_i) \\ Q_m^c(\tau, +\mu_i) \\ U_m^s(\tau, +\mu_i) \\ V_m^s(\tau, +\mu_i) \end{pmatrix}_{1\times 4N}; \tag{3.14}$$

with the source vectors of

$$\mathbf{S}_m^c(\tau, -\mu) = \begin{pmatrix} I_m^c(\tau, -\mu_i) \\ Q_m^c(\tau, -\mu_i) \\ U_m^s(\tau, -\mu_i) \\ V_m^s(\tau, -\mu_i) \end{pmatrix}_{1\times 4N}, \quad \mathbf{S}_m^c(\tau, \mu) = \begin{pmatrix} I_m^c(\tau, +\mu_i) \\ Q_m^c(\tau, +\mu_i) \\ U_m^s(\tau, +\mu_i) \\ V_m^s(\tau, +\mu_i) \end{pmatrix}_{1\times 4N}. \tag{3.15}$$

The amplitude of each Fourier component is a function of the zenith angle. Furthermore, the amplitude is discretized at a series of zenith angles (or streams) so that the combined Stokes vectors for each cosine and sinusoidal mode can be generalized as

$$\mu_i \frac{\mathrm{d}}{\mathrm{d}\tau} \begin{bmatrix} \mathbf{I}_m(\tau, \mu_i) \\ -\mathbf{I}_m(\tau, \mu_{-i}) \end{bmatrix} = \begin{bmatrix} \mathbf{I}_m(\tau, \mu_i) \\ \mathbf{I}_m(\tau, \mu_{-i}) \end{bmatrix} - \omega \sum_{j=1}^{N} \begin{bmatrix} \mathbf{M}_m(\mu_i, \mu_j) & \mathbf{M}_m(\mu_i, \mu_{-j}) \\ \mathbf{M}_m(\mu_{-i}, \mu_j) & \mathbf{M}_m(\mu_{-i}, \mu_{-j}) \end{bmatrix} \begin{bmatrix} \mathbf{I}_m(\tau, \mu_j) \\ \mathbf{I}_m(\tau, \mu_{-j}) \end{bmatrix} w_j - \begin{bmatrix} \mathbf{S}_m(\tau, \mu_i) \\ \mathbf{S}_m(\tau, \mu_{-i}) \end{bmatrix}, \tag{3.16}$$

where $w_j = a_j/4$ and $\mu_{-i} = -\mu_i$. The discretization to the integral term in the vector (or polarized) radiative transfer equation is fundamental toward the final solution of the radiative vector. The rest of the work is to solve a set of ordinary differential equations through finding the general and specific solution and then to determine the coefficients in the general solution through the internal and external boundary conditions. Eq. (3.16) can also be expressed as a one-vector equation as

$$\frac{d\mathbf{I}}{d\tau} = \mathbf{AI} - \mathbf{S}, \tag{3.17}$$

where

$$\mathbf{I} = [\mathbf{I}(\tau, \mu_1), \mathbf{I}(\tau, \mu_2), \ldots, \mathbf{I}(\tau, \mu_N), \mathbf{I}(\tau, \mu_{-1}), \mathbf{I}(\tau, \mu_{-2}), \ldots, \mathbf{I}(\tau, \mu_{-N})]^T, \tag{3.18}$$

and the composite phase matrix

$$\mathbf{A} = \begin{bmatrix} \mathbf{u}^{-1} & 0 \\ 0 & -\mathbf{u}^{-1} \end{bmatrix} \begin{bmatrix} \mathbf{E} - \omega\mathbf{M}(\mu, \mu) & \omega\mathbf{M}(\mu, -\mu) \\ \omega\mathbf{M}(-\mu, \mu) & \mathbf{E} - \omega\mathbf{M}(-\mu, -\mu) \end{bmatrix} = \begin{bmatrix} \boldsymbol{\alpha}_1 & \boldsymbol{\beta}_1 \\ -\boldsymbol{\beta}_2 & -\boldsymbol{\alpha}_2 \end{bmatrix}, \tag{3.19}$$

where $\boldsymbol{\alpha}$ and $\boldsymbol{\beta}$ are the $4N$ by $4N$ matrices and related to the elements of the phase matrices as

$$\boldsymbol{\alpha}_1(\mu_i, \mu_j) = [\mathbf{E} - \omega\mathbf{M}_m(\mu_i, \mu_j)]/\mu_i, \tag{3.20a}$$

$$\boldsymbol{\beta}_1(\mu_i, \mu_{-j}) = \omega\mathbf{M}_m(\mu_i, \mu_{-j})/\mu_i, \tag{3.20b}$$

$$\boldsymbol{\alpha}_2(\mu_{-i}, \mu_{-j}) = [\mathbf{E} - \omega\mathbf{M}_m(\mu_{-i},\mu_{-j})]/\mu_i, \tag{3.20c}$$

$$\boldsymbol{\beta}_2(\mu_{-i}, \mu_j) = \omega\mathbf{M}_m(\mu_{-i}, \mu_j)/\mu_i. \tag{3.20d}$$

In addition, $\mathbf{u}$ is a $4N$ by $4N$ matrix that has nonzero elements at its diagonal direction such that

$$\mathbf{u} = [\mu_1, \mu_1, \mu_1, \mu_1, \mu_2, \ldots\ldots, \mu_N, \mu_N, \mu_N, \mu_N]_{diagonal}. \tag{3.21}$$

Note that the source term for the cosine Stokes vector is

$$\mathbf{S} = (1 - \omega)B(T)\delta_{m0} \begin{bmatrix} \mathbf{u}^{-1} & 0 \\ 0 & -\mathbf{u}^{-1} \end{bmatrix} \boldsymbol{\Xi} + \frac{\omega F_0}{\pi} \exp(-\tau/\mu_0)\boldsymbol{\Psi}, \tag{3.22}$$

where $\boldsymbol{\Xi}$ and $\boldsymbol{\Psi}$ are the vectors that has $8N$ elements as

$$\boldsymbol{\Xi} = [1, 0, 0, 0, 1, 0, 0, 0, \ldots, 1, 0, 0, 0]^T \tag{3.23a}$$

and

$$\boldsymbol{\Psi} = [M_{11}(\mu_1, \mu_0)/\mu_1, M_{12}(\mu_1, \mu_0)/\mu_1, M_{13}(\mu_1, \mu_0)/\mu_1, M_{14}(\mu_1, \mu_0)/\mu_1,$$
$$M_{11}(\mu_2, \mu_0)/\mu_2, M_{12}(\mu_2, \mu_0)/\mu_2, M_{13}(\mu_2, \mu_0)/\mu_2, M_{14}(\mu_2, \mu_0)/\mu_2,$$

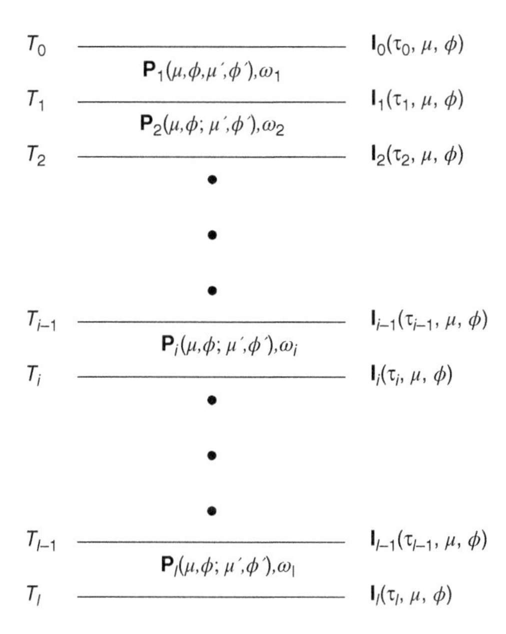

Figure 3.1 A schematic diagram of a multilayer medium for the vector radiative transfer calculation. The temperature at each level is specified as known; the phase matrix, single-scattering albedo, and optical thickness at each layer are calculated from the Mie theory. The radiative vector, including four radiative components, at each level, is calculated from the multilayer discrete-ordinate method.

$$\ldots, M_{11}(\mu_{-N}, \mu_0)/\mu_{-N}, M_{12}(\mu_{-N}, \mu_0)/\mu_{-N}, M_{13}(\mu_{-N}, \mu_0)/\mu_{-N},$$
$$M_{14}(\mu_{-N}, \mu_0)/\mu_{-N}]. \tag{3.23b}$$

Equation (3.17) is a linear differential system, and its solution within a homogeneous layer labeled as l (see Figure 3.1) can be written as

$$\mathbf{I} = \exp[\mathbf{A}(\tau - \tau_{l-1})]\mathbf{C}_l + \mathbf{S}_l, \tag{3.24}$$

where $\mathbf{S}_l$ is the matrix associated with the thermal and solar source terms and $\mathbf{C}_l$ is the matrix of the coefficients having $8N$ elements in each layer that can be further derived from the continuity conditions required for radiances at the internal boundaries

$$\mathbf{S}_l = \delta_{m0}\{B(\tau_{l-1})\Xi + \frac{B(\tau_{l-1}) - B(\tau_l)}{\tau_{l-1} - \tau_l}[\mathbf{A}_l^{-1}\Xi + (\tau - \tau_{l-1})\Xi]$$
$$+ \mu_0[\mu_0\mathbf{A}_l + \mathbf{E}]^{-1}\frac{\omega F_0}{\pi}\exp(-\tau/\mu_0)\mathbf{\Psi}\}, \tag{3.25}$$

where τ_l and τ_{l-1} are the optical depths at the top and bottom of the layer. Note that the coefficient vector $\mathbf{C}_l$ ($8N$ elements in each layer) in Eq. (3.24) can be determined from the continuity condition of the internal boundary.

However, the continuity conditions at the external boundaries are dependent on the mode. For example, for the cosine mode and at the top of the atmosphere,

$$\mathbf{I}_l(\tau_{l-1}) = \mathbf{I}_{l-1}(\tau_{l-1}), \tag{3.26a}$$

and at the top of atmosphere, we also have

$$\mathbf{I}_1(0) = \mathbf{I}_0, \tag{3.26b}$$

and on the surface

$$\mathbf{I}_L(\tau_L) = \boldsymbol{\varepsilon}B(T_s) + \mathbf{R}\mathbf{I}_L(\tau_L) + \mathbf{R}_0\frac{F_0}{\pi}\exp(-\tau_L/\mu_0)\overline{\Xi}, \tag{3.26c}$$

where

$$\overline{\Xi} = [1, 0, 0, 0, 1, 0, 0, 0, \ldots.., 1, 0, 0, 0]^T_{4N\times 1}, \tag{3.27}$$

and $\boldsymbol{\varepsilon}$ is the surface emissivity vector ($4N$), $\mathbf{R}$ is the surface reflection matrix ($4N$ by $4N$), $\mathbf{R}_0$ is the surface reflection vector ($4N$) at the sun zenith angle, and T_s is the surface temperature. Substituting Eq. (3.24) into (3.26) yields a set of algebraic equations for the determination of $\mathbf{C}$. This approach is generally referred to as the discrete-ordinate radiative transfer (DISORT) method and has been discussed in detail in many previous studies [46–48].

Substituting Eq. (3.24) into Eqs. (3.26a–c) results in a set of algebraic equations for solving the coefficient c_1 such that

$$\mathbf{PC} = \mathbf{V}, \tag{3.28}$$

where

$$\mathbf{P} = \begin{bmatrix} \overline{\mathbf{E}} & \overline{\mathbf{0}} & \overline{\mathbf{0}} & \overline{\mathbf{0}} & \overline{\mathbf{0}} & \overline{\mathbf{0}} & \overline{\mathbf{0}} & \overline{\mathbf{0}} & \overline{\mathbf{0}} \\ \mathbf{e}_1 & -\mathbf{E} & \mathbf{0} & \mathbf{0} & \mathbf{0} & \mathbf{0} & \mathbf{0} & \mathbf{0} & \mathbf{0} \\ \mathbf{0} & \mathbf{e}_2 & -\mathbf{E} & \mathbf{0} & \mathbf{0} & \mathbf{0} & \mathbf{0} & \mathbf{0} & \mathbf{0} \\ \mathbf{0} & \mathbf{0} & .. & .. & .. & .. & .. & .. & \mathbf{0} \\ \mathbf{0} & \mathbf{0} & .. & \mathbf{e}_{l-1} & -\mathbf{E} & .. & .. & .. & \mathbf{0} \\ \mathbf{0} & \mathbf{0} & .. & .. & \mathbf{e}_l & -\mathbf{E} & .. & .. & \mathbf{0} \\ \mathbf{0} & .. & .. & .. & .. & .. & .. & .. & \mathbf{0} \\ \mathbf{0} & \mathbf{0} & \mathbf{0} & .. & .. & .. & \mathbf{e}_{L-2} & -\mathbf{E} & \mathbf{0} \\ \mathbf{0} & \mathbf{0} & \mathbf{0} & \mathbf{0} & \mathbf{0} & \mathbf{0} & \mathbf{0} & \mathbf{e}_{L-1} & -\mathbf{E} \\ \overline{\mathbf{0}} & \overline{\mathbf{0}} & \overline{\mathbf{0}} & \overline{\mathbf{0}} & \overline{\mathbf{0}} & \overline{\mathbf{0}} & \overline{\mathbf{0}} & \overline{\mathbf{0}} & \overline{\overline{\mathbf{E}}}\mathbf{e}_L - \mathbf{R}\overline{\mathbf{E}}\mathbf{e}_L \end{bmatrix}, \tag{3.29}$$

$$\mathbf{C} = [\mathbf{c}_1, \mathbf{c}_2, \ldots\ldots, \mathbf{c}_{L-1}, \mathbf{c}_L]^T, \tag{3.30}$$

$$\mathbf{V} = [\overline{\mathbf{E}}\mathbf{s}_1(0), \mathbf{s}_2(\tau_1) - \mathbf{s}_1(\tau_1), \ldots, \mathbf{s}_L(\tau_{L-1}) - \mathbf{s}_{L-1}(\tau_{L-1}), \boldsymbol{\varepsilon}B(T_s)\delta_{om} + \mathbf{R}\overline{\mathbf{E}}\mathbf{s}_L(\tau_L) + \mathbf{R}_0\mu_0\frac{F_0}{\pi}\exp(-\tau_L/\mu_0)\overline{\Xi} - \overline{\overline{\mathbf{E}}}\mathbf{s}_L(\tau_L)]^T, \tag{3.31}$$

where $\mathbf{E}$ and $\mathbf{0}$ are unit and zero matrices ($8N \times 8N$), respectively; $\bar{\mathbf{0}}$ is a matrix with all elements of zero ($4N \times 8N$); $\bar{\mathbf{E}}$ and $\bar{\bar{\mathbf{E}}}$ are the matrices corresponding to the upper and lower $4N$ rows, respectively, in $\mathbf{E}$; and

$$\mathbf{e}_l = \exp[\mathbf{A}_l(\tau_l - \tau_{l-1})]. \tag{3.32}$$

For a vertically stratified atmosphere with a total of L layers, a total of $8N$ by L equations are coupled to derive all coefficients. Since $\mathbf{P}$ in Eq. (3.29) is a band matrix, the storage for its elements in computer memory can be optimally designed for speeding up the numerical calculations [45]. This formulation also allows for the derivation of the radiance gradient (or Jacobian) in an analytic form, as discussed in the next section.

3.4
Radiance Gradient or Jacobians

Since the radiance solution from the vector discrete radiative transfer (VDISORT) model is analytic in form, the Jacobian can also be explicitly obtained from Eq. (3.24). At the top of the atmosphere [$e^{\mathbf{A}(\tau-\tau_0)} = 1, l = 1$], the radiance gradient corresponding to a geophysical parameter (x_l) at the lth layer can be expressed as

$$\frac{\partial \mathbf{I}_1}{\partial x_l} = \frac{\partial \mathbf{c}_1}{\partial x_l} + \delta_{1l}\frac{\partial \mathbf{s}_1}{\partial x_l}, \tag{3.33}$$

where the second term on the right side of Eq. (3.33) can be derived directly from Eq. (3.25). Thus, the complexity of the Jacobians is largely dependent on the derivative of the coefficient matrix related to x_l. From Eq. (3.28),

$$\frac{\partial \mathbf{C}}{\partial x_l} = \mathbf{K}\left[\frac{\partial \mathbf{V}}{\partial x_l} - \frac{\partial \mathbf{P}}{\partial x_l}\mathbf{C}\right], \tag{3.34a}$$

where $\mathbf{K} = \mathbf{P}^{-1}$. Manipulating Eqs. (3.24) and (3.34a) results in

$$\frac{\partial \mathbf{I}_1(\mu)}{\partial x_l} = \sum_{k=l}^{L}\sum_{j=-4N}^{4N} \mathbf{K}_k(\mu,j)\left\{\frac{\partial\left(\mathbf{s}_k(\tau) - \mathbf{s}_{k-1}(\tau)\right)}{\partial x_l}|_{\tau=\tau_{l-1}}\right\}_j + \delta_{1l}\frac{\partial \mathbf{s}_1(\mu)}{\partial x_l}$$
$$- \sum_{j=-4N}^{4N} \mathbf{K}_l(\mu,j)\{\frac{\partial}{\partial x_l}\exp[\mathbf{A}_l(\tau - \tau_{l-1})]|_{\tau=\tau_l}\mathbf{c}_l\}_j, \tag{3.34b}$$

where the subscript index j outside the bracket {} represents the j-th element of the vector; $\mathbf{K}_l(\mu, j)$ is a vector at the lth layer. The first summation in the first term is the effect of the perturbation of the geophysical parameter (x_l) on the direct solar radiation. The derivatives on the right-hand side of Eq. (3.34a) can be directly derived using Eqs. (3.19) and (3.25). Note that Eq. (3.34a) is valid only for the layers above the surface. For the layer adjacent to the surface, the derivative includes more terms due to the surface reflection such that

$$\frac{\partial \mathbf{I}_1(\mu)}{\partial x_L} = \sum_{j=-4N}^{4N} \mathbf{K}_L(\mu, j)\{[\mathbf{R}\overline{\mathbf{E}} - \overline{\overline{\mathbf{E}}}]\frac{\partial \mathbf{s}_L(\tau)}{\partial x_L}|_{\tau=\tau_L}$$
$$- \mathbf{R}_0 \frac{F_0}{\pi} \exp(-\tau_L/\mu_0) \frac{\partial \tau}{\partial x_L}\bigg|_{\tau=\tau_L} \overline{\Xi}\}_j + \delta_{1L}\frac{\partial \mathbf{s}_1(\mu)}{\partial x_L}$$
$$- \sum_{j=-4N}^{4N} \mathbf{K}_L(\mu, j)\{[\overline{\overline{\mathbf{E}}} - \mathbf{R}\overline{\mathbf{E}}][\frac{\partial}{\partial x_L} \exp[\mathbf{A}_L(\tau - \tau_{L-1})]|_{\tau=\tau_L} \mathbf{c}_L\}_j. \tag{3.34c}$$

In Eq. (3.34c), the derivatives of the source term (**s**) and the composite phase matrix (**A**) on the right-hand side can be all analytically derived from Eqs. (3.24)–(3.25) if x_l is set for the optical thickness and single-scattering albedo. Thus, Jacobians including $\frac{\partial \mathbf{I}_1(\mu)}{\partial \tau_l}$ and $\frac{\partial \mathbf{I}_1(\mu)}{\partial \varpi_l}$ are analytic in form and can be computed very efficiently.

The Jacobian associated with the phase matrix variation may be derived if the angular dependence of all elements can be characterized in terms of the optical parameters. For example, in the phase function, the asymmetry parameter is introduced and used to characterize the angular distribution in the two-stream approximation. However, for a polarized two-stream approach, two additional parameters including phase polarization and asymmetry factors must be introduced so that the errors in the polarization forward modeling can be substantially reduced [51]. Thus, the Jacobian of the asymmetry parameter (g) can be directly computed from Eq. (3.34) (note that matrix A is analytically related to g). However, the derivation of the Jacobians related to the general phase matrix remains difficult.

The Jacobians related to the surface parameters (e.g., temperature and wind speed) can be derived as

$$\frac{\partial \mathbf{I}_1(\mu)}{\partial x_s} = \sum_{j=1}^{4N} \mathbf{K}_L(\mu, j)\{B(T_s)\frac{\partial \varepsilon}{\partial x_s} + \frac{\partial B(T_s)}{\partial x_s}\varepsilon + \frac{\partial \mathbf{R}}{\partial x_s}\overline{\mathbf{E}}\mathbf{s}_L(\tau_L)$$
$$+ \frac{\partial \mathbf{R}_0}{\partial x_s}\frac{F_0}{\pi} \exp(-\tau_L/\mu_0)\overline{\Xi}\}_j$$
$$+ \sum_{j=1}^{4N} \mathbf{K}_L(\mu, j)\left\{\frac{\partial \mathbf{R}}{\partial x_s}\overline{\mathbf{E}} \exp\left[\mathbf{A}_L\left(\tau_L - \tau_{L-1}\right)\right]\mathbf{c}_L\right\}_j. \tag{3.35}$$

Thus, the radiance gradient related to other physical parameters can be directly deduced from Eq. (3.35). For example, the Jacobian of water vapor mixing ratio is

$$\frac{\partial \mathbf{I}_1(\mu)}{\partial q_l} = \frac{\partial \tau_l}{\partial q_l}\frac{\partial \mathbf{I}_1(\mu)}{\partial \tau_l} + \frac{\partial \varpi_l}{\partial q_l}\frac{\partial \mathbf{I}_1(\mu)}{\partial \varpi_l} = \kappa_l^{abs}\left[\frac{\partial \mathbf{I}_1(\mu)}{\partial \tau_l} - \frac{\varpi_l}{\tau_l}\frac{\partial \mathbf{I}_1(\mu)}{\partial \varpi_l}\right], \tag{3.36}$$

where q_l and κ_l^{abs} are the integrated water vapor (kg/m^2) and the mass absorption coefficient (m^2/kg) of the water vapor at layer l, respectively. By the same token, the Jacobian of cloud liquid water can be derived as

$$\frac{\partial \mathbf{I}_1(\mu)}{\partial w_l} = \frac{\partial \tau_l}{\partial w_l}\frac{\partial \mathbf{I}_1(\mu)}{\partial \tau_l} + \frac{\partial \varpi_l}{\partial w_l}\frac{\partial \mathbf{I}_1(\mu)}{\partial \varpi_l} = \frac{\tau_l - \kappa_l^{abs} q_l}{w_l}\frac{\partial \mathbf{I}_1(\mu)}{\partial \tau_l} + \frac{\varpi_l \kappa_l^{abs} q_l}{w_l \tau_l}\frac{\partial \mathbf{I}_1(\mu)}{\partial \varpi_l}, \tag{3.37}$$

where w_l is the integrated cloud liquid water within layer l. Furthermore, the temperature Jacobian is

$$\frac{\partial \mathbf{I}_1(\mu)}{\partial T_l} = \frac{\partial \mathbf{I}_1(\mu)}{\partial T_l} + \frac{\partial \tau_l}{\partial T_l}\frac{\partial \mathbf{I}_1(\mu)}{\partial \tau_l} + \frac{\partial \varpi_l}{\partial T_l}\frac{\partial \mathbf{I}_1(\mu)}{\partial \varpi_l} = \frac{\partial \mathbf{I}_1(\mu)}{\partial T_l}$$
$$+ q_l \frac{\partial \kappa_l^{abs}}{\partial T_l}\left[\frac{\partial \mathbf{I}_1(\mu)}{\partial \tau_l} - \frac{\varpi_l}{\tau_l}\frac{\partial \mathbf{I}_1(\mu)}{\partial \varpi_l}\right], \tag{3.38}$$

where the derivative of the absorption coefficient related to temperature is generally negligible at visible wavelengths while that at thermal wavelengths can be either analytically derived or numerically evaluated. Thus, it is obvious that these Jacobians can be readily derived from a linear combination of those Jacobians related to the optical thickness and single-scattering albedo. Resulting computation efforts are optimally designed.

3.5 Benchmark Tests

Brightness temperatures computed from the VDISORT are compared with those from the doubling–adding polarized model [3] for a nonprecipitating atmosphere. The atmosphere ranges from 0 to 8 km and is divided into three layers, including a cloud layer between 3 and 4 km (see Table 3.1 for detailed parameters). In each layer, the thermal source term in terms of Planck's function linearly varies with the optical thickness. The parameters such as optical thickness, single-scattering albedo, and phase matrix within the cloudy layer are obtained from Mie calculations. Since the cloud droplets are spherical, the thermal source is essentially unpolarized. However, the underlying surface is in a fully polarized oceanic state and has an emissivity vector and a reflectivity matrix computed from an ocean polarimetric emissivity model. These parameters are computed at a surface wind speed of 10 m/s and a surface temperature of 300 K.

Table 3.1 also lists and compares the Stokes vectors computed from two models. The discrepancies between the two models are less than 0.01 K for the first two Stokes components, whereas the results are identical for the third and fourth components. Thus, the forward model calculations including beam and thermal

Table 3.1 Stokes vectors at the top of the atmosphere calculated from the VDISORT and compared with the results from the doubling–adding model for a microwave frequency of 37 GHz.

ϕ	Doubling–adding				VDISORT			
	I	Q	U	V	I	Q	U	V
0	228.599	32.371	0.000	0.000	228.604	32.367	0.000	0.000
15	228.385	32.088	−0.671	0.148	228.390	32.083	−0.671	0.148
30	227.846	31.343	−1.172	0.258	227.851	31.339	−1.172	0.258
45	227.231	30.403	−1.391	0.299	227.237	30.398	−1.391	0.299
60	226.799	29.575	−1.309	0.259	226.805	29.571	−1.309	0.259
75	226.678	29.080	−0.997	0.146	226.684	29.076	−0.997	0.146
90	226.802	28.976	−0.578	−0.010	226.808	28.972	−0.578	−0.010
105	226.966	29.161	−0.182	−0.170	226.971	29.157	−0.182	−0.170
120	226.950	29.458	0.102	−0.290	226.956	29.454	0.102	−0.290
135	226.663	29.712	0.238	−0.335	226.668	29.708	0.238	−0.335
150	226.193	29.853	0.239	−0.291	226.198	29.849	0.239	−0.291
165	225.759	29.900	0.143	−0.168	225.765	29.896	0.143	−0.168
180	225.584	29.906	−0.000	0.000	225.590	29.902	0.000	−0.000
195	225.759	29.900	−0.143	0.168	225.765	29.896	−0.143	0.168
210	226.193	29.853	−0.239	0.291	226.198	29.849	−0.239	0.291
225	226.663	29.712	−0.238	0.335	226.668	29.708	−0.238	0.335
240	226.950	29.458	−0.102	0.290	226.956	29.454	−0.102	0.290
255	226.966	29.161	0.182	0.170	226.971	29.157	0.182	0.170
270	226.802	28.976	0.578	0.010	226.808	28.972	0.578	0.010
285	226.678	29.080	0.997	−0.146	226.684	29.076	0.997	−0.146
300	226.799	29.575	1.309	−0.259	226.805	29.571	1.309	−0.259
315	227.231	30.403	1.391	−0.299	227.237	30.398	1.391	−0.299
330	227.846	31.343	1.172	−0.258	227.851	31.339	1.172	−0.258
345	228.385	32.088	0.671	−0.148	228.390	32.083	0.671	−0.148
360	228.599	32.371	−0.000	0.000	228.604	32.367	0.000	−0.000

Microwave problem

Frequency: 37 GHz; atmosphere stratification: three layers
Gamma size distribution of cloud droplets;
Effective radius: 10 mm; liquid water path (3 – 4 km): 0.5 mm;
Surface temperature: 300 K; surface wind speed: 10 m/s
Local zenith angle: 53°.

245 K ——— 8 km
Vapor
273 K ——— 4 km
Vapor and cloud
280 K ——— 3 km
Vapor
300 K ——— 0 km

The radiances are converted to brightness temperatures and shown as a function of the relative azimuthal angle (ϕ) between the viewing direction and wind direction with a fixed zenith angle of 53°. Ocean surface emissivity vector is calculated from the polarimetric model with a wind speed of 10 m/s.

sources, scattering, and surface polarization are reliable and accurate. The polarization signals arise primarily from the ocean emission that is fully polarimetric. Atmospheric gases and cloud hydrometeors in the second case mainly attenuate the surface polarization and thermally emit unpolarized radiation.

Radiance Jacobians corresponding to various geophysical parameters can be calculated using the finite differential method (FDM) that computes the radiance twice with one related to the basic state and the other corresponding to the perturbed condition. In this approach, a perturbation to the parameter within a layer requires new calculations of all the optical parameters at other layers; thus, the technique demands huge computational resources. Furthermore, the perturbation to each parameter should be small but large enough to produce a meaningful radiance perturbation. Strictly speaking, the ratio of radiance perturbation to the variable increment approaches the actual gradient when the increment approaches zero. In general, there is no criterion for selecting the perturbation magnitude. Thus, there is always an uncertainty in using the FDM for the radiance gradient calculation.

To illustrate how the FDM is converged to the VDISORT Jacobian model, we compute and compare the Jacobians relative to the various parameters under a cloudy atmosphere where the hydrometeors in various phases coexist. The atmospheric profiles including temperature, water vapor, cloud liquid, ice, and rain water contents are the outputs of the mesoscale model version 5 (MM5) simulations of Hurricane Bonnie (see Figure 3.2). At 37 GHz, both cloud liquid (nonprecipitating) and rain water (precipitating) contents are used because ice clouds are relatively transparent and their radiative effects can be neglected at this frequency. Table 3.2 displays the mean single-scattering albedo and optical thickness at 37 GHz from Mie calculations. A general gamma size distribution function

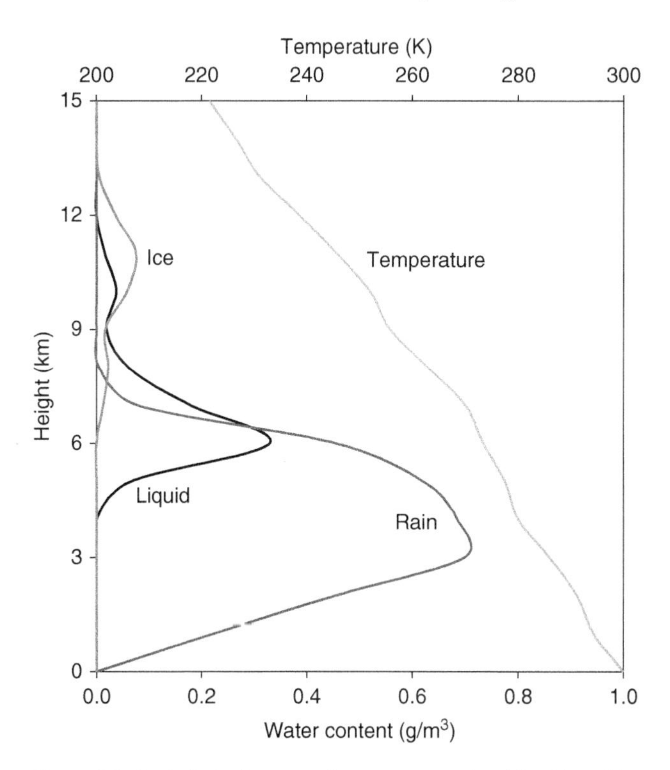

Figure 3.2 Vertical distributions of temperature, cloud liquid, and ice water content contributed from nonprecipitating hydrometers, rain water content from precipitating hydrometers. (Weng and Liu 2003 [46] http://journals.ametsoc.org/doi/pdf/10.1175/1520-0469%282003%29060%3C2633%3ASDAINW%3E2.0.CO%3B2.)

Table 3.2 Optical thickness and single-scattering albedo at 37 GHz for cloud liquid and rain water profiles.

Layer (km)	Single-scattering albedo	Optical thickness
14–15	0.00000	0.00040
13–14	0.00000	0.00060
12–13	0.00000	0.00070
11–12	0.00000	0.00100
10–11	0.00004	0.00890
9–10	0.00005	0.01870
8–9	0.00005	0.01120
7–8	0.00313	0.02870
6–7	0.09686	0.11930
5–6	0.28951	0.61180
4–5	0.35888	0.76450
3–4	0.37552	0.83850
2–3	0.37512	0.86530
1–2	0.32877	0.55410
0–1	0.29511	0.21940

is used with an effective radius of 10 μm for nonraining clouds and 500 μm for raining clouds. At 37 GHz, both the optical thickness and single-scattering albedo vary significantly with altitude and peak between 3 and 4 km where the rain water content is maximal. However, a maximum of single scattering albedo is only 0.38, which indicates that the scattering is not a predominate process.

Table 3.3 displays the ratio of the Jacobians of the optical thickness derived from the VDISORT to those derived from the FDM at 37 GHz. At a smaller $\Delta\tau$ of 0.01, the Jacobian ratio approaches unity for all four components throughout most of the atmosphere. At 37 GHz, $\Delta\tau$ of 0.001 may be needed for the FDM to obtain an accuracy approaching that obtained from the VDISORT. This again illustrates that the perturbation used in the FDM must be small but large enough relative to the basic state so that the radiance perturbation is computed to be numerically meaningful. Table 3.4 displays the ratio of the Jacobians of the single-scattering albedo derived from the VDISORT to those derived from the FDM at 37 GHz. It shows that the two methods agree well.

The VDISORT Jacobian model is further utilized to compute various Jacobian profiles. Figure 3.3a,b displays the Jacobians at 37 GHz relative to cloud liquid water and rain water content, respectively. The Jacobian in I-component is positive, whereas that in Q-component is slightly negative (see Figure 3.3a). This implies that nonprecipitating cloud emits the radiation at this frequency and results in an increase in the brightness temperatures as the cloud liquid content increases. However, for precipitating clouds (see Figure 3.3b), the Jacobian is initially positive between 7 and 9 km and then becomes largely negative between 3 and 6 km. This is due to the emission from a small amount of the raining droplets aloft and the scattering from the large raindrops at lower levels. The

Table 3.3 The ratio of the optical thickness Jacobian at 37 GHz computed from the doubling and adding method to that obtained from the finite differential method, as a function of an increment of the optical thickness.

Z (km)	$\Delta\tau = 0.1$		$\Delta\tau = 0.01$		$\Delta\tau = 0.001$	
	I	Q	I	Q	I	Q
15	0.988	0.994	0.994	0.997	0.993	0.994
14	1.003	0.995	0.996	0.997	0.994	0.994
13	1.035	0.997	1.000	0.998	0.994	0.994
12	1.114	0.999	1.010	0.998	0.995	0.994
11	1.416	1.001	1.047	0.998	0.999	0.994
10	−2.909	1.003	0.513	0.999	0.944	0.994
9	0.531	1.004	0.938	0.999	0.988	0.994
8	0.734	1.008	0.963	0.999	0.990	0.994
7	0.785	1.008	0.969	0.999	0.991	0.994
6	0.809	0.956	0.972	0.991	0.991	0.993
5	1.024	0.888	0.997	0.982	0.994	0.992
4	0.946	−5.059	0.988	0.346	0.993	0.928
3	0.945	0.972	0.988	0.991	0.993	0.993
2	0.941	0.955	0.988	0.989	0.993	0.993
1	0.955	0.954	0.989	0.989	0.993	0.993

Table 3.4 The ratio of the single-scattering albedo Jacobian at 37 GHz computed from the doubling and adding method to that obtained from the finite differential method, as a function of an increment of the single-scattering albedo.

Z (km)	$\Delta\tau = 0.1$				$\Delta\tau = 0.01$			
	I	Q	U	V	I	Q	U	V
15	1.060	0.634	1.372	0.907	1.007	0.959	1.041	0.991
14	1.061	0.658	2.228	0.900	1.007	0.962	1.136	0.990
13	1.030	1.091	0.979	1.000	1.004	1.010	0.996	1.000
12	1.016	1.001	1.062	1.000	1.002	0.996	0.996	1.000
11	0.999	1.041	1.000	1.000	1.000	1.000	1.000	1.000
10	0.999	1.035	1.000	1.000	1.000	1.000	1.000	1.000
9	0.999	1.038	1.000	1.000	1.000	1.000	1.000	1.000
8	0.999	1.031	1.000	1.000	1.000	1.000	1.000	1.000
7	0.999	1.030	1.000	1.000	1.000	1.000	1.000	1.000
6	0.999	1.030	1.000	1.000	1.000	1.000	1.000	1.000
5	0.999	1.030	1.000	1.000	1.000	1.000	1.000	1.000
4	1.354	1.000	1.000	1.000	1.041	1.000	1.000	1.000
3	0.982	1.036	1.000	1.000	0.998	0.998	1.000	1.000
2	0.996	1.031	1.000	1.000	1.000	1.000	1.000	1.000
1	0.997	1.031	1.000	1.000	1.000	1.000	1.000	1.000

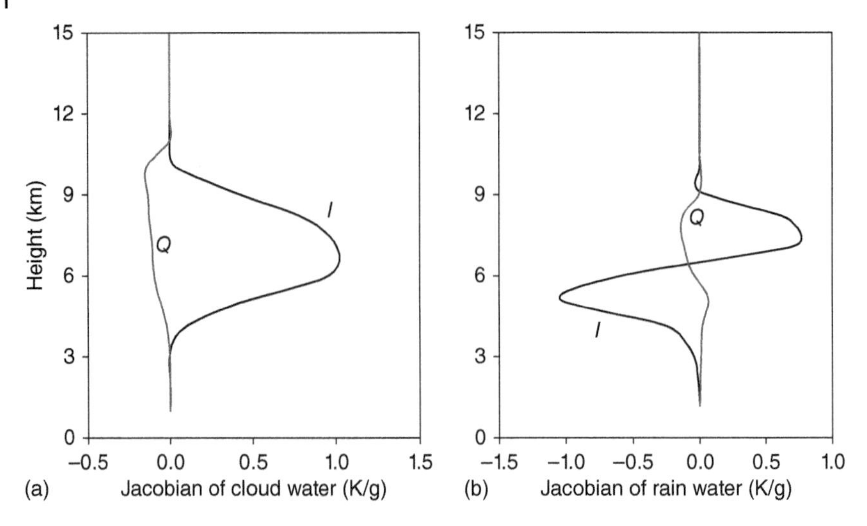

Figure 3.3 Vertical distributions of Jacobians of (a) cloud liquid water and (b) rain water. Note that both Jacobians are divided by a factor of 10. (Weng and Liu 2003 [46] http://journals.ametsoc.org/doi/pdf/10.1175/1520-0469%282003%29060%3C2633%3ASDAINW%3E2.0.CO%3B2.)

positive Jacobian in Q-component at the lower levels results from the scattering of larger raindrops.

3.6 The Zeroth-Order Approximation to Radiative Transfer Solution

Microwave radiative transfer can be simplified if single and multiple scattering terms are neglected and there are no azimuthally dependent terms. Thus, we can rewrite Eq. (3.12) in a scalar form as

$$\mu \frac{\mathrm{d}I(\tau,\mu)}{\mathrm{d}\tau} = I(\tau,\mu) - B(\tau), \tag{3.39}$$

where I is the zeroth-order term of radiance in the cosine mode in Eq. (3.11). For convenience, we neglect the subscript of Fourier zeroth component. When the terms from single and multiple scattering are neglected, Eq. (3.39) can be expressed in the following form:

$$\begin{aligned} I(\tau_0,\mu) &= I(\tau_s,\mu)\exp(-\tau_s/\mu) \\ &\quad + \int_0^1 r_s(\mu,\mu')\mathrm{d}\mu' \int_{\tau_0}^{\tau_s} B(\tau,T)\exp\left[-\frac{(\tau-\tau_0)}{\mu'}\right]\mathrm{d}\tau/\mu \\ &\quad + \int_{\tau_s}^{\tau_0} B(\tau,T)\exp\left[-\frac{(\tau_s-\tau)}{\mu}\right]\mathrm{d}\tau/\mu, \end{aligned} \tag{3.40}$$

or

$$I(\tau_0, \mu) = I(\tau_s, \mu)\exp(-\tau_s/\mu) + I_u + I_d, \quad (3.41)$$

$$I_u = \int_{\tau_s}^{\tau_0} B(\tau, T)\exp\left[-\frac{(\tau_s - \tau)}{\mu}\right] \mathrm{d}\tau/\mu, \quad (3.42)$$

$$I_d = \int_0^1 r_s(\mu, \mu')\mathrm{d}\mu' \int_{\tau_0}^{\tau_s} B(\tau, T)\exp\left[-\frac{(\tau - \tau_0)}{\mu'}\right] \mathrm{d}\tau/\mu, \quad (3.43)$$

where I_u and I_d are the downwelling and upwelling radiances, respectively; and r_s is the surface reflectivity. At the microwave frequencies, radiance is related to brightness temperature under Rayleigh–Jean approximation. In addition, we only consider the first Stokes component (i.e., intensity), which is the brightness temperature. After some manipulation, we can derive

$$T_b = \varepsilon T_s\exp(-\tau_s/\mu) + T_u + (1 - \varepsilon)(1 + \Omega)(T_d + T_c)\exp(-\tau_s/\mu), \quad (3.44)$$

$$T_u = \int_{\tau_s}^{\tau_0} B(\tau, T)\exp\left(-\frac{(\tau_s - \tau)}{\mu}\right) \mathrm{d}\tau/\mu, \quad (3.45)$$

$$T_d = \int_{\tau_0}^{\tau_s} B(\tau, T)\exp\left(-\frac{(\tau - \tau_0)}{\mu}\right) \mathrm{d}\tau/\mu, \quad (3.46)$$

where ε is the surface emissivity, T_s is the surface temperature, and T_c is the cosmic background brightness temperature. The parameter Ω is introduced for nonspecular effect of surface reflection and varies with surface roughness, sea surface wind speed, frequency, and atmospheric transmittance [52]. For an isothermal atmosphere, upwelling and downwelling components in terms of brightness temperatures can be approximated as

$$T_u \approx T_d = (1 - \Upsilon)T_m, \quad (3.47)$$

where $\Upsilon = \exp\left(-\frac{(\tau_s - \tau_0)}{\mu}\right)$ and T_m is the atmospheric temperature. Thus,

$$T_b = T_s[1 - (1 - \varepsilon)\Upsilon^2] - \Delta T(1 - \Upsilon)[1 + (1 - \varepsilon)\Upsilon], \quad (3.48)$$

where $\Delta T = T_s - T_m$. It is apparent that brightness temperatures under these approximations are directly related to the layer mean temperature and atmospheric transmittance. Under a low emissivity condition, the brightness temperature increases as the atmospheric transmittance decreases. This physical principle drives the microwave remote sensing of clouds over oceans.

3.7
The First-Order Approximation to Radiative Transfer Solution

For a scattering and absorbing medium, the radiance may be considered azimuthally independent so that the radiative transfer equation is given as

$$\mu \frac{dI(\tau, \mu)}{d\tau} = I(\tau, \mu) - \frac{\omega(\tau)}{2} \int_{-1}^{1} P(\mu, \mu')I(\tau, \mu')d\mu' - (1 - \omega(\tau))B(T), \quad (3.49)$$

where I is the radiance; $\omega(\tau)$ the single-scattering albedo; $P(\mu, \mu')$ the phase function; $B(T)$ the Planck function; T the thermal temperature; τ the optical thickness; μ the cosine of incident zenith angle; and μ' the cosine of scattering zenith angle. A solution for Eq. (3.49) is derived at a viewing angle with a two-stream approximation [53] as follows:

$$\mu \frac{dI(\tau, \mu)}{d\tau} = [1 - \omega(1 - b)]I(\tau, \mu) - \omega b I(\tau, -\mu) - (1 - \omega)B, \quad (3.50a)$$

$$-\mu \frac{dI(\tau, -\mu)}{d\tau} = [1 - \omega(1 - b)]I(\tau, -\mu) - \omega b I(\tau, \mu) - (1 - \omega)B, \quad (3.50b)$$

where b and $1 - b$ is the ratio of the integrated scattering energy in the backward and forward directions, respectively. For an isotropic scattering, $b = 1/2$ so that the scattered energy is the same in both directions. Since $b = (1 - a)/2$ and is generally less than 1/2 for the Mie scattering case, the forward scattering is much stronger than backward scattering, and the resulting upwelling radiation is reduced. Equation (3.50a,b) can be combined into two second-order differential equations for the final solution [53], assuming that ω, b, and B are independent of τ.

3.8
Ocean Emissivity Model

3.8.1
Ocean Roughness Phenomena

The roughness over oceans is the result of winds blowing over the surface, transferring energy to the surface and generating waves. The energy transfer between near-surface winds and oceans results in a directional surface wave height spectrum. The directionality is dependent on the local wind direction. Wind waves, capillary waves, and foam result from a momentum balance between wind input and dissipative processes such as viscous damping and wave breaking. The initial ocean response to wind forcing is the formation of capillary waves. Under continued wind forcing, wind waves can grow continuously until the water surface becomes unstable and breaks. This process dissipates the excess energy provided by the wind and entrains air bubbles. The process is completed when the air bubbles rise to the surface forming whitecaps and foam.

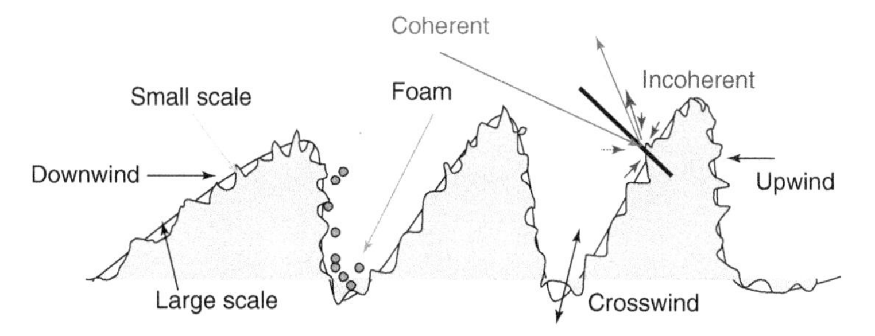

Figure 3.4 Sea surface roughness producing the coherent and incoherent reflections.

Microwave radiation is affected by the scattering and emission from the Earth's atmosphere and surface. In radiative transfer modeling, the contribution from the surface to the upwelling radiance or brightness temperature is expressed as a product of surface emissivity and radiance as well as the surface reflection term. It is generally dependent on the frequency and surface conditions. The hydrodynamic processes affect the sea surface brightness temperatures through three primary scattering mechanisms. First, capillary waves having a wavelength comparable to the radiation wavelength affect the brightness temperature through Bragg scattering. Second, tilting effects, caused by the waves with wavelengths longer than the radiation wavelength, change the effective incidence angle and rotate the polarization states and, as a result, the observed brightness temperature. Third, whitecaps and foam significantly increase the emissivity of the sea surface, thus altering the observed brightness temperature. Thus, the emissivity at microwave frequencies over oceans depends on polarization, surface roughness, foam, and water dielectric constant. A novel ocean wave model approximates roughness with two scales, as shown in Figure 3.4. In this model, small-scale (or capillary) waves are superimposed onto large-scale (or gravity) waves that are called local facets.

For a rough oceanic surface whose slopes have a normal distribution, the angular distribution of diffusely reflected radiation can be well approximated if the surface is considered to be composed of many small facets, each reflecting independently and specularly [54–57]. For fast computations of ocean reflectivity or emissivity, a model called fast microwave emissivity model (FASTEM) was developed [58]. In the latest version of FASTEM, the reflection coefficients in describing the large-scale wave effects are derived by a regression, that is, fitting the results generated from the geometric optics (GO) theory [54]. The GO model is a first-order approximation to the full-scale reflectivity or emissivity model, and the accuracy of the derived emissivity is not adequate for microwave remote sensing applications at low frequencies. In particular, the GO model lacks coherent and incoherent interactions among the bistatic scattering coefficients and does not produce the fourth Stokes component, besides significantly underestimating the wind speed dependence of the horizontally polarized radiation. The wind-induced sea surface emissivity was also derived from the two-scale ocean roughness theory. All the coefficients in the surface emissivity model are derived through a regression against the satellite observations [59] and the model function is referenced to the five frequencies and 49° incidence angle at which the satellite observations operate. In general, the sea surface emissivity is accurately represented by a composite model in which the emission from the rough and foam-free water is given by two-scale scattering theory, and the foam emission is given by a layered dielectric theory [60]. The two-scale model takes the same facet treatment as the GO model for the large-scale wave but uses the bistatic scattering coefficients instead of the earlier approach through which the Fresnel reflection coefficients are multiplied by an exponential function of the height variance for small ripples. The coherent part of the bistatic scattering coefficients is the sum of the Fresnel reflection coefficients and the correction of the specular reflection coefficients caused by the small surface perturbation, which depend on the shortwave part of the surface roughness spectrum. The incoherent scattering coefficients are proportional to the shortwave part of the surface roughness spectrum.

In calculating the Fresnel reflectivity of the small facets, the dielectric constant is based on the single Debye relaxation law [61] and is a function of salinity, surface temperature, and frequency [62]. With all the updates to the water dielectric models from laboratory measurements [60, 63, 64], the model frequencies are now valid between 1.4 and 410 GHz. The FASTEM is further improved with the two-scale sea surface roughness model by Durden and Vesecky [65, 66]. In the following subsections, the theoretical basis for the FASTEM is described in detail.

3.8.2 Approximation of Water Dielectric Constant

When an electric field is applied in water, charge separation and molecular rearrangement within water occur, causing the phenomenon of polarization. The magnitude of the polarization can be measured by the water dielectric constant. This macroscopic property is related to the microstructure through the molecular polarizability and dipole moment. The dipole orientation takes place as the molecule attempts to align with the electric field to adopt a low-energy configuration. By neglecting intermolecular interaction, Debye [61] proposed an approximate equation to describe the dielectric constant (also referred to as permittivity) as follows:

$$e = e_\infty + \frac{e_s - e_\infty}{1 + j2\pi f\tau}, \tag{3.51}$$

where e_s and e_∞ are the dielectric constant at zero (static) and infinite frequencies, respectively; τ is the relaxation time constant; and f is the electromagnetic frequency in gigahertz. All the three parameters are a function of water temperature. For seawater, dissolved salts render it a good conductor and contribute to the imaginary part of the dielectric constant as

$$e = e_\infty + \frac{e_s - e_\infty}{1 + j2\pi f\tau} + j\frac{\alpha}{2\pi f e_0}, \tag{3.52}$$

where α is the ionic conductivity, and e_0 is a parameter. The conductivity of seawater increases with the number of ions, and the extent of polarization due to the displacement of bound charges in seawater depends on its salinity [67]. Therefore, the static dielectric constant, relaxation time, and ionic conductivity depend on the salinity of seawater. Below 20 GHz, the effect of salinity on the permittivity is not negligible. Klein and Swift [62] presented the permittivity as a function of frequency and water temperature as well as salinity, based on laboratory measurements. Klein and Swift used measurements at 1.43 and 2.653 GHz to derive the coefficients of their permittivity model. Their model has sufficient accuracy for the very low frequency range of microwaves and has been used in numerous applications for many years. Guillou *et al.* [68] published a new permittivity model based on the measurements performed at the Laboratoire de Physique des Interactions Onde Matière in France. Seawater samples were collected from different regions to cover the natural salinity variations over the globe. Samples from the Mediterranean Sea exhibited salinities at 38‰ and 38.9‰. Medium-salinity samples (35.7‰) were found in the mid-North Atlantic, while low-salinity samples were taken from the Atlantic–Gironde estuary, rich in sediment content (23.2‰, 28‰), and from polar seas (30.2‰). The measurements were performed between −2 and 30 °C with two vector network analyzers. In the same year (1998) and in the same journal, Radio Science, Ellison *et al.* [69] published another new permittivity model based on the measurements over the frequency range of 3–20 GHz in 0.1 GHz steps and over the temperature range between −2 and 30 °C in 1° steps. The measurements at 23.8, 36.5, and 89 GHz and at selected temperatures between −2 and 30 °C are also applied in their model derivation. Later work published by Ellison *et al.* [60] shows the large discrepancy on the permittivity at high frequencies between measurements and model simulations using the Debye formula. The work may evidence that intermolecular interaction, which is ignored in the Debye approximation, needs to be taken into account for high frequencies. The intermolecular interaction may be the second polarization that can be considered in a double Debye model [60]:

$$e = e_\infty + \frac{e_s - e_1}{1 + j2\pi f \tau_1} + \frac{e_1 - e_\infty}{1 + j2\pi f \tau_2} + j\frac{\alpha}{2\pi f e_0}, \tag{3.53}$$

where e_1 is the permittivity at an intermediate frequency and f is the frequency in Hz. Barthel *et al.* [70] have pointed out that the double Debye form is necessary and sufficient for the permittivity at high frequencies. For lower alcohols, a triple Debye form is needed. The permittivity at the infinite frequency is a function of the water temperature. The permittivity at zero and intermediate frequencies as well as relaxation times is a function of salinity and temperature. The function is empirical. Most of permittivity models use a polynomial function [68, 71]. Meissner and Wentz [72] used combined functions: polynomial functions, fraction with polynomials in both numerator and denominator, and exponential function. All those empirical functions rely on the coefficients that fit the measurement data.

The permittivity formulation was also derived by Ellison *et al.* [60] and was used in the FASTEM model [58]. It is a double Debye model with the coefficients determined by fitting the permittivity measurement data. The permittivity of synthetic seawater with a constant salinity of 35‰ in about 2 GHz steps from 30 to 105 GHz and at temperatures of −2, 5, 10, 15, 20, 25, and 30 °C was obtained from a free-air propagation method by using the ABMM measuring system [58]. The permittivity model by Ellison *et al.* [60] fits the measurements very well, but it does not include sensitivity to salinity. The Soil Moisture and Ocean Salinity (SMOS) satellite has carried the L-band microwave radiometer from space to measure ocean salinity and soil moisture. The Y-shape that carries 69 separate antenna receivers measures radiation at 1.4 GHz band (1.400–1.427 GHz) to derive the information on soil moisture and ocean salinity [73]. The new capability drives this study to develop an upgraded permittivity model in the FASTEM for studying ocean salinity. The measurement data from 1.4 to 410 GHz are used to determine the fitting coefficients in a new permittivity model. The permittivities at 23.8 and 36.5 GHz and 89 GHz are measured at a constant salinity of 38.89‰ and at water temperatures of −2, 12, 20, and 30 °C [69]. The dielectric properties of seawater are derived at 1.43 and 2.653 GHz [74, 75], and the permittivity of synthetic seawater is also observed at constant salinity of 35‰ from 30 to 105 GHz and at temperatures of −2, 5, 10, 15, 20, 25, and 30 °C. The pure water permittivity is obtained from 1.7 to 410 GHz [70, 71]. The wide variability of the permittivity measurement data with temperature, salinity, and frequency is unique for determining the fitting coefficients for the permittivity model. The double Debye model is applied, and its coefficients are derived through polynomial fitting using water temperature, T, in Celsius and water salinity, S, in parts per thousand, according to the following expressions:

$$e_\infty = a_0 + a_1 T, \tag{3.54a}$$

$$e_s = (a_2 + a_3 T + a_4 T^2 + a_5 T^2)(1 + a_6 S + a_7 S^2 + a_8 TS), \tag{3.54b}$$

$$e_1 = (a_9 + a_{10} T + a_{11} T^2)(1 + a_{12} S + a_{13} S^2 + a_{14} TS), \tag{3.54c}$$

$$e_0 = 8.8429 \times 10 \ \ \mathrm{F/m}, \tag{3.54d}$$

$$\tau_1 = (a_{15} + a_{16} T + a_{17} T^2 + a_{18} T^3)(1 + a_{19} S + a_{20} ST + a_{21} ST^2), \tag{3.54e}$$

$$\tau_2 = (a_{22} + a_{23} T + a_{24} T^2 + a_{25} T^3)(1 + a_{26} S + a_{27} ST + a_{28} S^3), \tag{3.54f}$$

$$\alpha = \alpha_{25} \exp(-\beta\delta), \tag{3.54g}$$

where

$$\delta = 25 - T, \tag{3.54h}$$

$$\beta = a_{29} + a_{30}\delta + a_{31}\delta^2 + S(a_{32} + a_{33}\delta + a_{34}\delta^2), \tag{3.54i}$$

and the ionic conductivity at a temperature of 25 °C

$$\alpha_{25} = S(a_{35} + a_{36} S + a_{37} S^2 + a_{38} S^3). \tag{3.54j}$$

The coefficients a_i are determined by fitting the measurement data.

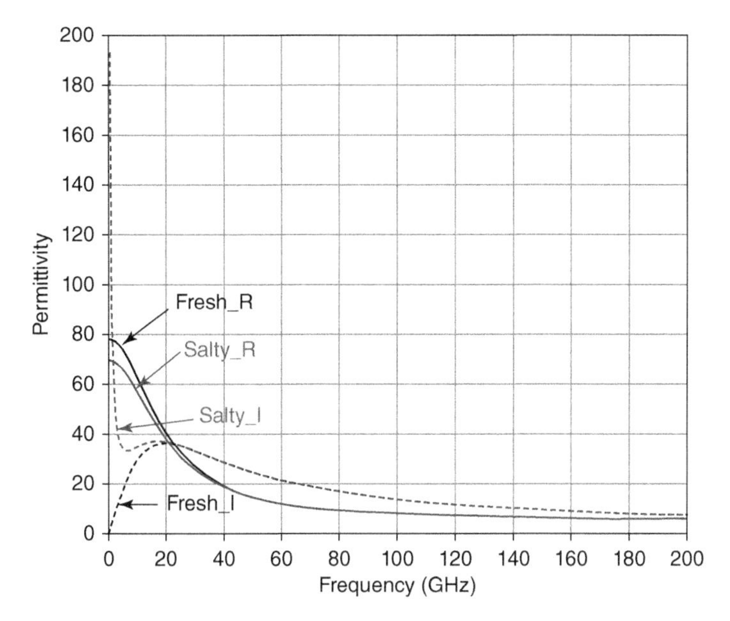

Figure 3.5 Water permittivity corresponding to freshwater and salty water at 25 °C and salinity of 35‰. The solid and dashed lines represent the permittivity in real and imaginary parts, respectively.

Figure 3.5 shows the comparison of the permittivity in both real and imaginary parts between measurements and our model calculations. The black line denotes pure water at a temperature of 25 °C. The red line denotes seawater at a temperature of 25 °C and at a salinity of 35‰. In general, the difference between the measured and the modeled permittivity is less than 3%, within the uncertainty of the measurements. The root-mean-square errors (rms) of the real and imaginary parts between the measurements and the new model simulations are 0.91 and 0.50, respectively. The slightly large error in the real part of the permittivity for salted water may be due to a bias in the measurement. The permittivity real parts above 30 GHz show a large difference between freshwater and salt water, contrary to the common knowledge that both freshwater and salted water should have the same permittivity at high frequencies.

3.8.3
Ocean Roughness Heights and Spectrum

Sea surface roughness is mainly driven by the surface wind vector [76], but wind vector alone cannot fully resolve the surface roughness spectra since the boundary stability and wave development stage also affect the roughness. In the infrared and visible surface emissivity and reflectivity model, the surface slope variance, σ^2, is often used for a surface roughness of a Gaussian distribution. The slope variance measured by Cox and Munk [56] for a clean surface is a linear function of wind speed $U_{12.5}$ (in m/s) at 12.5 m above the surface, namely,

$$\sigma_u^2 = 0.003 + 0.00316 \times U_{12.5}, \tag{3.55a}$$

for an upwind direction and

$$\sigma_c^2 = 0.00192 \times U_{12.5}, \tag{3.55b}$$

for a crosswind direction.

Wind vector is usually accepted to be a unique vector describing the sea surface state. Wu and Fung [77] applied a simple function (BK^{-4}) in terms of a wave number, K, and a wind-dependent constant, B, to calculate the microwave surface emissivity. The full-wave models developed by Bjerkaas and Riedel [78], Durden and Vesecky [65] (hereinafter referred to as DV), Donelana and Pierson [79], Apel [80], and Elfouhaily *et al.* [81] are widely used in microwave sea surface emissivity models. Dinnat *et al.* [82] have investigated the effect of the surface roughness models by Durden and Vesecky and Elfouhaily *et al.* Their results suggested that the DV surface roughness spectrum is more adequate for microwave emissivity calculation at the L-band. Using the surface roughness spectrum by Elfouhaily *et al.*, the sensitivity of the brightness temperature to wind speed is obviously problematic for surface wind speeds below 7 m/s. The spectrum model by Bjerkaas and Riedel [78] was used in a Monte Carlo emissivity model [83] and was able to simulate the brightness temperatures comparable to satellite measurements. But the spectrum is sophisticated and divided into four subspectra. The spectrum does not have a smooth and continuous derivative to the wave number, which may cause problems in tangent-linear and adjoint calculation in the data assimilation. Thus, we chose the DV surface spectrum for this study. The DV surface roughness spectrum can be written as

$$W(K, \phi) = \frac{1}{2\pi K} S(K)\Phi(K, \phi), \tag{3.56a}$$

where $S(K)$ is an omnidirectional spectrum, $\Phi(K, \phi)$ is the angular portion of the spectrum, and ϕ is the wave direction relative to the wind. The DV omnidirectional part is described by the Pierson–Moskowitz spectrum,

$$S(k) = \frac{a}{2\pi} K^{-3} \exp[-0.74(K_0/K)^2], \tag{3.56b}$$

for the wave number $K < K_j$ and $K_0 = g/U_{19.5}^2$, where the gravity acceleration constant $g = 9.8\ \text{ms}^{-2}$, and $U_{19.5}$ is the wind speed at 19.5 m above the surface.

Using a dimensional analysis, Durden and Vesecky [65] have proposed the spectrum for $K \geq K_j$ as

$$S(K) = \frac{a}{2\pi} K^{-3} \left(\frac{bKu_*^2}{g + \gamma K^2} \right)^{c \log_{10}(K/K_j)} \exp[-0.74(K_0/K_j)^2], \tag{3.56c}$$

where $\gamma = 7.25 \times 10^{-5}$. The angular portion of the spectrum is written as

$$\Phi(K, \phi) = 1 + d[1 - \exp(-sK^2)] \cos 2\phi. \tag{3.56d}$$

Obviously, the aforementioned spectrum depends on the empirical parameters, K_j, a, b, c, d, and s. Here, the values for these parameters are set as

$$K_j = 2\,\text{m}^{-1},\ b = 1.75,\ c = 0.25,\ s = 0.00015\,\text{m}^2.$$

The friction velocity of the surface wind u_* can be calculated from the surface wind at a given elevation. Conversely, the wind speed at any elevation z can be calculated from the friction velocity by

$$U(z) = \frac{u_*}{0.4} \log\left(\frac{z}{Z_0}\right), \tag{3.57a}$$

where

$$Z_0 = 0.0000684/u_* + 0.00428 \times u_*^2 - 0.000443. \tag{3.57b}$$

From Eqs. (3.57a,b), the friction velocity of the surface wind, u_*, can be calculated from the surface wind speed at a height of z, and vice versa. It is found that $a = 0.008$ as suggested by Yueh *et al.* [84], who compared the model simulations with the measurements. In our revised ocean two-scale model, $a = 0.016$ is applied, which is referred to as the DV2 model.

The parameter d is an important parameter used to describe the angular portion of the spectrum (see Eq. (3.56d)). For a symmetric distribution (i.e., $d = 0$) of the ocean roughness spectrum, the wind direction cannot be resolved from the remote sensing data. In the existing spectrum models, the parameter d is positive [85]. Thus, the upward slope variance is mathematically larger than the crosswind slope variance. For the DV model, the parameter d is determined by forcing the ratio of the slopes between crosswind and upwind to equal the ratio given by the Cox and Munk model [84], which is

$$d = \frac{2}{1 - N}\left(\frac{1 - R}{1 + R}\right), \tag{3.57c}$$

where

$$N = \int_0^\infty K^2 S(K) \exp(-sK^2) dK \Big/ \int_0^\infty K^2 S(K) dK. \tag{3.57d}$$

For the parameter R, an offset of 0.003 is added to avoid a zero value in the denominator and to ensure that the ratio is less than 1, that is,

$$R = \frac{0.003 + 0.00192 U_{12.5}}{0.003 + 0.00316 U_{12.5}}. \tag{3.57e}$$

It is worth emphasizing that the slope variance calculated from the spectrum is generally different from the slope variance value obtained from the Cox and Munk model, although the ratio of the slope variance between crosswind and upwind is adopted from the Cox and Munk model. In fact, for the full spectrum, the mean slope variance of a rough sea surface can be calculated from its roughness spectrum by

$$\sigma^2 = \int_0^\infty \int_0^{2\pi} K^3 W(K,\phi) \mathrm{d}K \; \mathrm{d}\phi, \tag{3.58a}$$

The slope variance along the up- and downwind direction is then given by

$$\sigma_u^2 = \int_0^\infty \int_0^{2\pi} K^3 W(K,\phi) \mathrm{d}K \cos^2\phi \; \mathrm{d}\phi = \frac{\sigma^2}{1+R}, \tag{3.58b}$$

and the slope variance along the crosswind direction is given by

$$\sigma_c^2 = \int_0^\infty \int_0^{2\pi} K^3 W(K,\phi) \mathrm{d}K \sin^2\phi \; \mathrm{d}\phi = \frac{R\sigma^2}{1+R}, \tag{3.58c}$$

and the height variance (i.e., displacement variance) is given by

$$\xi^2 = \int_0^\infty \int_0^{2\pi} K W(K,\phi) \mathrm{d}K \; \mathrm{d}\phi. \tag{3.58d}$$

Using Eq. (3.56a) and integrating over the direction component, the height variance for a cutoff wave number K_c can be written as

$$\xi_c^2 = \int_{K_c}^\infty S(K) dK. \tag{3.58e}$$

The DV2 spectrum model is applied for computing the surface slope variance. The slope variance obtained from the original DV model is just half of the slope variance obtained from the DV2 model. Figure 3.6 shows the dependence of the total slope variance on the wind speed at 10 m above the surface for the DV2 model (dark black), the DV model (light gray), and the Cox–Munk model (moderate gray). The Cox–Munk slope variance is a linear function of the wind speed. The DV2 and DV total slope variances display a slightly parabolic relationship. Both the original DV and the Cox–Munk model agree, but the variance is far lower than the DV2 total slope variance. For the microwave radiation, the surface effective slope variance is smaller than the total slope variance obtained using Eq. (3.58a).

Figure 3.6 shows that the DV2 model is adequate for computing the microwave surface emissivity because of the significant small-scale slope variance to the electromagnetic wavelength [86]. However, for an infrared or a visible spectrum, K_c is very large, and the small-scale slope variance is negligible. The DV model should be used because its slope variance agrees with the Cox–Munk model that consists

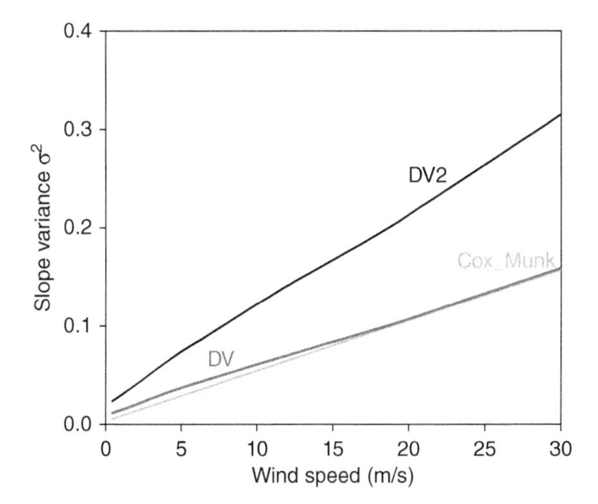

Figure 3.6 The surface slope variance versus wind speed at 37 GHz for three ocean roughness models.

in photographing the Sun's glitter pattern on the sea surface from a plane, for determining the statistics of the surface slope distribution.

Figure 3.7 shows the omnidirectional spectra for three wind speeds. A high wind speed corresponds to a large spectrum. The spectrum part for a large wave number is important for determining the capillary wave, which is used to correct a small-scale of the microwave reflectance.

It can be seen from Figure 3.8 that K_c increases as the frequency or surface wind speed increases. For a given frequency and a given surface wind speed, the cutoff wave number can be calculated from Eq. (3.58c). Figure 3.8 shows the variation of K_c with frequencies for surface wind speeds of 3, 7.5, and 15 m/s.

For microwave remote sensing applications, the electromagnetic wavelength can be comparable to the small irregularities, and both gravity and capillary waves need to be taken into account in the surface microwave emissivity calculation. The large-scale surface roughness is governed by the gravity force (gravity waves), while the small-scale surface roughness (capillary waves) is mainly driven by the surface tension. Several criteria to separate the small-scale surface from the large-scale surface have been suggested in the literatures [87, 88]. The criteria for the small-scale surface were derived by Guissad and Sobieski [89]:

$$k\xi_c \ll 1, \tag{3.59a}$$

and

$$\frac{K_c}{k} \ll 1, \tag{3.59b}$$

with k ($k = \frac{2\pi f}{c}$, c is the speed of light, and f is the frequency) the wave number of the incident electromagnetic wave, K_c the cutoff wave number separating the

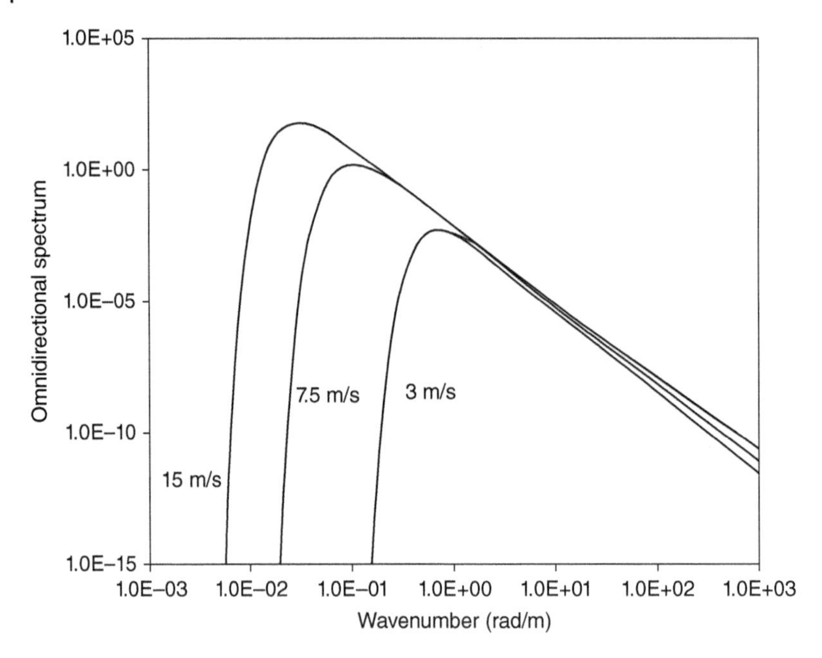

Figure 3.7 Variation of omnidirectional spectrum with wave numbers for three wind speeds at 3, 7.5, and 15 m/s, respectively.

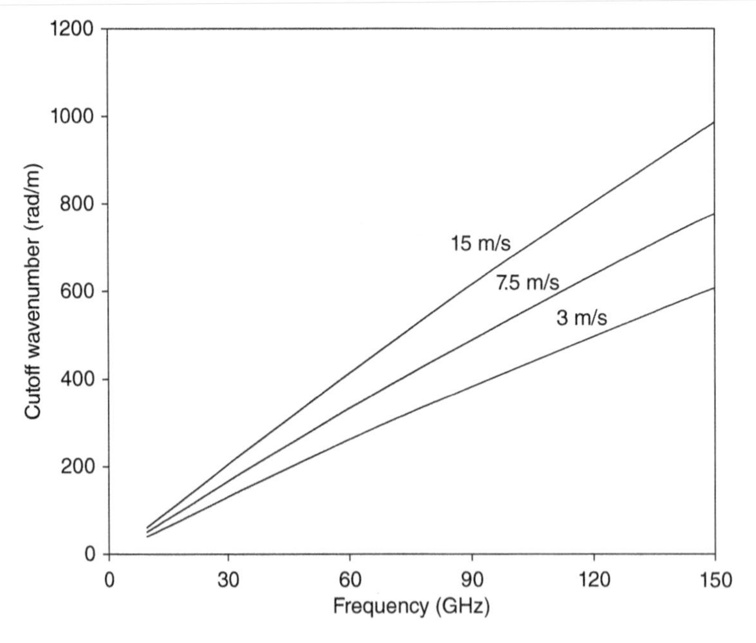

Figure 3.8 Variation of cutoff wave number with frequency for three wind speeds at 3, 7.5, and 15 m/s, respectively.

small-scale surface roughness from the large-scale surface roughness, and ξ_c the height variance for small ripples. Using Eqs. (3.58a,b) and (3.57e) by postulating $k\xi_c = \frac{K_c}{k}$, the cutoff wave number K_c is obtained from

$$\frac{K_c^2}{k^4} = \xi_c^2 = \int_{K_c}^{\infty} S(K)dK. \tag{3.59c}$$

Obviously, the cutoff wave number K_c of the surface roughness to a large-scale or a small-scale wave depends on the wind speed and the electromagnetic wavelength acting on the surface [83]. For microwave frequencies, the cutoff wave number is between 10 and 300 m^{-1}, depending the surface wind speed and electromagnetic wavelength. The cutoff value will affect the slope variance for the large-scale wave, which is used in the GO and two-scale models [84]. The slope variance for the large-scale wave and for a given K_c can be adjusted by changing the integration upper limit of (Eq. 3.58a–c) from ∞ to K_c. For an infrared or visible wavelength, K_c is large enough, and adjustment is not necessary. However, the adjustment to the large-scale slope variance for microwave sensors is significant.

3.8.4
Foam Coverage

For a wind speed larger than several meters per second, foam starts to affect the surface emissivity. Foam is often a mixture of air and water. The air volume fraction in the foam can be very high and is greater than 0.95. The foam coverage was expressed by Tang [90]:

$$f_c = 7.75 \times 10^{-6} \left(\frac{V}{V_0} \right)^{3.231}, \tag{3.60}$$

where V is the wind speed in m/s at 10 m above the sea surface and V_0 is a constant of 1 m/s. The total reflectivity is calculated from the sum of the foam reflectivity weighted with the foam coverage (f_c) and the reflectivity of water weighted with the water coverage ($1 - f_c$). Both foam emissivity and coverage affect the surface emissivity, and the two parameters may also depend on the stability in a lower boundary layer. Schrader [91] explicitly treated sea foam as the third scale in his microwave ocean emissivity model. Stogryn [92] derived an empirical sea foam emissivity model as a function of frequency, incidence angle, and sea surface temperature from radiometric measurements. However, a recent study by Rose *et al.* [93] shows that the measurements of foam emissivity at 10.8 and 36.5 GHz are greater than the foam emissivity obtained using the Stogryn model. According to their measurements, the measured foam emissivity is greater than 0.9 over a range of incidence angles between 30° to 60° for the vertical polarization. Kazumori *et al.* [94] further improved the foam emissivity model that depends on polarization, angle, and frequency.

3.8.5 Surface Emissivity Vector

The two-scale modeling of the ocean surface emissivity [77, 84, 88] requires integrations over the ocean wave spectrum and over all facet directions, which demand huge computational resources. The Monte Carlo two-scale microwave emissivity model [83] demands even more computational time. A new approach has been developed for effective calculations for various applications such as satellite radiance assimilation and in the retrieval of geophysical parameters. The first version of the fast emissivity model (FASTEM) is an empirical and parameterized model that is based on the two-scale accurate simulations. It is composed of small-scale corrected Fresnel reflection coefficient, large-scale adjustment, and foam. FASTEM-1 is found to be problematic when it is used for simulating the special sensor microwave/imager (SSM/I) brightness temperatures. In the revised FASTEM (version 2), the total atmospheric transmittance is used in a correction factor to the reflectivity to account for angular-dependent downward radiation. The FASTEM-1 and -2 do not include the Stokes third and fourth emissivity components. To enable the processing of a polarimetric sensor, the FASTEM is extended to its version 3 by including the polarimetric components.

The theoretical basis for the FASTEM starts from a modified GO model in which a small-scale correction is applied to the Fresnel reflectivity. The modified GO model for the vertically and horizontally polarized reflectance can be written as [83]

$$\begin{bmatrix} \Gamma_v(\theta) \\ \Gamma_h(\theta) \end{bmatrix} = \iint Rot_1 \begin{bmatrix} F_v^m & 0 \\ 0 & F_h^m \end{bmatrix} Rot_2 \begin{bmatrix} 1 \\ 1 \end{bmatrix} P \sin\theta_s \mathrm{d}\theta_s \mathrm{d}\varphi_s, \tag{3.61}$$

where θ, φ and θ_s, φ_s are the zenith and azimuth angles of the incident and scattering direction, respectively; and P is the probability function of facets that depends on the surface slope variance. The modified Fresnel reflectance, F_v^m and F_h^m as shown in eq. (3.61), are evaluated at a local plane of the facet. The rotation matrix Rot_2 first rotates the incident radiation vector into the local plane of the facet, and the rotation matrix Rot_1 rotates from the facet reflected radiation vector to the meridian plane, which contains the surface normal direction and the outgoing radiation direction [83, 89], showing that the modified GO model achieves a good accuracy for the vertically and horizontally polarized brightness temperatures in comparison to the Hollinger's measurements and the satellite measurements. The model was also validated between the model calculation and the satellite measurements. The calculated brightness temperatures agree in general with the satellite measurements. The rms error between the modeled and the measured microwave brightness temperature is less than 2 K for 19.35, 22.235, and 37 GHz.

Assuming that the ocean surface reflection is a quasi-specular, the product of the rotation matrices and the modified Fresnel reflectance around the incident direction in Eq. (3.61) can be expanded as

$$\begin{bmatrix}\Gamma_v(\theta)\\ \Gamma_h(\theta)\end{bmatrix} = \iint \begin{bmatrix}F_v^m(\theta) & 0\\ 0 & F_h^m(\theta)\end{bmatrix}\begin{bmatrix}1\\1\end{bmatrix} P\sin\theta_s \mathrm{d}\theta_s \mathrm{d}\varphi_s$$
$$+ \iint L \begin{bmatrix}1\\1\end{bmatrix} P\sin\theta_s \mathrm{d}\theta_s \mathrm{d}\varphi_s$$
$$\approx \begin{bmatrix}F_v^m(\theta)\\ F_h^m(\theta)\end{bmatrix} + \begin{bmatrix}R_{Lv}\\ R_{Lh}\end{bmatrix}. \tag{3.62}$$

The second term on the right side of Eq. (3.62) represents the large-scale correction. The surface emissivity also depends on the relative azimuth angle between the wind direction and the sensor azimuth angle. Here, three cosine harmonic functions are used for the vertical and horizontal polarization and three sinusoidal harmonic functions for the third and the fourth Stokes components [95], as suggested by St Germain and Poe [96]. The emissivity at a zenith angle, θ, and a relative azimuth angle, φ_R, thus can be summarized as follows:

$$\varepsilon_v = (1 - f_c)[1 - F_v \exp(-y\cos^2\theta)] + \text{Large\_cor}_v]$$
$$+ f_c \times \varepsilon_{foam\_v} + \sum_{m=1}^{3} c_m \cos(m\varphi_R), \tag{3.63a}$$

$$\varepsilon_h = (1 - f_c)[1 - F_h \exp[-y\cos^2\theta)] + \text{Large\_cor}_h]$$
$$+ f_c \times \varepsilon_{foam\_h} + \sum_{m=1}^{3} d_m \cos(m\varphi_R), \tag{3.63b}$$

$$\varepsilon_3 = \sum_{m=1}^{3} e_m \sin(m\varphi_R), \tag{3.63c}$$

$$\varepsilon_4 = \sum_{m=1}^{3} g_m \sin(m\varphi_R), \tag{3.63d}$$

where the small-scale correction parameter $y = h(k\xi_c)^2$, and the large-scale correction terms, Large_cor_v and Large_cor_h, are determined by fitting the rigorous two-scale model calculations. The coefficients c_m, d_m, e_m, g_m are obtained by fitting both surface measurement data and the rigorous two-scale model calculations, which enable us to apply the azimuthal part to various zenith angles and frequencies. The effects of sea foam on the emissivity for the third (ε_3) and fourth (ε_4) Stokes components are not explicit and may be partially included by fitting to the measurement data.

In Eq. (3.63), the small-scale correction parameter, y, is also parameterized according to surface wind speed by fitting it with the results derived from the small-scale correction parameter:

$$y = h_1 \times W \times f + h_2 \times W \times f^2 + h_3 \times W^2 \times f + h_4 \times W^2 \times f^2 + h_5 \times W^2/f + h_6 \times W^2/f^2 + h_7 \times W + h_8 \times W^2, \tag{3.64}$$

where W is the wind speed with a unit of meter per second, and f is the frequency in GHz. The coefficients h_i are obtained by fitting the height variance computed from the surface roughness.

The large-scale correction parts are written in the following regression equations:

$$\begin{aligned}\text{Large\_cor}_v = {} & z_1 + z_2 f + z_3 f^2 + (z_4 + z_5 f + z_6 f^2)/\cos\theta \\ & + (z_7 + z_8 f + z_9 f^2)/\cos^2\theta + (z_{10} + z_{11} f + z_{12} f^2) \\ & \times W + (z_{13} + z_{14} f + z_{15} f^2) \times W^2 + (z_{16} + z_{17} f + z_{18} f^2) \\ & \times W/\cos\theta,\end{aligned} \tag{3.65a}$$

$$\begin{aligned}\text{Large\_cor}_h = {} & z_{19} + z_{20} f + z_{21} f^2 + (z_{22} + z_{23} f + z_{24} f^2)/\cos\theta \\ & + (z_{25} + z_{26} f + z_{27} f^2)/\cos^2\theta + (z_{28} + z_{29} f + z_{30} f^2) \\ & \times W + (z_{31} + z_{32} f + z_{33} f^2) \times W^2 + (z_{34} + z_{35} f + z_{36} f^2) \\ & \times W/\cos\theta.\end{aligned} \tag{3.65b}$$

The regression coefficients z_i are evaluated using the large-scale contribution extracted from the rigorous two-scale model calculations with a constraint to

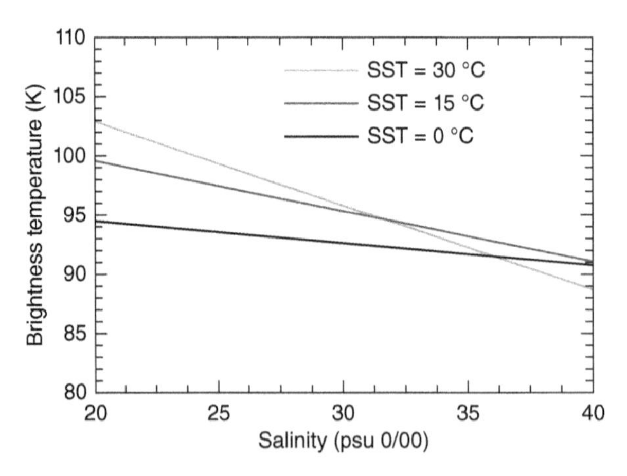

Figure 3.9 Variation of the surface brightness temperature with the seawater salinity for the sea surface temperatures at 0, 15, and 30 °C.

ensure the same emissivity value for both vertical and horizontal polarization at nadir.

Figure 3.9 shows brightness temperatures at 1.4 GHz as a function of seawater salinity. In general, brightness temperatures linearly decrease as the salinity increases. Under higher temperature conditions, the brightness temperatures are sensitive to salinity. For salinity less than 30 per thousand in volume, the brightness temperatures are more sensitive to temperature. Note that these computations are performed for a wind speed of 7 m/s and under the nadir condition.

Figure 3.10 shows the variation of surface emission at the third and fourth Stokes components from the ocean Stokes emissivity model calculated from the radiative transfer model. The variation of Stokes vector to the wind direction is generally less than 3 K. The amplitude of the variation for the fourth component of the

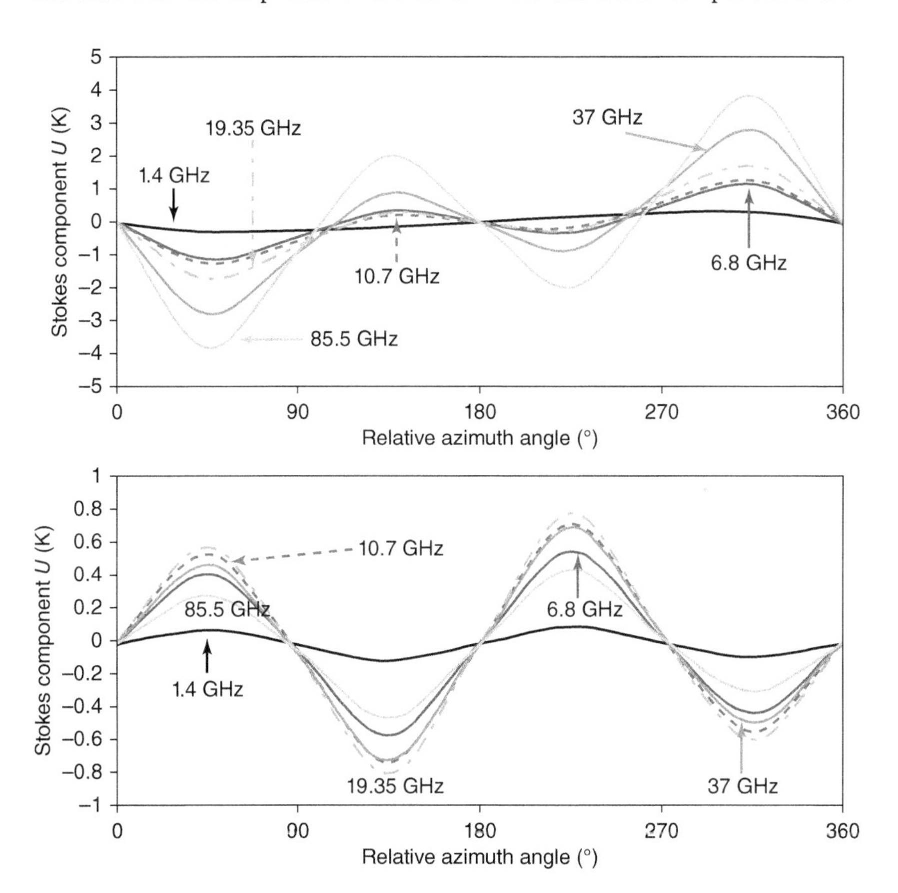

Figure 3.10 Variation of third and fourth Stokes components at 1.4, 6.8, 10.7, 19.35, 37, and 85.5 GHz for wind of 10 m/s above 19.5 m with relative azimuthal angle simulated from the two-scale emissivity model for a wind speed of 10 m/s at a height of 19.5 m and a surface temperature of 300 K. (Weng and Liu 2003 [46] http://journals.ametsoc.org/doi/pdf/10.1175/1520-0469%282003%29060%3C2633%3ASDAINW%3E2.0.CO%3B2.)

Stokes vector is only about 0.5 K and depends on the frequency, although the envelope is similar. Thus, under nonprecipitation conditions, the measurements at higher microwave frequencies such as 37 GHz may be very valuable for light wind retrievals because of the increasing sensitivity. The measurements at 85 GHz are also sensitive to wind direction, but the noises from atmospheric absorption and scattering limit the high frequency applications for wind vector retrievals.

3.9 Land Emissivity Model

Microwave emissivity spectra over land were simulated and measured over a limited range of frequencies and surface conditions. For bare soil, the emissivity was derived as a function of soil moisture, soil textural components (e.g., clay and sand), and surface roughness [97, 98]. However, the bare soil emissivity was only modeled at lower frequencies where the soil dielectric constants are empirically adjusted using the ground-based measurements. The emissivity of the vegetation canopy was also simulated, using a simplified radiative transfer model, and the canopy optical parameters in the model were derived from Rayleigh's approximation [99], which is only valid for low frequencies. More sophisticated radiative transfer schemes were proposed to simulate the effects of the vegetation canopy on the microwave emissivity [100–102]. For a snow-covered surface where ice particles have a high fractional volume, the optical parameters were approximated using an effective wave propagation constant for the medium [103]. Alternatively, the Mie phase matrix must be modified to account for scattering interaction of closely spaced scatterers [101] and then used in the radiative transfer equation. In addition to modeling the emissivity, the techniques were developed to retrieve the microwave emissivity over land using the data from satellite window channels together with some auxiliary data [104–108].

In this section, the latest improvements on microwave land emissivity model are discussed. A general radiative transfer scheme and its approximated solution used for simulating land emissivity of scattering and emitting medium are proposed.

3.9.1 Theoretical Approach for Land Emission

The microwave land emissivity over various surfaces such as snow, deserts, and vegetation is computed from a radiative transfer scheme [106]. The reflected and emitted components occurring at the interfaces above and below the scattering layer are taken into account, and the cross-polarization and attenuation due to surface roughness are parameterized as a function of roughness height and frequency (see Figure 3.11). For the vegetation canopy, the optical parameters are derived using GO. For a medium with a higher fractional volume of particles such as snow and deserts, the scattering and absorption coefficients are approximated using the dense medium theory. The model takes satellite zenith angle, frequency,

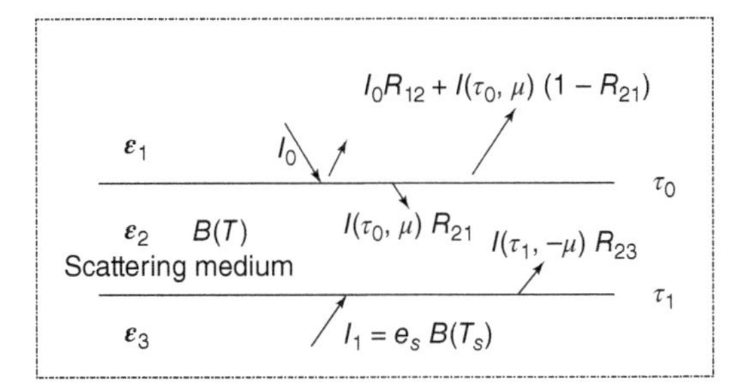

Figure 3.11 A schematic diagram illustrating the radiative transfer process through a three-layer medium. The boundary layer is generalized to include the interface reflection caused by the dense particles. (Weng *et al.* 2001 [106]. Reproduced with permission of Wiley.)

soil moisture content, vegetation fraction, soil temperature, land surface temperature, and snow depth as inputs and computes surface emissivity at both vertical and horizontal polarization states.

In general, the upwelling and downwelling radiances can be derived as

$$I(\tau, \mu) = \frac{I_0'[\gamma_1 e^{\kappa(\tau-\tau_1)} - \gamma_2 e^{-\kappa(\tau-\tau_1)}] - I_1'[\beta_3 e^{\kappa(\tau-\tau_0)} - \beta_4 e^{-\kappa(\tau-\tau_0)}]}{\beta_1\gamma_4 e^{-\kappa(\tau_1-\tau_0)} - \beta_2\gamma_3 e^{\kappa(\tau_1-\tau_0)}} + B, \quad (3.66a)$$

$$I(\tau, -\mu) = \frac{I_0'[\gamma_4 e^{\kappa(\tau-\tau_1)} - \gamma_3 e^{-\kappa(\tau-\tau_1)}] - I_1'[\beta_2 e^{\kappa(\tau-\tau_0)} - \beta_1 e^{-\kappa(\tau-\tau_0)}]}{\beta_1\gamma_4 e^{-\kappa(\tau_1-\tau_0)} - \beta_2\gamma_3 e^{\kappa(\tau_1-\tau_0)}} + B, \quad (3.66b)$$

where μ is the cosine value of the local zenith angle, κ is the eigenvalue for solving the differential equations and is related to the particle optical parameters, and B is the Planck function of the scattering medium.

The parameters I_1' and I_0' are defined as

$$\begin{aligned} I_1' &= I_1 - B(1 - R_{23}), \\ I_0' &= I_0(1 - R_{12}) - B(1 - R_{21}), \end{aligned} \quad (3.66c)$$

where I_1 is the upwelling radiance at $\tau = \tau_1$ from the bottom layer, I_0 is the downwelling radiance at $\tau = \tau_0$ from the top layer, and R_{ij} is the reflectivity at the interface between the two layers. The coefficients (γ_i, β_i) are derived as

$$\begin{aligned} \gamma_1 &= 1 - R_{23}/\beta, \\ \gamma_2 &= 1 - R_{23}\beta, \\ \gamma_3 &= 1/\beta - R_{23}, \\ \gamma_4 &= \beta - R_{23}, \\ \beta &= (1 - a)/(1 + a), \end{aligned} \quad (3.66d)$$

and

$$\beta_1 = 1 - R_{21}/\beta,$$
$$\beta_2 = 1 - R_{21}\beta,$$
$$\beta_3 = 1/\beta - R_{21},$$
$$\beta_4 = \beta - R_{21}, \tag{3.66e}$$

respectively, where

$$\beta = (1-a)/(1+a),$$
$$a = \sqrt{(1-\omega)(1-\omega g)},$$
$$\kappa = \sqrt{(1-\omega)(1-\omega g)}/\mu. \tag{3.66f}$$

where ω and g are the single-scattering albedo and the asymmetry factor, respectively. Note that we have introduced the generalized conditions at upper and lower boundaries so that the solutions can be used for both surface and atmospheric scattering cases. If $I_1' = 0$, the upwelling radiance emanating from the second layer is

$$I(\tau_0, \mu) = \frac{I_0'[\gamma_1 e^{\kappa(\tau_0-\tau_1)} - \gamma_2 e^{-\kappa(\tau_0-\tau_1)}]}{\beta_1\gamma_4 e^{-\kappa(\tau_1-\tau_0)} - \beta_2\gamma_3 e^{\kappa(\tau_1-\tau_0)}} + B. \tag{3.67}$$

The interface between layers 1 and 2 can also cause an additional reflection to the incident radiation. Thus, the downwelling radiance from the first layer is reflected and the upwelling radiation from the second layer is internally reflected at the interface so that the total radiance is given by

$$I_t(\tau_0, \mu) = I_0 R_{12}(\mu) + I(\tau_0, \mu_t)[1 - R_{21}(\mu_t)], \tag{3.68}$$

where μ_t is the cosine of the upwelling angle being related to μ through Snell's law.

The emissivity of the three-layer medium is defined as a ratio of the total radiance emanating from the medium to the blackbody radiance calculated using the Planck function, that is, $\varepsilon = I_t/B$. As a result,

$$\varepsilon = \alpha R_{12} + (1 - R_{21})\left\{\frac{(1-\beta)\,[1+\gamma e^{-2\kappa(\tau_1-\tau_0)}]}{(1-\beta R_{21}) - (\beta - R_{21})\gamma e^{-2\kappa(\tau_1-\tau_0)}},\right.$$
$$\left. + \frac{\alpha\left(1-R_{12}\right)[\beta - \gamma e^{-2\kappa(\tau_1-\tau_0)}]}{(1-\beta R_{21}) - (\beta - R_{21})\gamma e^{-2\kappa(\tau_1-\tau_0)}}\right\}, \tag{3.69}$$

where $\alpha = \frac{I_0}{B}$, $\beta = \frac{1-a}{1+a}$, and $\gamma = \frac{\beta - R_{23}}{1-\beta R_{23}}$. Since the reflectivity at the interface depends on polarization, the emissivity derived from Eq. (3.69) is also a function of polarization.

3.9.2
Optical Parameters of Vegetation Canopy

According to Eq. (3.69), the most important parameters affecting the emissivity are the optical thickness (τ_1, τ_0) and the interface reflection coefficients (R_{ij}). The scatterers within layer 2 could have very complicated shapes. For instance, a deciduous forest consists of leaves, small twigs, branches, and even trunks, which are disk- and needle-shaped and cylindrical. The computation of the optical thickness for different vegetation shapes was discussed elsewhere [109–116]. However, most of these results are only valid at low frequencies (e.g. <10 GHz) and therefore have limited applications. Recently, Wegmüller *et al.* [102] applied the GO for canopy leaves to compute the optical parameters at a wide frequency range (1–100 GHz) and concluded that the results were much improved in comparison with the commonly used Rayleigh approximation [99]. This might have an important application for the satellite microwave sensors because of the simplicity of the computational procedure. Thus, the approach proposed by Wegmüller *et al.* [102] is briefly discussed here and used together with this emissivity model.

The canopy leaf is approximated through a slab with a thickness and a dielectric constant as shown in Figure 3.12. If a leaf size is much larger than the EM wavelength, the geometrical optics is applied to compute the reflectivity and transmissivity from the Fresnel equation. The reflectivity, transmissivity, and absorptivity for the leaf are given as follows [88, 102]:

$$R_p = \left| \frac{r_p \left(1 - e^{-i2\kappa_{z1}d}\right)}{1 - r_p^2 e^{-i2\kappa_{z1}d}} \right|^2, \tag{3.70}$$

$$T_h = \left| \frac{4\kappa_{z0}\kappa_{z1} e^{i(\kappa_{z0}-\kappa_{z1})d}}{(\kappa_{z1} + \kappa_{z0})^2 (1 - r_h^2 e^{-i2\kappa_{z1}d})} \right|^2, \tag{3.71a}$$

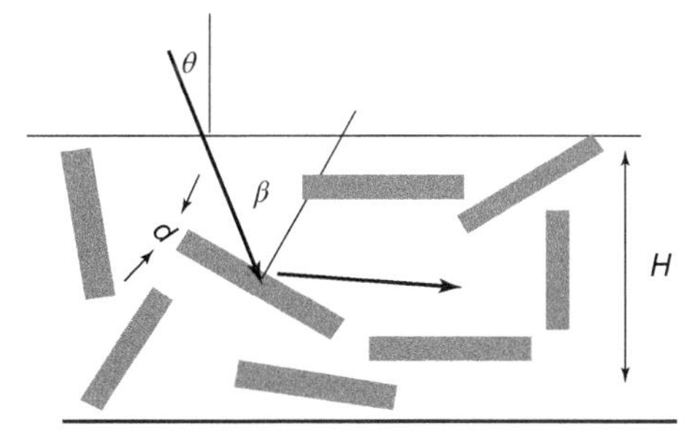

Figure 3.12 Vegetation canopy model used for microwave optical parameter calculation.

$$T_v = \left| \frac{4\kappa_{z0}\kappa_{z1} e^{i(\kappa_{z0}-\kappa_{z1})d}}{(\kappa_{z1}+\kappa_{z0})^2(1-r_v^2 e^{-i2\kappa_{z1}d})} \right|^2, \tag{3.71b}$$

$$A_p = 1 - R_p - T_p, \tag{3.72}$$

respectively, where $i = \sqrt{-1}; p = v, h$ and stands for vertical and horizontal polarization; d is the leaf thickness and

$$\kappa_{z0} = \kappa_0 cos\xi, \tag{3.73}$$

$$\kappa_{z1} = \kappa_0 (e_{veg} - sin^2\xi)^{\frac{1}{2}}, \tag{3.74}$$

$$r_h = \frac{\kappa_{z0} - \kappa_{z1}}{\kappa_{z0} + \kappa_{z1}}, \tag{3.75}$$

$$r_v = \frac{e_{veg}\kappa_{z0} - \kappa_{z1}}{\kappa_{z0} + \kappa_{z1}}, \tag{3.76}$$

where ξ is the leaf orientation angle. The leaf dielectric constant, e_{veg}, is obtained using a formula that treats the canopy leaf as a matrix of saline water, bound water, and dry matter [117].

$$\begin{aligned} e_{veg} &= 1.7 - (0.74 - 6.16m_g)m_g + m_g \times (0.55m_g - 0.076) \\ &\times [4.9 + 75.0/(1 + y_i) - y_i] + 4.64m_g^2/(1 + 7.36m_g^2) \\ &\times [2.9 + 55.0/(1.0 + \sqrt{y_i})], \end{aligned} \tag{3.77}$$

where y_i is a complex value given as

$$y_i = iv/18.0,$$

and m_g is the gravimetric water content (g/kg), and v is the frequency in GHz. Other mixing formulas were also derived and validated for leaves [118] having a higher gravimetric water content (e.g., > 0.5), which is

$$e_{veg} = (0.52 - 0.69m_d)e_w + 3.84m_d + 0.51, \tag{3.78}$$

where m_d is the dry matter content and e_w is the dielectric of water.

If canopy leaves are considered as independent scatterers, the absorption (K_{ap}) and scattering (K_{sp}) coefficients are calculated by integrating the individual optical parameters according to the leaf orientation function and density distribution function, namely,

$$K_{ap} = \frac{\text{LAI}}{H} \int_0^{\pi/2} A^* \cos\xi n(\xi) \mathrm{d}\xi, \tag{3.79}$$

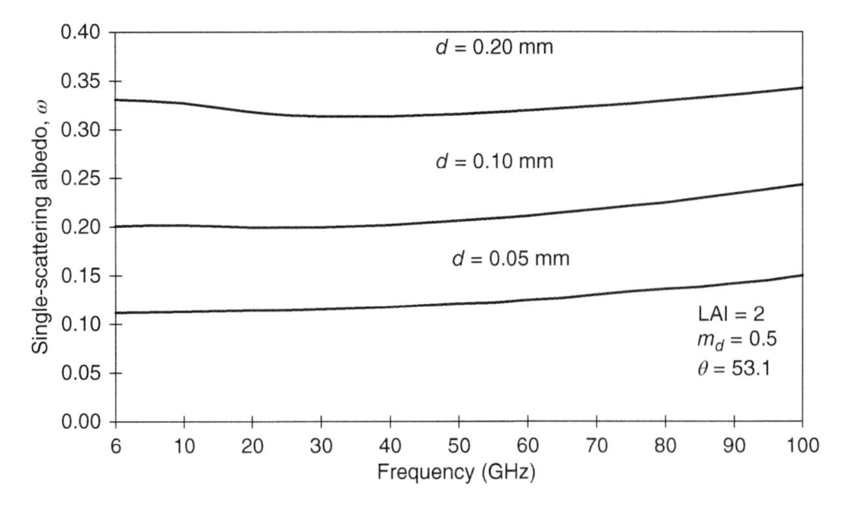

Figure 3.13 Single-scattering albedo versus frequency as a function of leaf thickness.

and

$$K_{sp} = \frac{\text{LAI}}{H} \int_0^{\pi/2} R^* \cos \xi n(\xi) \mathrm{d}\xi, \tag{3.80}$$

where LAI is the leaf area index, n is the leaf orientation function, H is the canopy depth, and A^* and R^* are the absorption and reflectivity, respectively, of each leaf averaged between two polarizations.

Vegetation leaf thickness has a significant impact on its bulk absorption and scattering properties. Figure 3.13 shows the leaf albedo as a function of frequency. The albedo first decreases with frequency and then increases. Larger leaf thickness results in higher albedo. The computation of this albedo result is derived for a leaf area index of 2 and a volumetric water content of 0.5.

3.9.3
Optical Parameters of Snow

In addition to their complex shapes, the surface scatterers may occupy a large fractional volume. Various procedures have been developed to handle the coherent scattering effects resulting from closely spaced particles. In the early development of the radiative transfer theory, only diffuse media containing sparsely populated scatterers were treated. Here, the particles are located in the far-field zone of one another, and it was sufficient to define the optical properties and phase function in terms of the scattered properties of an isolated particle without accounting for the effects of neighboring particles.

For snow, the scatterers are ice needles, which are closely spaced, occupying typically 20% to 40% of the volume. Thus, the adjacent scatterers are not always in the far field of one another, and it is necessary to use the exact scattered field without the far zone approximation in the calculation of the optical parameters. A modified Mie phase function was introduced to account for the interference effects of closely spaced neighboring particles having arbitrary sizes [101, 119]. This procedure further complicates the traditional Mie calculations of the optical parameters, although it offers a valuable tool for future research. An alternative method for analyzing the scattering in dense media is to use the perturbation theory, as proposed by Tsang *et al.* [103]. Since the optical parameters can be derived analytically, this methodology offers a very good physical understanding and is computationally efficient. However, the theory is only valid when the particle size is smaller than the wavelength.

Dense media are represented by an inhomogeneous dielectric constant whose effective propagation constant is determined by applying the perturbation theory to Maxwell equations [103]. For scattering particles of radius, r, having a dielectric constant, e_s, immersed in a uniform background dielectric content, e_0, the effective propagation constant of the medium, K, is [103]

$$K^2 = \kappa_0^2 + \frac{3v_a\kappa_0^2 y}{1 - v_a y}\left[1 + i\frac{2}{3}\frac{\left(\kappa_0 r\right)^3 y(1 - v_a)^4}{(1 - v_a y)(1 + 2v_a)^2}\right], \tag{3.81}$$

where v_a is the fraction volume and y is a complex variable given by

$$y = y_r + iy_i = \frac{e_s + e_0}{e_s + 2e_0}, \tag{3.82}$$

where y_r and y_i are the real and imaginary components, respectively.

It should be noted that the aforementioned derivations are strictly valid when $\kappa_0 r < 1$, which limits the range of particle radii for which the theory can be applied. Further decomposing the wave propagation constant in Eq. (3.81) into real and imaginary parts yields

$$K_r = \kappa_0\left[\frac{1 + 2v_a y_r}{1 - v_a y_r} - \frac{4v_a y_r y_i\left(\kappa_0 r\right)^3(1 - v_a)^4}{(1 + 2v_a)^2(1 - v_a y_r)^3}\right]^{\frac{1}{2}}, \tag{3.83a}$$

and

$$K_i = \frac{\kappa_0^2}{2K_r}\left[\frac{3v_a y_i}{\left(1 - v_a y_r\right)^2} + \frac{2v_a(\kappa_0 r)^3(1 - v_a)^4 y_r^2}{(1 + 2v_a)^2(1 - v_a y_r)^2}\right], \tag{3.83b}$$

respectively, where the imaginary part, K_i, represents the total attenuation caused by the dense medium particles and can be further decomposed into scattering and absorption coefficients assuming that $K_r^2 \gg K_i^2$ [103]:

$$K_s = \frac{\kappa_0 v_a y_r^2(\kappa_0 r)^3(1 - v_a)^4}{(1 + 2v_a)^2(1 - v_a y_r)^2}\left(\frac{1 - v_a y_r}{1 + 2v_a y_r}\right)^{\frac{1}{2}}, \tag{3.84}$$

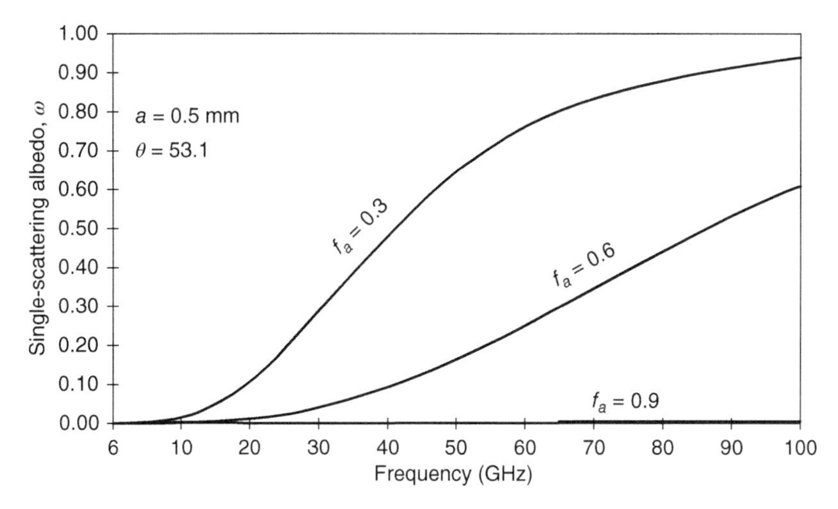

Figure 3.14 Snow optical parameter versus frequency as a function of volume fraction of particle.

$$K_a = K_i - K_s. \tag{3.85}$$

The characteristic of snow single-scattering albedo calculated from dense medium is different than that from the diffuse, especially for a high volume fraction of particles. As shown in Figure 3.14, the albedo is small or close to zero, and thus the medium is more or less an interface problem.

After snow experiences a metamorphosis process, it forms stratification and snow grains become sticky and form clusters. The aged snow requires a dense media radiative transfer (DMRT) theory with the quasicrystalline approximation (QCA), which provides more accurate results when compared to emissions determined by a homogeneous snowpack and other scattering models [120]. The DMRT model accounts for adhesive aggregate effects, which lead to dense media Mie scattering by using a sticky particle model. With the multilayer model, both the frequency and polarization dependencies of the brightness temperatures from representative snowpacks are derived and compared to the results from a single-layer model. It is found that the multilayer model predicts higher polarization differences, twice as much, and weaker frequency dependence [120].

3.9.4
Surface Reflection at Layer Interfaces

As shown in Eq. (3.69), the emissivity is also affected by the surface reflection. It is found that the reflective energy from a rough surface at small angles of incidence can be approximated by the GO approximation [121], whereas the reflectivity at large angles is obtained using the perturbation solution [101]. This leads to the development of a two-scale model, which is particularly useful for modeling the sea surface emissivity [83, 86, 88, 122, 123]. Over land, however, the two-scale models remain impractical because of the unpredictable relationship between surface roughness and geophysical parameters. Instead, various empirical relationships, though being limited by frequency, were derived to relate the surface reflectivity with roughness parameters [97].

For a smooth interface, the reflectivity is obtained from the Fresnel equations, which is a function of the incident angle and the dielectric constant of the layers. An empirical dielectric constant was derived for bare soil using a dielectric soil–water mixing model [124]. The dielectric constant is expressed as a function of the soil volumetric moisture content with soil temperature and sand and clay fractions as parameters. Of particular interest is the dielectric constant for dry and wet snow [125]. These surfaces are composed of scatterers having a high fractional volume so that the dielectric constant of the dense medium is derived from the effective propagation constant given by Eq. (3.81) using $K^2 = \kappa_0^2 e$ and is expressed as

$$e = \frac{1 + 2v_a y}{1 - v_a y} + i\frac{2v_a y^2 (\kappa_0 r)^3 (1 - v_a)^4}{(1 - v_a y)^2 (1 + 2v_a)^2}. \tag{3.86}$$

Surface roughness modifies the Fresnel reflection coefficients [97]. For example, the dynamic range of reflectivity of wet and dry soil decreases as the surface roughness increases. There is also a cross-polarization contribution, which must be taken into account, namely,

$$R_h' = [(1 - Q)R_h + QR_v]P, \tag{3.87a}$$

$$R_v' = [(1 - Q)R_v + QR_h]P, \tag{3.87b}$$

where

$$P = e^{-4\kappa_0^2 \sigma^2 cos^2(\theta)}, \tag{3.88}$$

$$Q = 0.35(1 - e^{-0.61f\sigma^2}). \tag{3.89}$$

where P and Q depend on the frequency, incident angle, and standard deviation of the surface roughness height, σ (in unit of centimeter).

The cross-polarization factor derived from Eq. (3.89) is close to zero as the frequency is near 20 GHz. Thus, its applications are very limited. Thus, a general hyperbolic tangent function is proposed with

$$P = \frac{1}{2}\left[A + B\tanh\left(\frac{x - x_0}{w_0}\right) + C\tanh\left(\frac{x - x_0}{w_1}\right)\right], \tag{3.90}$$

$$Q = 0.35(1.0 - e^{-15.0x}), \tag{3.91}$$

where $x = 4\pi f \cos(\theta)/30.0$. This revision renders much better performance in terms of roughness and cross-polarization correction. The reflectivity data over bare soil from various studies can be fit well, as shown in Figure 3.15. In addition,

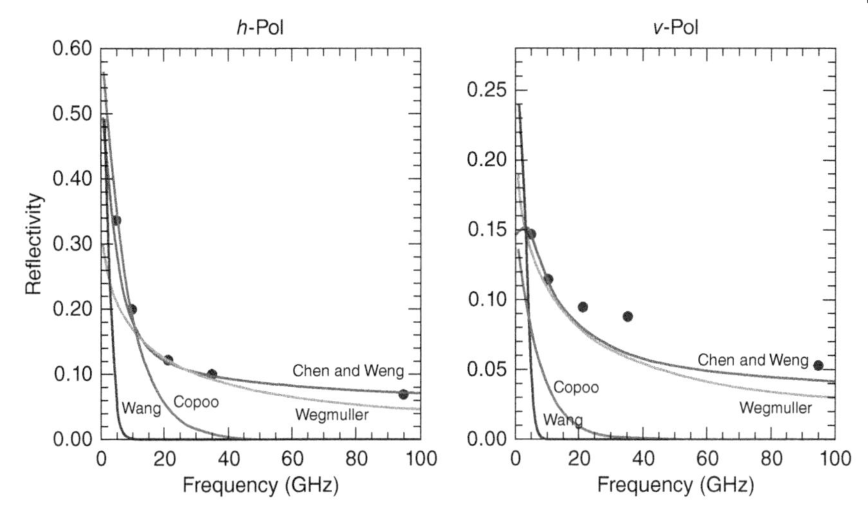

Figure 3.15 Microwave land reflectivity of bare soil derived from new roughness model, compared with the measurements [126–128].

the dependence of the reflectivity on roughness height and angle is significantly improved, compared to the measurements.

3.9.5 Soil Dielectric Constant

Dobson *et al.* [124] developed a mixing rule for soil dielectric constant [106], which is

$$e_m^{\alpha} = 1 + \frac{\rho_b}{\rho_s}(e_s^{\alpha} - 1) + m_v^{\beta} e_w^{\alpha} - m_v, \tag{3.92}$$

where m_v is the soil volumetric moisture, e_s is the dielectric constant of solids, and ρ_b is the density of soil, ρ_s is the density of solids, which are calculated from sand and clay fractions. The exponents α, β depend on the soil type as:

$$\alpha = 0.65, \tag{3.93a}$$

$$\beta = 1.09 - 0.11S + 0.18C, \tag{3.93b}$$

$$e_s = (1.01 + 0.44 \times \rho_s)^2 - 0.062, \tag{3.93c}$$

3.9.6 Simulated Surface Emissivity Spectra

From this new emissivity model, land microwave emissivity spectra are calculated for these four surface conditions. For a wetland, its volumetric moisture content is 0.3. The fractions of sand and clay are 0.2 and 0.8, respectively. For a corn field,

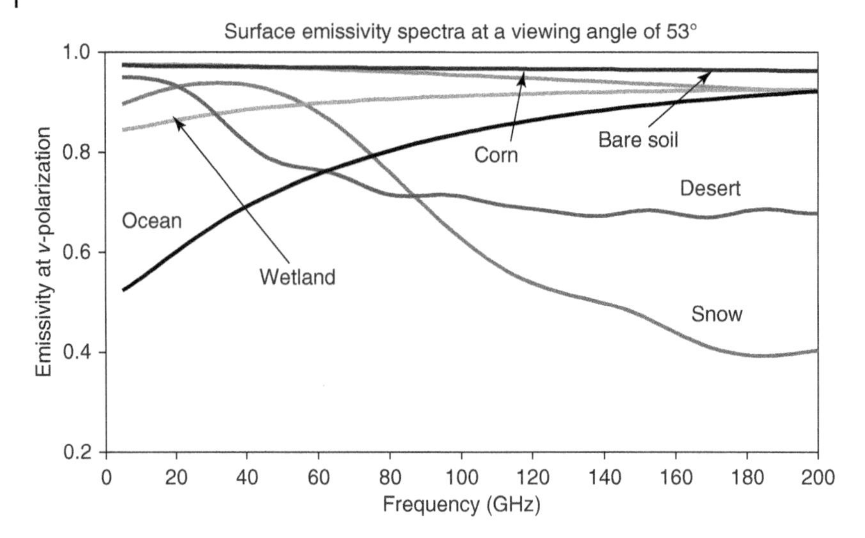

Figure 3.16 Microwave emissivity spectra corresponding to various Earth surface conditions derived from land and ocean models as discussed in the following sections. Shown are the values of two polarization states at a local zenith angle of 53°. (Weng and Liu 2003 [46] http://journals.ametsoc.org/doi/pdf/10.1175/1520-0469%282003%29060%3C2633%3ASDAINW%3E2.0.CO%3B2.)

the leaf area index is 4. For a snow and desert scattering medium, the particle size is 0.5 mm and its volumetric fraction is 0.3 for snow and 0.8 for deserts, as shown in Figure 3.16. The emissivity of snow decreases as the frequency increases due to the scattering of snow particles. The characteristics of wetland emissivity are similar to those over oceans in that the emissivity increases and the polarization decreases as the frequency increases. However, the emissivity spectra of other land surfaces do not exhibit a large variability in most of the frequency range. Thus, the uncertainty in simulating land emissivity is the largest over snow-covered and desert regions where the dense medium scattering may still be problematic.

3.10 Summary and Conclusions

The discrete-ordinate method is a fundamental tool for solving the atmospheric radiative transfer problem and derives an analytic form of radiance in mathematical elegancy. In principle, it is an idea for remote sensing applications since the Jacobians (radiance gradient) can also be derived accurately with low cost of computational resources. The vector form of solution for the scattering medium that contains spherical or nonspherical particles is also entirely analytic. Its research codes have been updated several times in history and linked to more applications in radiance data assimilation systems and remote sensing applications. However, knowledge of surface emissivity vector or reflectivity matrix is further required.

The dependence of surface emission or reflection on azimuthal angle must be in good harmony with atmospheric particle scattering phase matrix so that the sinusoidal and cosine modes of Fourier harmonics in radiances can be fully decoupled, as discussed in this chapter.

For computing the surface emissivity over oceans, the coefficients in the double Debye water dielectric constant or permittivity model are derived with the laboratory measurements collected from various studies, and the new model can be applied for a wide range of microwave frequencies from 1.4 to 200 GHz. This update is critical for use in simulations of surface reflectivity with the Fresnel formula. The two-scale roughness model is used to compute the surface Stokes emissivity vector. It is shown that the first two emissivity components at horizontal and vertical polarization states are dependent on the azimuthal angle and can be uniquely expanded as a series of Fourier cosine harmonics, whereas the third and fourth components are expanded as a series of sinusoidal harmonics. This property allows for a solution of the vector radiative transfer when its lower boundary is specified as the oceanic condition.

A general approach is also developed for simulating the land emissivity over bare soil, vegetated land, and snow. The inputs to the emissivity model are primarily based on the available information from the current land data assimilation system. There are many parameters set as fixed values (e.g., leaf thickness, snow grain size). More studies are needed for the medium that is more vertically stratified, and, as a result, more sophisticated radiative transfer schemes are more appropriate.

4
Microwave Radiance Simulations

4.1
Introduction

Clouds and precipitation affect microwave radiative transfer process through scattering and emission. A simple rain cloud model was used by Wilheit *et al.* [41] to simulate the relationship between brightness temperature and rain rate. The model has two free parameters: the surface rain rate and height of the freezing level. Other rain cloud properties, such as temperature, humidity, and precipitation rate profiles, were "hard-wired" in terms of these parameters, and other properties, such as column-integrated cloud liquid water, were assumed constant. The rain intensity was assumed constant between the freezing level and the surface, and the drop size distribution assumed was that used by Marshall and Palmer [129]. Later, Spencer [130] found out that frozen precipitation aloft in rain clouds causes scattering at a frequency of 37 GHz or higher. This scattering signature is much less directly related to the surface rain rate. At higher rain rates, large precipitation size in liquid phase can also result in scattering; thus, a relationship between brightness temperature and rain rate over the ocean is nonlinear, and thus, there is an ambiguity in deriving the surface rain rate from a single-channel microwave measurement, especially under heavy precipitation conditions. Mugnai and Smith [131] and Smith and Mugnai [132] further studied the multifrequency responses to precipitation using a 2D numerical cumulus model output as input to a plane-parallel radiative transfer model. Their results demonstrated that the cloud water content can alter and obscure the relationship between microwave brightness temperature and rain rate, and its influence is dependent on the frequency. To utilize multichannel information for precipitation retrieval, we need sophisticated models of the relationship between rain cloud properties and observed radiances: for example, if the low frequencies of a microwave sensor can be viewed as primarily responding to the rain water path below the freezing level and high frequencies as responding primarily to scattering by frozen precipitation aloft. Some of the better known current methods for utilizing multichannel information in physical retrievals are based on the output of three-dimensional numerical cloud models using bulk microphysical parameterization schemes to generate ensembles of realistic rain

Passive Microwave Remote Sensing of the Earth: For Meteorological Applications, First Edition. Fuzhong Weng.

cloud structures that are then fed to a suitable radiative transfer model. Surface rain rates and hydrometeor profiles are retrieved by identifying the cloud model structure(s) whose computed brightness temperatures most closely match the multichannel observations [133–135]. Olson *et al.* [135] also addressed some of the complexities and potential uncertainties associated with modeling microwave radiative transfer in rain clouds.

While the simulations of microwave imager radiance under various clouds and precipitation conditions and the relationships between the brightness temperature and hydrometeor contents are well characterized, the impacts of clouds and precipitation on sounder channel radiances are less understood. In this chapter, we develop a method to characterize the uncertainty of simulated radiances with respect to observations. First, various modules used in Community Radiative Transfer Model (CRTM) are introduced in Section 4.2. The algorithm for the calculation of antenna brightness temperature is presented in Section 4.3. In Sections 4.4–4.6, simulations of the radiances of an ATMS and characterization of the errors of forward models are carried out using the global forecast model outputs, the level 2 data from Global Positioning Satellite (GPS) Radio Occultation (RO) data and Tropical Rainfall Measuring Mission's (TRMM) Microwave Imager (TMI) collocated with the Advanced Microwave Sounding Technology (ATMS) measurements as the input to CRTM. Advanced radiative transfer simulation is described in Section 4.7. In Section 4.8, the bias between ATMS simulations and observations is presented.

4.2
Fast Radiative Transfer Simulations

In data assimilation systems or satellite retrieval algorithms, simulations of satellite radiances and/or brightness temperatures need to be performed at a high speed and with a high accuracy. The concept of CRTM was conceived when the Joint Center for Satellite Data Assimilation (JCSDA) was established in 2002. The CRTM is a software library for computing the satellite instrument radiances and their gradients with respect to various atmospheric and surface state variables required in the data assimilation systems. The CRTM is based on a design framework that emphasizes modularity and code reuse (see Figure 4.1). From its initial conception, the CRTM library was designed to be both platform independent and assimilation system independent for ease of implementation for all of the JCSDA partners. The CRTM has progressed into a system for use primarily at infrared and microwave frequencies, with development opportunities at visible wavelengths.

The gaseous absorption module in CRTM is developed for each individual instrument. As discussed in Chapter 2, the gas absorption coefficient profiles in the module are directly predicted using atmospheric state variables or their integrated products. In addition, the oxygen absorption at microwave frequencies associated with Zeeman splitting is parameterized in terms of the atmospheric temperature, geomagnetic field strength, and angle between the geomagnetic field vector and the electromagnetic wave propagation direction. The Zeeman

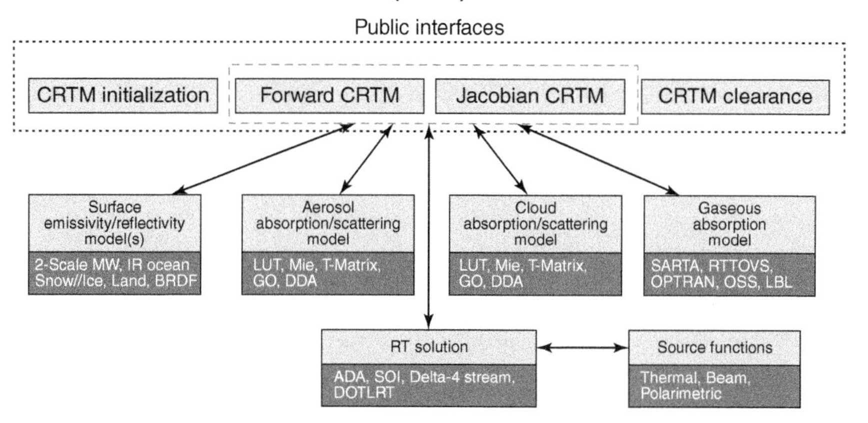

Figure 4.1 The CRTM for fast and accurate calculations of satellite radiances and Jacobians at the top of atmospheres. (Chen *et al.* 2008 [158]. Reproduced with permission of Wiley.)

splitting absorption model, Special Sensor Microwave Imager/Sounder (SSMIS) on board the Defense Meteorology Satellite Program (DMSP) F-16 satellite probes the atmospheric temperature from the surface to 100 km. SSMIS channels 19–22 are significantly affected by Zeeman splitting, which is dependent on the Earth's magnetic field. Thus, in satellite data assimilation or retrieval systems, SSMIS brightness temperatures and their Jacobians (or gradient with respect to temperature) must be computed with a fast radiative transfer scheme that takes into account the Zeeman splitting effect. In CRTM, an averaged transmittance within the channel frequency passbands is parameterized and predicted with the atmospheric temperature, geomagnetic field strength, and angle between the geomagnetic field vector and the electromagnetic wave propagation direction [136]. As discussed in Chapter 2, the coefficients of the predictors are trained with a line-by-line radiative transfer model that accurately computes the monochromatic transmittances at fine frequency steps within each passband. Parameterization of gas absorption near 14 μm for the Stratospheric Sounding Unit (SSU) on the early NOAA satellites takes into account CO_2 cell pressure leakage. The absorption models of historical satellite instruments such as the 29-year-old SSU satellite data records are required for numerical weather model reanalysis and climate studies. The SSU measures temperatures in the middle and upper stratosphere. However, the SSU is a complicated infrared sensor because of its additional spectral response to a CO_2 cell pressure. The CO_2 cell pressure leaking during the mission life complicates the use of the data. An algorithm is developed to parameterize the gas absorption coefficients at each optical level as a function of the cell pressure. The fast SSU algorithm achieves accuracy better than 0.1 K in comparison to a detailed line-by-line calculation. A good agreement is obtained between the SSU measurements and simulations using the microwave limb sounder (MLS)-derived temperatures.

Cloud and precipitation scattering and absorption are computed for liquid-phase cloud (liquid water and rain clouds) and solid-phase cloud (ice, snow, graupel, and hail clouds). Spherical scatters are used for liquid clouds in the infrared and microwave ranges. The size distribution for these clouds is described by the modified Gamma distribution. Various effective radii of the size distribution for a fixed effective variance are chosen for these clouds. For any given effective radius, the layer water content is used to determine the total number of cloud particles and to scale the optical quantities for the normalized size distribution. A lookup table contains the extinction coefficients, single-scattering albedo, asymmetry factor, delta truncation factors for removing the forward peaks, and expansion coefficients from the Lorentz–Mie calculations. For ice clouds, nonsphericity of ice crystals is accounted for in the development of the lookup tables for the extinction coefficients, single-scattering albedo, and asymmetry factors. A database of nonspherical ice particles based on the discrete dipole approximation (DDA) is also included in CRTM [33].

Cloud and aerosol optical properties at visible and infrared wavelengths also constitute a fast lookup table. All the particles are assumed to be spherical and distributed with a bi-mode lognormal function. The scattering and absorption coefficients, asymmetry factor, and the Legendre expansion coefficients for scattering-phase matrices are precalculated using the Lorentz–Mie theory and stored in a lookup table. The expansion coefficients can be used to produce the phase function according to the number of streams and the number of the components in the Stokes vector in the radiative transfer solver. Dust particles can also be spheroids with an aspect ratio of 1.7. To compute the single-scattering properties of these particles, we applied a combination of the T-matrix method [137] and an improved geometric optics method that includes the edge effects [138]. It is found that the optical properties of nonspherical (spheroidal) particles are quite different from their spherical counterparts in the backscattering directions. For ice clouds, the habit (shape) distribution of ice crystals is assumed to be consistent with that used in the MODIS operational (Collection 5 version) cloud retrieval [139]; that is, ice crystals are assumed to be 100% droxtals for size bins less than 60 μm; 35% plates, 15% bullet rosettes, 50% solid columns, for size bins between 60 and 1000 μm; 45% solid columns, 45% hollow columns, 10% aggregates for size bins between 1000 and 2000 μm; and 3% aggregates and 97% bullet rosettes for size bins larger than 2000 μm. In total, 1117 size distributions from *in situ* measurements are used to derive the bulk optical properties of ice clouds on the basis of the single-scattering properties of individual ice crystals [140]. Furthermore, we plan to improve the aerosol and cloud lookup tables, considering more realistic particle morphologies such as surface roughness for ice crystals and triaxial ellipsoidal model for dust particles.

A fast emissivity model (FASTEM) [58, 66] is implemented into CRTM to compute the ocean emissivity at microwave wavelengths. This model had been updated several times in the past. Its sixth release (FASTEM-6) primarily corrects the larger biases for microwave imager radiance simulations. An empirical correction was made to adjust the coefficients in FASTEM-5 [141]. The microwave land emissivity model is generally designed for a two-layer scattering and emission medium with the top layer representing scatters related to vegetation and snow and the bottom layer for soil. The soil reflectivities at both polarization states are first calculated from the Fresnel formula and weighted with roughness and cross-polarization factors [106]. The emissivity from vegetation or snow is then further derived from the two-stream radiative transfer solutions. The recent improvements in land surface roughness model allows for much better simulations of microwave imager radiances within a wide frequency range [142].

The infrared sea surface emissivity model utilizes a lookup table (LUT) of sea surface emissivities derived from the emissivity model for a wind-roughened sea surface [143, 144]. The sea surface is modeled by numerous small facets whose slopes approximately follow the normal and isotropic distribution [56]. Each of the facets is treated as a specular surface, and emission at the observation angle is computed with the geometrical optics, taking into account wave shadowing effects and surface reflection of surface emission taken. The LUT variables are zenith angle, frequency, and wind speed. Currently, linear interpolation is performed between LUT values. The infrared surface emissivities for snow-free land, snow, and sea ice are provided as a database [145]. The database contains surface reflectance measurements as a function of wavelength in both visible and IR spectral regions for the 24 surface types. The emissivity is calculated as 1 minus the reflectance under the assumption of a Lambertian surface in the IR spectral region.

At present, the advanced doubling–adding (ADA) method is used as CRTM radiative transfer solver [44]. The new study strictly derived an analytical expression replacing the most complicated thermal source terms in the doubling–adding method. The solution within any layer is given by the optical properties and the Planck source function of the layer, and the radiance at the top of the atmosphere is integrated from the surface and vertical layers using a stack procedure. However, only the scalar ADA is implemented into CRTM. A general polarimetric doubling and adding [3] is considered for future CRTM implementation. In addition, VDISORT [47] is an alternative scheme for full vector radiative transfer simulations.

Snow and sea ice emissivities are also developed for microwave instruments such as Microwave Humidity Sounder (MHS) on NOAA-18 and METOP-A satellites and SSMIS on DMSP F-16 and F-17 satellites, which are new sensors, and the snow and sea emissivity calculations are updated with new coefficients in the empirical part [146]. The empirical snow and sea ice microwave emissivity models are derived primarily using satellite brightness temperatures at window channels and static databases from the early measurements at the discrete microwave frequencies and a fixed viewing angle. This data set is utilized to produce a variety of empirical emissivity spectra according to the snow/sea ice types, which are applicable for a frequency range from 5 to 200 GHz. The emissivity at a zenith angle of 50° is calculated from one of empirical emissivity spectra specified using a snow/sea ice type identification algorithm associated with window channels of satellite brightness temperatures, while angular dependence of emissivity relies on the microwave land emissivity model developed previously by Weng *et al.* [106]. This so-derived emissivity results in much improved accuracy in simulating the brightness temperatures at the top of the atmosphere and improves the uses of satellite microwave data in numerical weather prediction models.

4.3
Calculations of Antenna Brightness Temperatures

For a cross-track scanning microwave radiometer such as ATMS, the quasi-vertical and quasi-horizontal antenna brightness temperatures (TDR), T_a^{Qv} and T_a^{Qh}, are derived as follows [5]:

$$T_a^{Qv} = \eta_{me}^{vv} T_b^{Qv} + \eta_{me}^{hv} T_b^{Qh} + \eta_{se}^{vv} T_{b,se}^{Qv} + \eta_{se}^{hv} T_{b,se}^{Qh} + (\eta_{sc}^{vv} + \eta_{sc}^{hv}) T_{c,RJ} + S_a^{Qv}, \tag{4.1a}$$

$$T_a^{Qh} = \eta_{me}^{hh} T_b^{Qh} + \eta_{me}^{vh} T_b^{Qv} + \eta_{se}^{hh} T_{b,se}^{Qh} + \eta_{se}^{vh} T_{b,se}^{Qv} + (\eta_{sc}^{hh} + \eta_{sc}^{vh}) T_{c,RJ} + S_a^{Qh}, \tag{4.1b}$$

where η_{me}^{vv} and η_{me}^{hh} are the co-polarized and cross-polarized antenna main beam efficiencies; η_{me}^{vh} and $\eta_{me,}^{hv}$ are the cross-polarized ones; η_{se}^{vv}, η_{se}^{hh}, η_{se}^{hv}, and η_{se}^{vh} are the co-polarized and cross-polarized antenna side-lobe beam efficiencies; and η_{sc}^{vv}, η_{sc}^{hh}, η_{sc}^{hv}, and η_{sc}^{vh} are the co-polarized antenna and cross-polarized side-lobe cold-space beam efficiencies. It is reminded that for ATMS, each frequency is only measured at a single polarization state, that is, either horizontal or vertical (see Table 4.1). Therefore, there are only one co-polarization efficiency and one cross-polarization antenna beam efficiency for the main beam, side lobes, and cold-space side lobes. Values of the antenna beam efficiencies for the Earth-scene main beam, Earth-scene side lobes, and cold-space side lobes are also required (see [5]).

The last term in Eq. (4.1a) or (4.1b) (S_a^{Qv} or S_a^{Qh}) come from the radiation contributed by the antenna near-field side lobe or other effects such as the emitted radiation from ATMS flat reflector, and its net effect was estimated from the ATMS pitch maneuver data [147]. The quasi-vertical and quasi-horizontal sensor brightness temperatures (SDR), T_b^{Qv} and T_b^{Qh}, are related to the pure vertically and horizontally polarized brightness temperatures, T_b^v and T_b^h, through Eq. (1.18). Neglecting the third component of the Stokes vector results in

$$T_b^{Qh} = T_b^v \sin^2\theta + T_b^h \cos^2\theta, \tag{4.2a}$$

$$T_b^{Qv} = T_b^v \cos^2\theta + T_b^h \sin^2\theta, \tag{4.2b}$$

where θ is the scan angle.

Table 4.1 Requirements and characteristics of ATMS and their weighting function peak for a US standard atmospheric condition.

Channel	Center frequency (GHz)	Total bandpass (GHz)	Polarization	Accuracy (K)	NEΔT (K)	EFOV cross-track(°)	EFOV along-track(°)	Peak of weight function (hPa)
1	23.8	0.27	*Qv*	1.00	0.50	6.3	5.2	Window
2	31.4	0.18	*Qv*	1.00	0.60	6.3	5.2	Window
3	50.3	0.18	*Qh*	0.75	0.70	3.3	2.2	Window
4	51.76	0.40	*Qh*	0.75	0.50	3.3	2.2	950
5	52.8	0.40	*Qh*	0.75	0.50	3.3	2.2	850
6	53.596 ± 0.115	0.17	*Qh*	0.75	0.50	3.3	2.2	700
7	54.4	0.40	*Qh*	0.75	0.50	3.3	2.2	400
8	54.94	0.40	*Qh*	0.75	0.50	3.3	2.2	250
9	55.5	0.33	*Qh*	0.75	0.50	3.3	2.2	200
10	57.29	0.33	*Qh*	0.75	0.75	3.3	2.2	100
11	57.29 ± 0.217	0.078	*Qh*	0.75	1.20	3.3	2.2	50
12	57.29 ± 0.322 ± 0.048	0.036	*Qh*	0.75	1.20	3.3	2.2	25
13	57.29 ± 0.322 ± 0.022	0.016	*Qh*	0.75	1.50	3.3	2.2	10
14	57.29 ± 0.322 ± 0.010	0.008	*Qh*	0.75	2.40	3.3	2.2	5
15	57.29 ± 0.322 ± 0.0045	0.003	*Qh*	0.75	3.60	3.3	2.2	2
16	88.2	3	*Qv*	1.00	0.30	2.2	2.2	Window
17	165.5	3	*Qh*	1.00	0.60	2.2	1.1	Window
18	183.31 ± 7.0	2	*Qh*	1.00	0.80	2.2	1.1	800
19	183.31 ± 4.5	2	*Qh*	1.00	0.80	2.2	1.1	700
20	183.31 ± 3.0	1	*Qh*	1.00	0.80	2.2	1.1	500
21	183.31 ± 1.8	1	*Qh*	1.00	0.80	2.2	1.1	400
22	183.31 ± 1.0	0.5	*Qh*	1.00	0.90	2.2	1.1	300

Figure 4.2 shows the brightness temperature with pure horizontal and vertical polarization states as well as quasi-horizontal and vertical polarization. The polarization differences at ATMS channels 1–2 are the largest and decrease with increasing frequency. The difference between the pure vertically and horizontally polarized brightness temperatures, that is, $T_b^v - T_b^h$, is zero at the nadir and increases with the scan angle. The difference between the quasi-vertical and quasi-horizontal sensor brightness temperatures (SDR), that is, $T_b^{Qv} - T_b^{Qh}$, are zero at both nadir and 45° scan angle. The polarization differences of the brightness temperatures between quasi-vertical and quasi-horizontal polarization states are smaller than those at the pure polarization states.

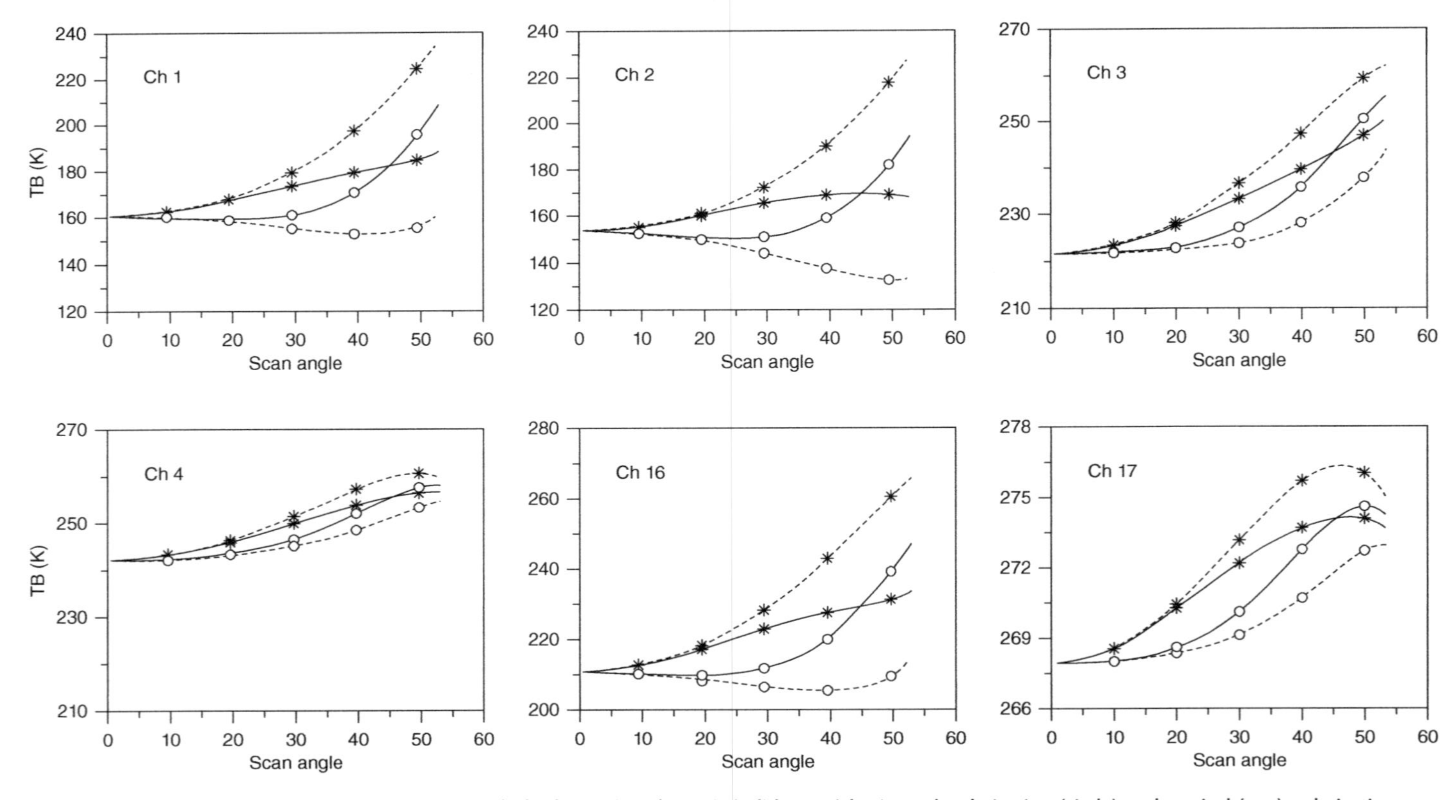

Figure 4.2 The brightness temperature with pure (dashed curve) and quasi- (solid curve) horizontal polarization (circle) and vertical (star) polarization states using the US standard atmospheric profile with sea surface wind speed being 5 m/s and sea surface temperature being 290 K. (Weng *et al.* 2013 [5]. Reproduced with permission of Wiley.)

4.4 Simulations of ATMS Sounding Channels Using Global Forecast Model Outputs

An important application of ATMS data is for improving the numerical weather prediction (NWP) forecast skill through data assimilation. All data assimilation methods employ either a maximum likelihood estimate or a minimum variance estimate under the assumption that both observations and models are unbiased. Any bias related to the instrument and forward modeling must be quantified and removed in satellite data assimilation. Since the weighted differences between observations and model simulations, $O - B$, are minimized in satellite data assimilation, the observation bias (μ^o) and model bias (μ^b) can be lumped together as follows:

$$(O - \mu^o) - (B - \mu^b) = O - B - (\mu^o + \mu^b). \tag{4.3}$$

Therefore, $O - B$ statistics can be used to estimate the sum of observation and model biases.

An assessment of the ATMS data biases requires a forward radiative transfer model for calculating the microwave radiation at 22 ATMS frequencies at the top of the atmosphere for any given atmospheric state (e.g., temperature and water vapor profiles) and the Earth's surface properties (e.g., surface temperature, surface emissivity, and surface wind speed). In this study, the CRTM and National Center for Environmental Prediction (NCEP) global forecast system (GFS) 6 h forecasts are used for bias characterization. The NCEP GFS 6 h forecast fields have a horizontal resolution of 0.3125° × 0.3125° and 64 vertical levels. The highest vertical level is around 0.01 hPa.

Brightness temperatures simulated by CRTM using NWP analysis/forecast fields can be used to evaluate the performance of ATMS, especially for the sounding channels under clear-sky conditions over oceans. Here, ATMS observations under clear-sky conditions during December 20–27, 2011, are used for characterizing the performance of the ATMS temperature sounding channels 5–15. To detect a cloud-affected ATMS field of view (FOV) measurement, an algorithm similar to that developed by Weng *et al.* [47] for AMSU-A is used for retrieving atmospheric cloud liquid water path (LWP) from ATMS channels 1 and 2 measurements. As demonstrated, microwave measurements at lower frequency window channels can be directly related to LWP and water vapor path (WVP) through an emission-based radiative transfer model [47, 106, 119, 148–151]. The effects of surface parameters such as emissivity and temperature on the measurements at two ATMS channels are taken into account from GFS forecast fields.

Figure 4.3 presents the global distributions of brightness temperatures at channels 1 and 2 from ATMS and AMSU-A (Figure 4.3a–d) as well as the LWP retrievals derived from these two channels for the ascending nodes on December 20, 2011. The sensitivity of these two window channels to the Earth's surface (e.g., surface emissivity and surface skin temperature) yields a sharp contrast between land and ocean. Due to much large surface emission, the brightness temperatures

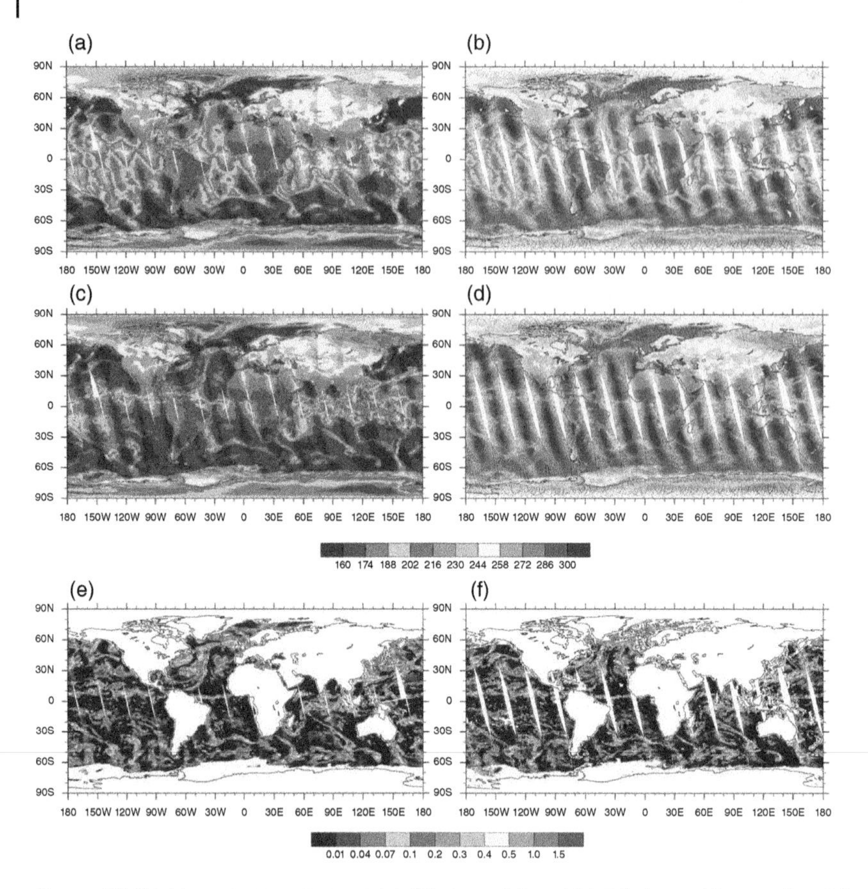

Figure 4.3 Brightness temperatures at (a), (b) channel 1 and (c), (d) channel 2, as well as (e), (f) LWP retrievals over ocean from ATMS (left panels) and AMSU-A (right panels) from the ascending nodes on December 20, 2011. (Weng *et al.* 2012 [147]. Reproduced with permission of American Geophysical Union.)

over land are higher than those than over ocean. The relative contribution of the atmospheric absorption to the total radiance over ocean is thus higher than that over land, leading to a stronger scan dependence of the brightness temperatures over ocean than over land. The global LWP distribution deduced from ATMS (Figure 4.3e) compares favorably with the AMSU-A-derived LWP (Figure 4.3f). The ATMS provides a nearly continuous distribution of global LWP while AMSU-A has large orbital gaps in low latitudes. Spatial features of large LWP (Figure 4.3e,f) can be seen in the global distribution of brightness temperature of channel 2 (Figure 4.3c,d), which is the primary channel for the LWP retrieval. Channel 1 is the most sensitive to atmospheric WVP, which is usually high over cloudy areas.

An LWP of 0.05 kg/m^2 is used as a threshold for detecting cloud-affected ATMS sounding channels. An ATMS sounding channel corresponding to the LWP less

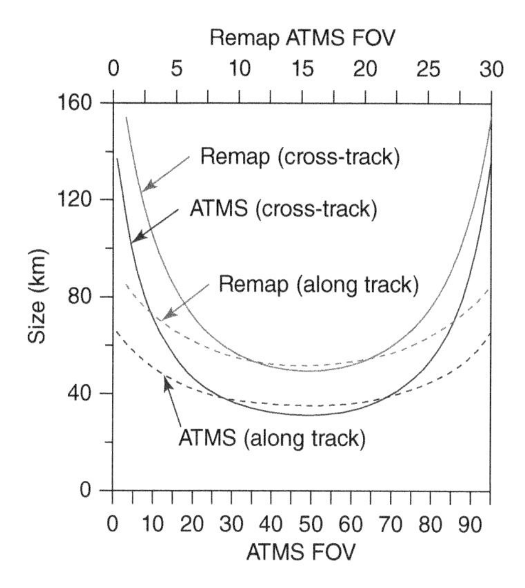

Figure 4.4 Cross-track (solid) and along-track (dashed) FOV size of ATMS and ATMS Remap. (Weng *et al.* 2012 [147]. Reproduced with permission of American Geophysical Union.)

than this threshold is treated as clear FOV [150]. During the study period, there are more than 250 data counts within any $1° \times 1°$ grid boxes over the globe.

Differences between ATMS data and its predecessor AMSU-A could be inferred from the differences between ATMS raw and remapped data. The differences in the observation resolutions between ATMS raw and remapped data are first examined (Figure 4.4). The ATMS FOV diameter at the nadir is 31.6, while that of the remapped data FOV is 48.6 km. The cross-track FOV size increases much rapidly than that in the along-track direction. The differences in the along-track FOV size between the raw and remapped ATMS TDR remain nearly constant with respect to the scan angle, while the size differences in the cross-track FOVs between the raw and remapped data decrease with the increase in the scan angle. The cross-track and along-track FOV sizes of ATMS at the largest scan angle (e.g., ±52.77°) are 136.7 and 60 km, respectively, while those of the remapped one at the largest scan angle (e.g., ±48.33°) are 155.2 and 85.6 km, respectively.

ATMS remapped data is a weighted average of the ATMS raw data. Differences in the observational resolutions between ATMS raw and remapped data change the dynamic ranges and standard deviations of the differences between observations and model simulations (O−B). The scatter plots of the temperature dependence of O−B for ATMS channel 6 is illustrated in Figure 4.5. It is shown that the original ATMS has a larger spread compared to the remapped data. This is partially due to higher channel noise and partially due to the fact that small-scale features of the real atmosphere, which vary rapidly in time, are not captured by the GFS fields, and the averaging improves the agreement between satellite data and model simulations. It is noticed that the O−B data points within the same FOV number appear to increase with respect to the observed brightness temperature value

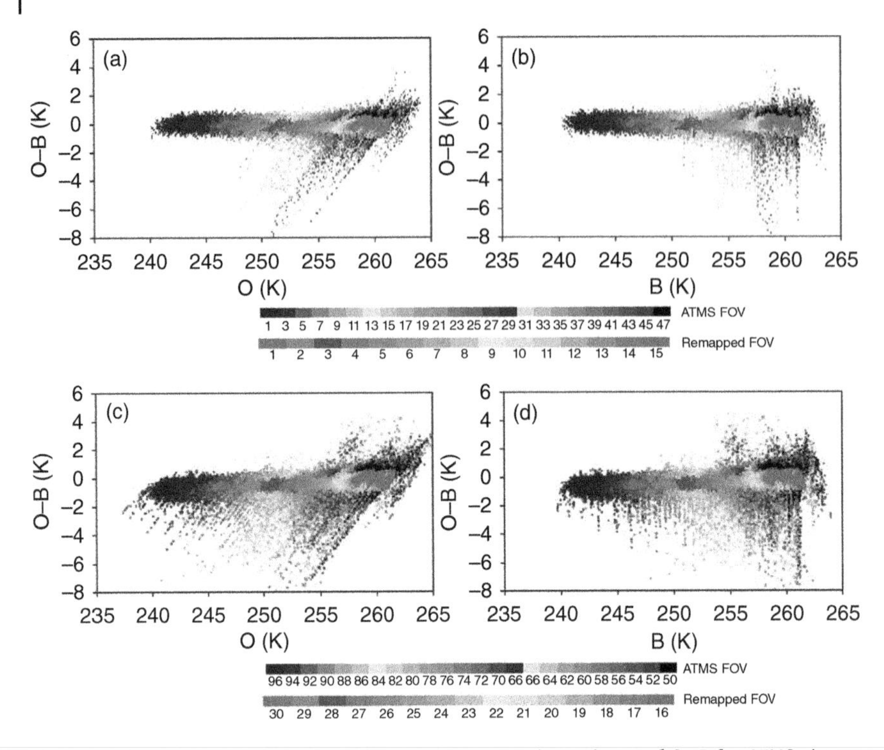

Figure 4.5 (a) and (b) Scatter plots of the temperature dependence of O–B for ATMS channel 6 with respect to the observed (left panels) and modeled (right panels) brightness temperatures at ATMS FOVs 1–48 (upper color bar) and ATMS remapped FOVs 1–15 (lower color bar) for all the data within 10S–10N on December 20, 2011. (c), (d) Same as (a), (b) except for ATMS FOVs 49–96 and remapped FOVs 16–30. (Weng *et al.* 2012 [147]. Reproduced with permission of American Geophysical Union.)

(Figure 4.5a,c) but not with respect to the simulated brightness temperature, especially near the nadir. This is due to a larger variability in the observations compared to that in model simulations, especially near the nadir where the peak WF altitude is the lowest for ATMS channel 6. The observed temperature range for the same FOV (Figure 4.5a,c) is larger than that for the model simulation (Figure 4.5b,d). It is also noticed that the observations for the ending half of the scan line (FOVs 49–96, Figure 4.5c,d) are more negatively biased compared to those for the beginning half of the scan line (FOVs 1–48, Figure 4.5a,b).

Figure 4.6 shows the biases and standard deviations of brightness temperatures for ATMS temperature sounding channels and the remapped data within [60S, 60N] under clear-sky conditions over ocean during December 20–27, 2011. It is reminded that the biases as shown in Figure 4.6a are not in the absolute sense but are relative to the GFS model fields. Negative biases are found for ATMS channels 5–9 located within the troposphere and low stratosphere, and positive biases are found for ATMS channels 10–14 in the stratosphere and higher. Channel 15 peaking at the last level has a negative bias. Impacts of remap on data biases are rather

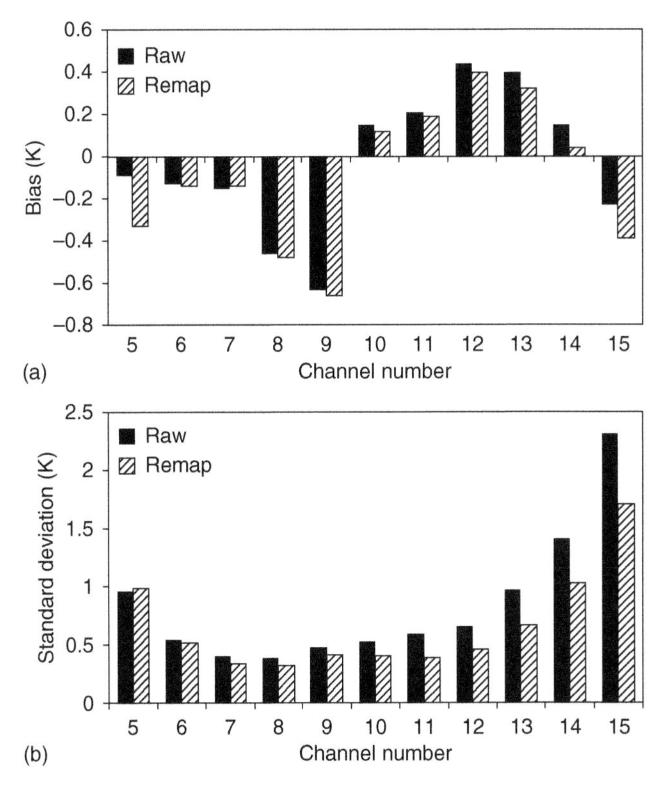

Figure 4.6 (a) Biases and (b) standard deviations of O–B brightness temperatures for ATMS temperature sounding channels with (dashed bar) and without (solid bar) remap for all the data within [60S, 60N] under clear-sky conditions over ocean during December 20–27, 2011. (Weng *et al.* 2012 [147]. Reproduced with permission of American Geophysical Union.)

small (e.g., ≤ 0.1 K) except for channels 5 and 15 (~0.2–0.3 K). The remap does not change the sign of the biases. The standard deviations of the remapped data are smaller than those of the ATMS raw data as expected, with a larger reduction of standard deviations for higher level channels.

The latitudinal dependence of bias and standard deviation is shown in Figure 4.7 and reveals that the biases of ATMS data in the middle and low troposphere (e.g., channels 5–7) are slightly higher at the high latitudes than at the middle and low latitudes, and the reverse is true for the remaining upper-level sounding channels except for channel 15. The standard deviation is larger for channels 14–15 at all latitudes and channel 5 at the middle latitudes with high terrain areas. The standard deviation is reduced at all latitudes after remapping.

A unique feature of a cross-track scanning radiometer instrument is the so-called limb effect, which arises from the variation of the optical path length with the scan angle. This limb effect is modeled through CRTM. Therefore, an *a priori* limb adjustment is not required for ATMS data assimilation. However,

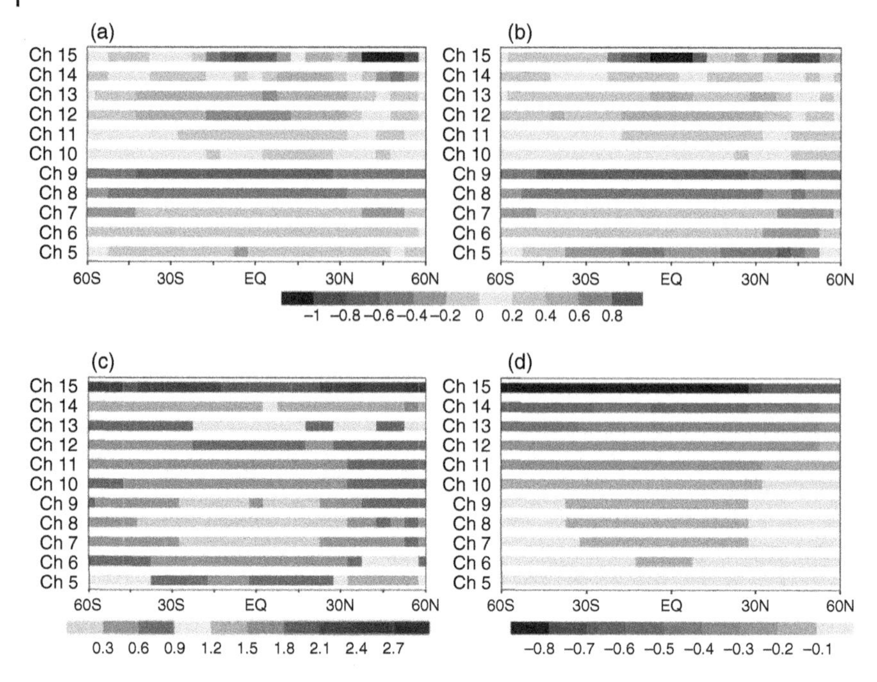

Figure 4.7 Latitudinal distributions of O−B biases for (a) ATMS raw data and (b) ATMS remap data, (c) standard deviations for ATMS channels 5–15 (σ^2_{ATMS}), and (d) differences of σ^2_{ATMS} and the standard deviation for ATMS remap data (σ^2_{remap}), that is, $\sigma^2_{ATMS} - \sigma^2_{remap}$. (Weng *et al.* 2012 [147]. Reproduced with permission of American Geophysical Union.)

the atmospheric inhomogeneity increases with the scan angle, which may not be explicitly simulated in radiative transfer models. An obstruction to satellite observations by the spacecraft radiation may occur at large scan angles, which is usually difficult to be taken into account in the forward model and calibration process. Therefore, scan-angle-dependent biases of both observed brightness temperatures and those simulated from radiative transfer models are anticipated for cross-track scanning radiometer instruments. In many applications such as NWP radiance assimilation, angular-dependent biases between the observed brightness temperatures and those simulated from radiative transfer models must be quantified and be removed from the data [8, 47, 152].

Figure 4.8 presents scan-dependent biases of ATMS channels 5–15 estimated separately for ascending and descending nodes. If the atmospheric inhomogeneity is the only source of biases, a symmetric bias distribution is expected. However, an asymmetric scan bias pattern is noticed for all ATMS channels examined. Channels 5–12 are more negatively biased near the ends of the ATMS scan line, and channels 13–15 are more negatively biased at the beginning of the ATMS scan line toward a cold temperature. A temperature dependence of scan biases is observed, evidenced by the different bias magnitudes for ascending and descending nodes of the same channel (e.g., channels 10–14) and the different bias

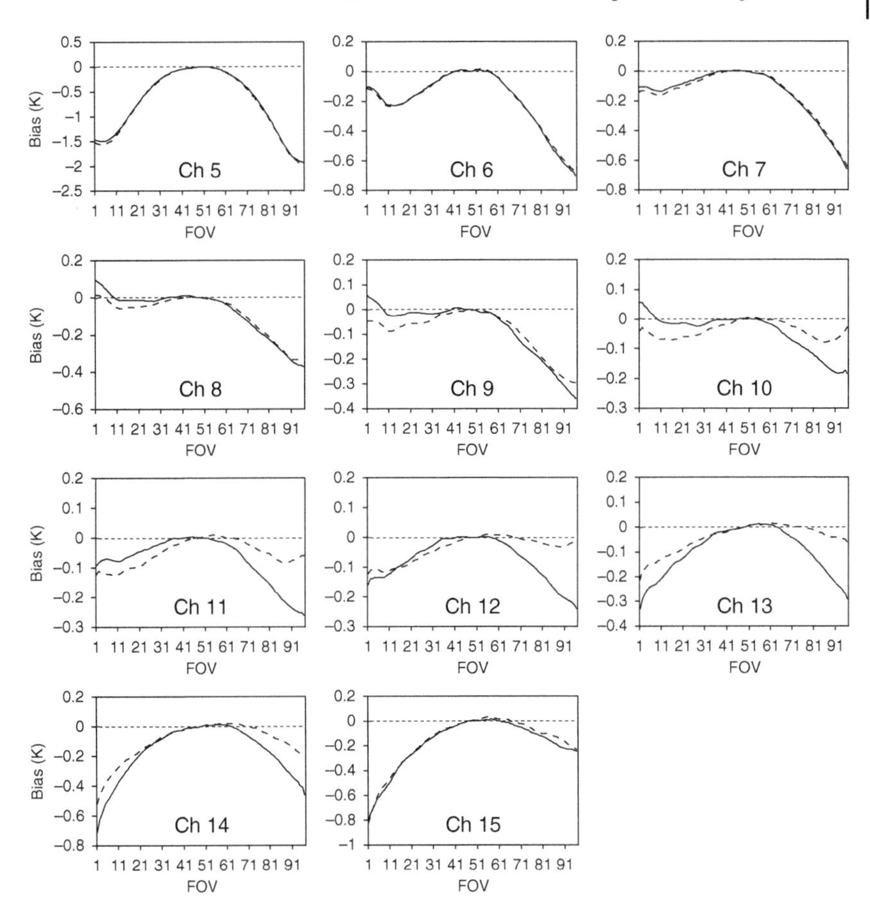

Figure 4.8 Scan-dependent biases of ATMS channels 5–15 at ascending (solid) and descending nodes (dashed) within [60S, 60N]. (Weng *et al.* 2012 [147]. Reproduced with permission of American Geophysical Union.)

magnitudes for different channels. This probably arises from the contributions from its near- (e.g., spacecraft) and far-field (e.g., Earth view) side lobes. Further studies are needed for finding and confirming the root causes of the asymmetric bias pattern found for ATMS antenna temperatures using pitch maneuver data.

4.5
Simulations of ATMS Sounding Channels Using GPSRO Data

4.5.1
Collocation of GPS RO and ATMS Data

The GPS RO is a limb-sounding technique that makes use of radio signals emitted from the GPS satellites for sounding Earth's atmosphere. Under the assumption of the spherical symmetry of the atmospheric refractive index, vertical profiles of bending angle and refractivity can be derived from the raw RO measurements of the excess Doppler shift of the radio signals transmitted by GPS satellites. The profiles of refractivity are then used to generate profiles of the temperature and water vapor retrieval using a one-dimensional variational data assimilation (1D-Var) algorithm [153, 154].

The COSMIC satellite system consists of a constellation of six low-Earth-orbit (LEO) microsatellites and was launched on April 15, 2006. Each LEO follows a circular orbit 512 km above the Earth surface, with an inclination angle of 72°. Currently, there are about 1000 soundings daily. The vertical resolution is 0.1 km from the surface to 39.9 km, and each GPS RO measurement quantifies an integrated atmospheric refraction effect over a few hundred kilometers along a ray path centered at the tangent point. The global mean differences between COSMIC and high-quality reanalysis within the height range between 8 and 30 km are estimated to be ~0.65 K [155]. The precision of COSMIC GPS RO soundings, estimated by comparison of closely collocated COSMIC soundings, is approximately 0.05 K in the upper troposphere and lower stratosphere [156]. In the water-vapor-abundant region in the lower troposphere or the ionosphere regions, GPS profiles become less accurate.

The COSMIC RO soundings are collocated with ATMS measurements for assessing the accuracy of ATMS measurements. The collocation criteria are set by a time difference of not more than 3 h and a horizontal spatial separation of less than 50 km at the altitude of peak weight function. If there are more than one ATMS pixel measurement satisfying these collocation criteria, the one that is closest to the related COSMIC sounding is chosen and the others are discarded. Because surface state variables and parameters are not provided by COSMIC ROs, only upper-level temperature sounding channels are simulated using COSMIC GPS RO data. The global biases and the angular dependence of biases are estimated.

As the GPS radio signal passes through the atmosphere, its ray path is bent over due to variations of the atmospheric refraction. Therefore, the geolocation of the perigee point (also called tangent point) of a single RO profile varies with altitude. On the other hand, a satellite measurement at a specific frequency represents a weighted average of radiation emitted from different layers of the atmosphere. The magnitude of such a weighting is determined by a channel-dependent weighting function (WF). The measured radiation is the most sensitive to the atmospheric temperature at the altitude where the WF reaches a maximum. The level of the peak WF also varies with the scan angle (see Figure 4.9 for ATMS channel 8). For each channel, the altitude of the peak WF is the lowest at the nadir and increases with the scan angle. Considering the geolocation change of the perigee point of a GPS RO profile with altitude, the geolocation of a GPS RO at the altitude where the WF for each collocated ATMS FOV of a particular sounding channel reaches the maximum is first considered, and then, the spatial distance between the GPS RO and ATMS channel must be less than 50 km. The altitude of the maximum

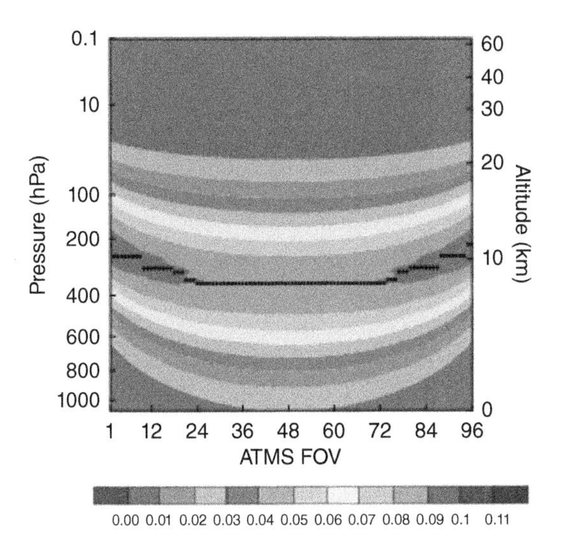

Figure 4.9 Angular-dependence of ATMS weighting function profiles for ATMS channel 8 (shaded color) calculated by CRTM using the U.S. standard atmospheric profile. The maximum weighting function altitudes at all FOVs are indicated in black dots. (Zou *et al.* 2014 [8]. Reproduced with permission of IEEE.)

WF used in the collocation study is precalculated by the use of CRTM with the US standard atmospheric profile (see [8]).

Since ATMS brightness temperatures at lower peaking sounding channels are affected by clouds, thus a cloud detection algorithm [147] is applied to separate the data under clear-sky conditions over ocean from the total ATMS measurements. Figure 4.10 presents the spatial distribution of the ATMS observations that are collocated with the COSMIC data under clear-sky conditions over ocean and between 60S and 60N from December 10, 2011 to June 30, 2012. Most collocated data are located in the subtropical regions in the Northern Hemisphere and middle latitudes in the Southern Hemisphere.

4.5.2 ATMS Bias with Respect to GPS RO Data

Simulations of the ATMS brightness temperatures using GPS RO profiles as input to CRTM are denoted B^{GPS} hereafter. The altitudes of the maximum WF of ATMS channels 14–15 are above 40 km, which is the top of COSMIC RO data. Therefore, biases of ATMS channels 14–15 are not included in this study. Figure 4.11 presents scatter plots of brightness temperature from ATMS observations (O) and GPS RO simulations (B^{GPS}) for all collocated data points under clear-sky conditions over ocean between 60S and 60N from December 10, 2011 to June 30, 2012.

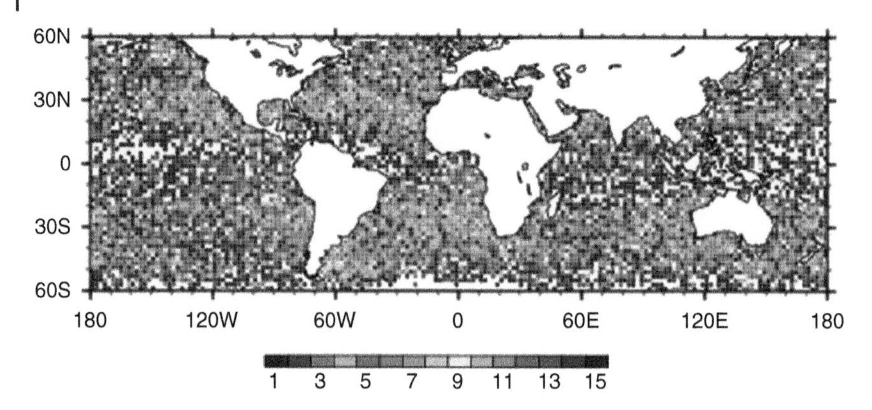

Figure 4.10 Spatial distribution of ATMS measurements that are collocated with COSMIC ROs in a 2° × 2° latitude and longitude grid for channel 6. (Zou *et al.* 2014 [8]. Reproduced with permission of IEEE.)

In general, CRTM simulations with GPS RO input profiles correlate quite well with ATMS observations. A noticeable temperature dependence of the differences ($O-B^{GPS}$) between observations and simulations is seen in channel 6. Global biases estimated by the mean differences ($O-B^{GPS}$) are positive for channels 6 and 10–13 with values smaller than 0.5 K and negative for channels 5 and 7–9 with values greater than −0.7 K (Figure 4.12). The standard deviation is smallest for channel 8 (~0.25 K) and increases with the channel number to about 2.0 K at channel 13 (see Figure 4.12). The standard deviations for channels 5 and 6 are 0.4 and 0.6 K, respectively.

For a cross-track scanning radiometer such as ATMS, the variation of the optical path length with the scan angle is modeled through CRTM. However, the variation of atmospheric inhomogeneity with the scan angle has not been explicitly simulated in CRTM. In addition, effects of the spacecraft radiation on brightness temperatures can also vary with the scan angle. Therefore, a scan-angle-dependent bias is expected for the cross-track scanning radiometer. In many applications such as weather predictions through radiance data assimilation, angular-dependent biases must be properly quantified so that they could be removed before data applications.

Figure 4.13 presents scan-dependent biases and standard deviations of ATMS channels 5–13 estimated by using GPS RO data. Variations of GPS RO profile numbers collocated with ATMS data are also shown in Figure 4.13. As expected, the total number of collocated GPS ROs increases with scan angle since the size of FOV increases with scan angle. An asymmetric scan bias pattern is noticed for ATMS channels 5–10. ATMS temperature sounding channels are more negatively biased near the end of ATMS scan lines. The standard deviations of ATMS channels 6, 12, and 13 are much larger than those of the remaining channels. The root causes of the asymmetric bias pattern came from the contributions from the near-field (e.g., spacecraft) side lobes, which were confirmed by using the ATMS pitch maneuver data [9].

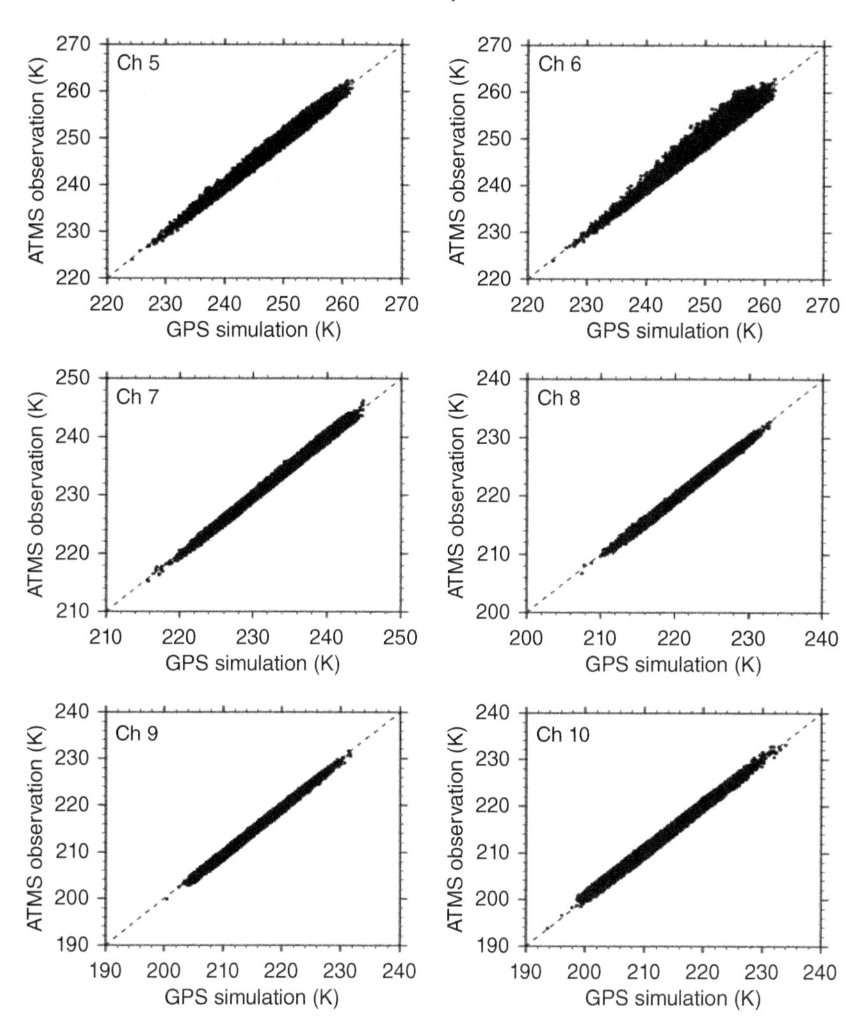

Figure 4.11 Scatter plots of brightness temperature from ATMS observations and CRTM simulations at six channels with the inputs from collocated COSMIC data under clear-sky over ocean between 60S and 60N from December 10, 2011 to June 30, 2012. (Zou *et al.* 2014 [8]. Reproduced with permission of IEEE.)

4.6 Uses of TRMM-Derived Hydrometeor Data in Radiative Transfer Simulations

4.6.1 Collocation of ATMS and TRMM Data

Tropical Rainfall Measurement Mission (TRMM) generated the vertical hydrometeor profile products from its microwave imager and radar. The TRMM Microwave Imager (TMI) is a multichannel dual-polarized passive microwave

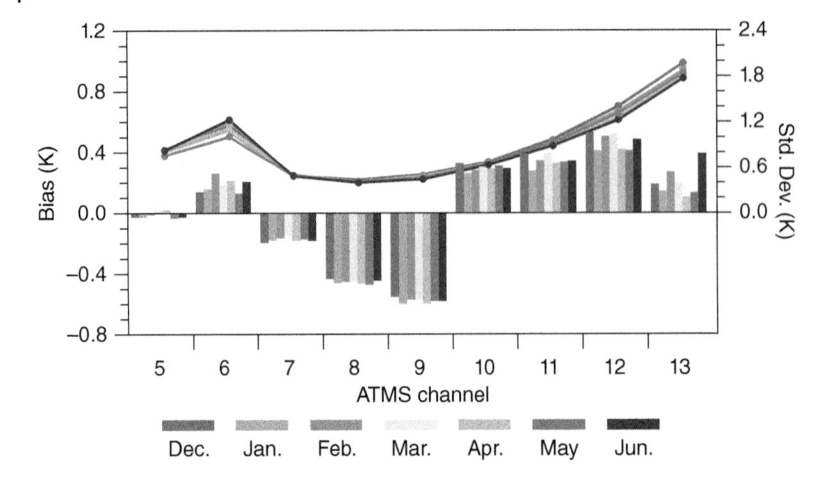

Figure 4.12 Biases (bars) and standard deviations (lines) of the differences between ATMS observations and GPS RO simulations (O−B^{GPS}) for all collocated data under clear-sky over ocean and between 60S and 60N from December 10, 2011 to June 30, 2012. (Zou *et al.* 2014 [8]. Reproduced with permission of IEEE.)

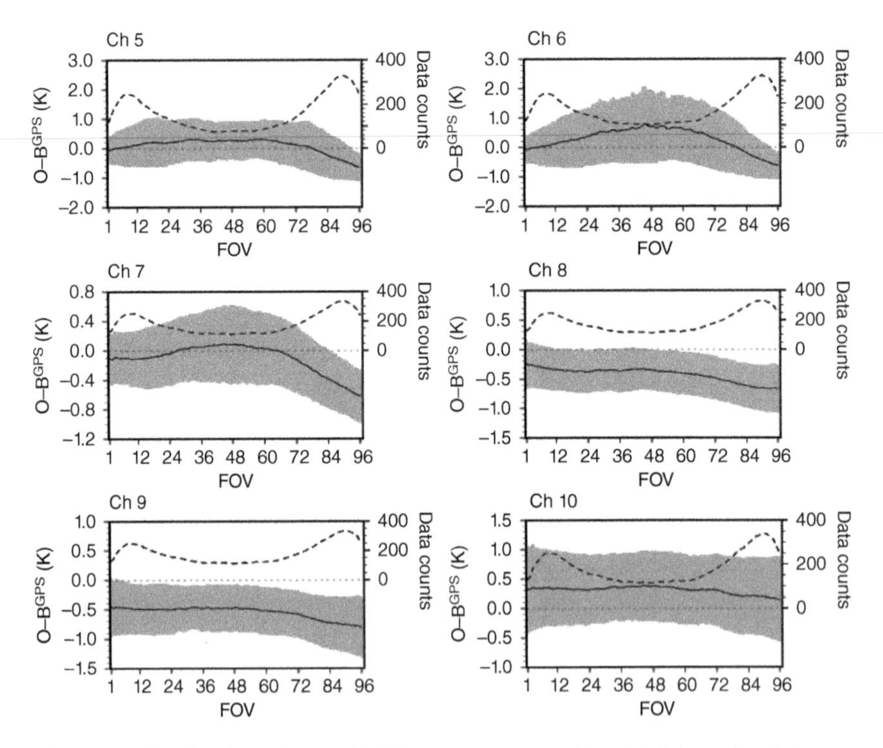

Figure 4.13 Angular-dependence of brightness temperature biases (solid curve) and standard deviations (shaded area) estimated by the differences between ATMS observations and GPS RO simulations (O−B^{GPS}) for collocated data under-clear sky over ocean between 60S and 60N. The GPS RO profiles numbers collocated with ATMS data are also shown (dashed curve). (Zou *et al.* 2014 [8]. Reproduced with permission of IEEE.)

radiometer with nine channels at four frequencies: 10.65, 19.35, 21.3, and 85.5 GHz. The TRMM 2A12 products include surface rainfall and vertical hydrometeor profiles generated from Goddard Profiling algorithm, GPROF2010. Atmospheric hydrometeor types include cloud liquid water, rain water, cloud ice, snow, and graupel. The TMI hydrometeor profiles are distributed in 28 levels from the surface to 18 km in height. In forward calculations, temperature and humidity profiles are generated from the European Center for Medium range Weather Forecasting (ECMWF) 6-h forecast fields. In addition, atmosphere pressure, water vapor, ozone, surface temperature, wind speed and direction, and surface type are also provided in the ECMWF forecast field. Whereas ECMWF forecast fields have 91 vertical levels, ECMWF output data are interpolated into the TRMM 28 levels.

Brightness temperatures at each ATMS channel are calculated using CRTM with the TRMM hydrometeor profiles as inputs. In addition, the brightness temperatures at both horizontal and vertical polarization states are computed for each set of TRMM profiles, and within each ATMS FOV, they are combined to produce the quasi-polarization brightness temperatures. The quasi-vertical and -horizontal brightness temperatures are then averaged with the antenna gain function to obtain the antenna brightness temperature (TDR). Figure 4.14 is a schematic diagram of ATMS channels 1 and 2 FOV coverage (elliptical shape) and TMI measurements (small dots). Essentially, ATMS FOV can be described in an elliptical equation as

$$\frac{(x-x_0)^2}{a^2} + \frac{(y-y_0)^2}{b^2} = 1, \quad (4.4)$$

where $(x-x_0)$ and $(y-y_0)$ are the distances between the ATMS and TMI centers along the ATMS FOV major (cross-track) and minor (along-track) axes. The lengths of the semimajor and -minor axes are represented by a and b. If a value on the left side of Eq. (4.4) is less than 1, then TMI pixel is within ATMS FOV. The spatial considering as well as the temporal considering will be included. The interval between ATMS and TRMM observations less than 15 min is acceptable for brightness temperature calculations.

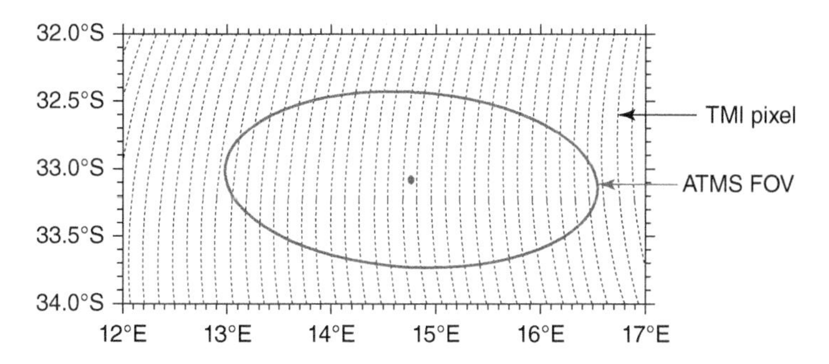

Figure 4.14 A spatial coverage of ATMS field of view (FOV) at channels 1 and 2 and TMI pixels within and near the ATMS FOV.

4.6.2 ATMS Biases With Respect to TRMM-Derived Simulations

Direct and accurate information on atmospheric hydrometeors can increase the accuracy of radiance simulation over cloudy regions. For the simulation of cloudy radiance, the inputs to CRTM consist of three major parts: atmospheric state (including surface parameters, vertical profiles of temperature/water vapor, the pressure level, and also absorption gases), vertical profiles of hydrometeor, and cloud optical properties. CRTM can deal with six cloud types: liquid cloud, ice cloud, rain, snow, graupel, and hail. In order to speed up the cloud radiance simulations, cloud optical properties are computed offline and stored as LUTs. For microwave, both liquid and ice cloud particles are assumed to be spherical, and following the modified gamma size distribution, the scattering properties are calculated based on the Mie theory. Table 4.2 lists the cloud size parameters used for calculating the size distributions of spherical particles in CRTM.

In this study, the DDA scattering database is tested when ice cloud particles are assumed to be nonspherical. This database contains single-scattering properties at frequencies ranging from 15 to 340 GHz, with temperatures from 0 to 40 °C, or particle size from 50 to 12 500 mm, and for 11 particle shapes. In the current experiment, the shapes of 6-bullet rosettes, dendrite snowflakes, thick plate and sector snowflakes are selected for ice cloud, snow, hail, and graupel, respectively. In deriving the single-scattering properties, all ice particles are assumed to be randomly oriented in space [33].

In this study, vertical hydrometeor profiles from the TRMM 2A12 product are used as inputs to CRTM to simulate the ATMS antenna brightness temperatures at 22 channels. Our case studies focus on simulations of Typhoon Neoguri on July 10, 2014, in the Northwestern Pacific ocean. ATMS and TRMM satellites observe the storm as shown in Figure 4.15.

Since TRMM 2A12 product only includes five hydrometeor contents, the particle size and the bulk volume density are set for each type of hydrometeor as summarized in Table 4.2. In CRTM, these values are required for searching the scattering parameters from Mie or DDA lookup tables. In future, when GPM data are available, the particle size for each hydrometeor will be retrieved through its dual polarization radar, and thus, it is deemed to improve the simulation accuracy.

Figure 4.16a–d shows the difference between ATMS observations and simulations at four selected ATMS channels 2, 3 16, and 17 as a function of ice water path (IWP), which is integrated from TRMM 2A12 products. Mie scattering lookup table is used in the forward simulations. As the IWP is less than 0.1 kg/m^2, ATMS

Table 4.2 Cloud hydrometeor size and bulk volume density used in CRTM simulations.

Hydrometer type	Cloud water	Cloud ice	Rain water	Snow	Graupel
Effective radius (μm)	10	50	500	500	500
Volume density (kg/m^3)	1000	920	1000	100	400

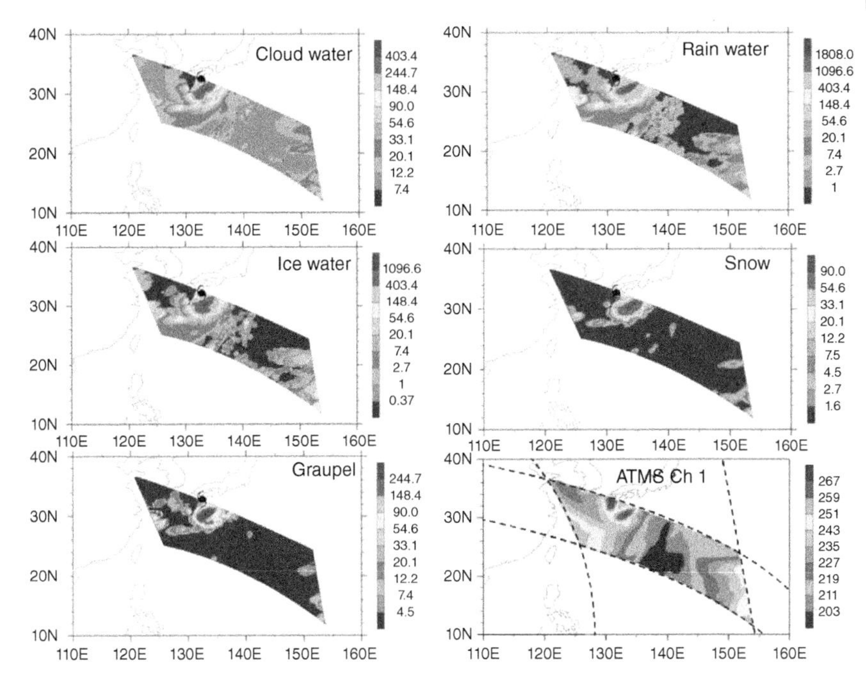

Figure 4.15 Typhoon Neoguri observed by ATMS and TMI over Pacific region at 0351 to 0401 UTC 10 July 2014. Hydrometeor profiles including cloud water, rain water, ice water, graupel, and snow derived from TMI 2A12 products are vertically integrated.

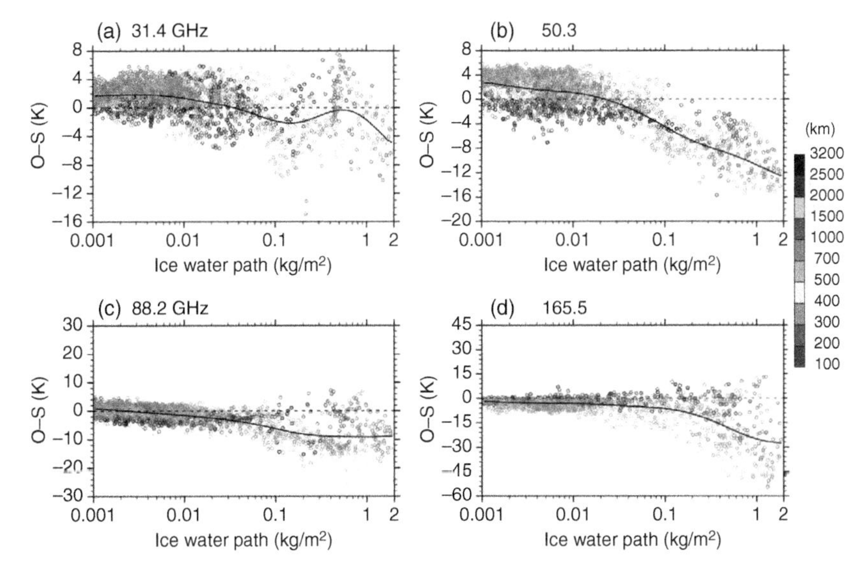

Figure 4.16 A difference between observation (O) and simulation (S) versus ice water path at four ATMS channels (a) 31.4 GHz, (b) 50.3 GHz, (c) 88.2 GHz, and (d) 165.5 GHz. The mean differences of brightness temperature (O–S) are indicated by black solid lines in each panel. Simulations are made through uses of Mie scattering table. The color bar indicates the distance of the data point to the storm center in km.

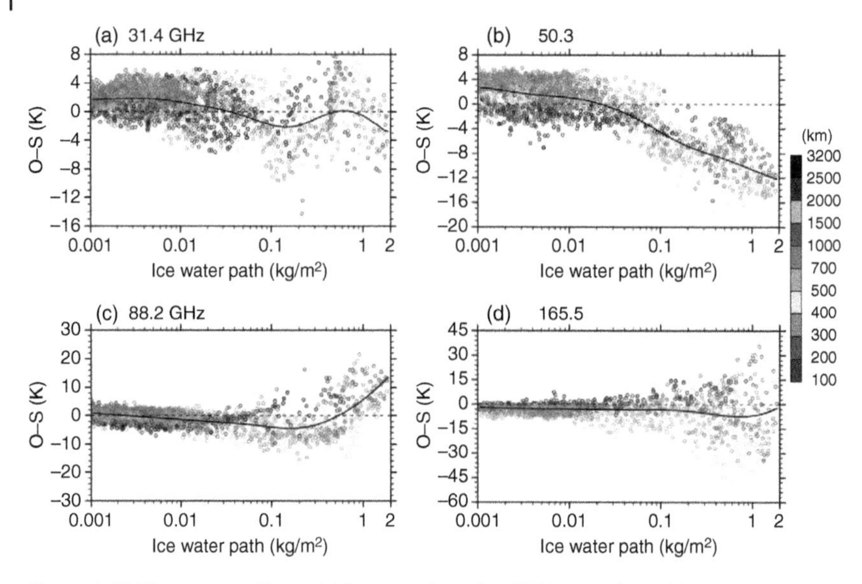

Figure 4.17 The same as Figure 4.16, except for using DDA scattering table.

observations have smaller biases. As the IWP is greater than 0.5 kg/m^2, the bias becomes larger, while the biases at higher frequencies become negative. Thus, the current Mie table results in warmer brightness temperature simulations in storm regions. Compared to the simulations with the Mie table, the brightness temperatures simulated from the DDA scattering table reduces the overall bias, especially at 165.5 GHz, as shown in Figure 4.17a–d. However, the O – S from DDA shows a similar scattering as IWP is greater than 1 kg/m^2, compared to Mie computations. Overall, O – S at 31.4 GHz does not show any specific dependence on IWP.

Figure 4.18a–d further illustrates O – S at four ATMS upper-air temperature sounding channels. In general, for IWP less than 0.7 kg/m^2, an inclusion of cloud and precipitation scattering in the simulation results in smaller biases for most of these temperature channels. Overall, the impact of clouds and precipitation on the ATMS temperature sounding channel diminishes as the WF becomes higher. For example, O – S at channel 7 (54.4 GHz) does not vary with IWP.

Figure 4.19a–d further illustrates O – S at four ATMS upper-air water-vapor sounding channels. Except for ATMS channel 18 at 183.31 ± 7 GHz, O – S does not significantly vary with IWP. Overall, the impact of clouds and precipitation on the ATMS water-vapor sounding channel also diminishes as the WF becomes higher. For example, O – S at channel 21 (183.31 ± 1.8 GHz) does not vary with IWP.

Figure 4.20 shows the dependence of O – S bias on IWP for the ATMS channels included in Figures 4.17–4.19 when the simulations are performed through the use of Mie and DDA scattering tables. For lower frequency window channels 1–2, the main cause of O – S is the water-phase cloud, and the scattering has some impact on the lower air temperature channels but not on the higher temperature channels. For water vapor channels, the difference between Mie and DDA is more

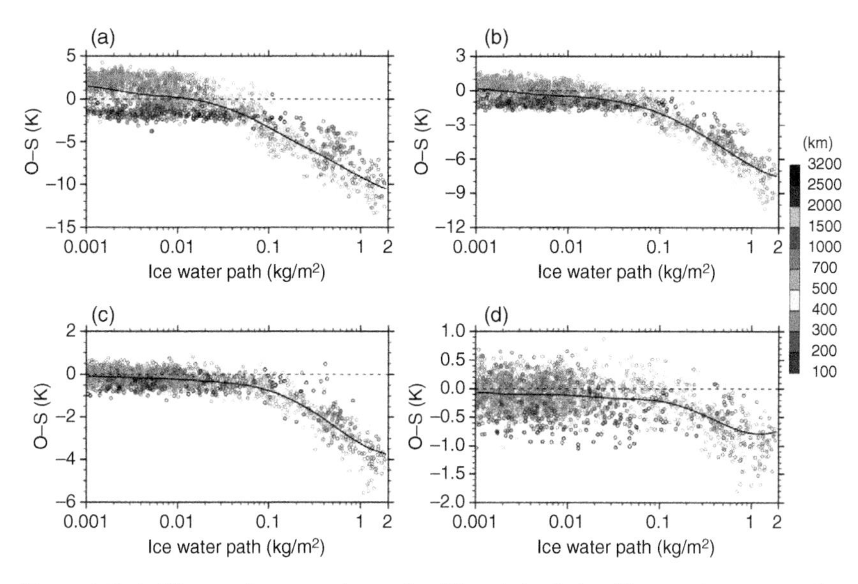

Figure 4.18 A difference between observation (O) and simulation (S) versus ice water path at four ATMS channels at (a) 51.76 GHz, (b) 52.8 GHz, (c) 53.6 GHz, and (d) 54.4 GHz. Simulations are made through uses of DDA scattering table.

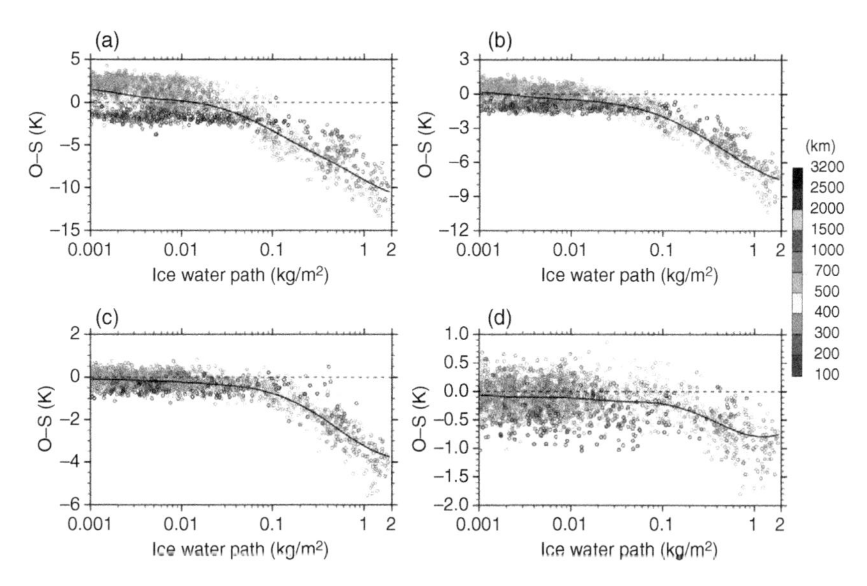

Figure 4.19 A difference between observation (O) and simulation (S) versus total water path at four ATMS channels (a) 183.31 ± 7 GHz, (b) 183.31 ± 4.5 GHz, (c) 183.31 ± 3 GHz, and (d) 183.31 ± 1.8 GHz. Simulations are made through uses of DDA scattering table.

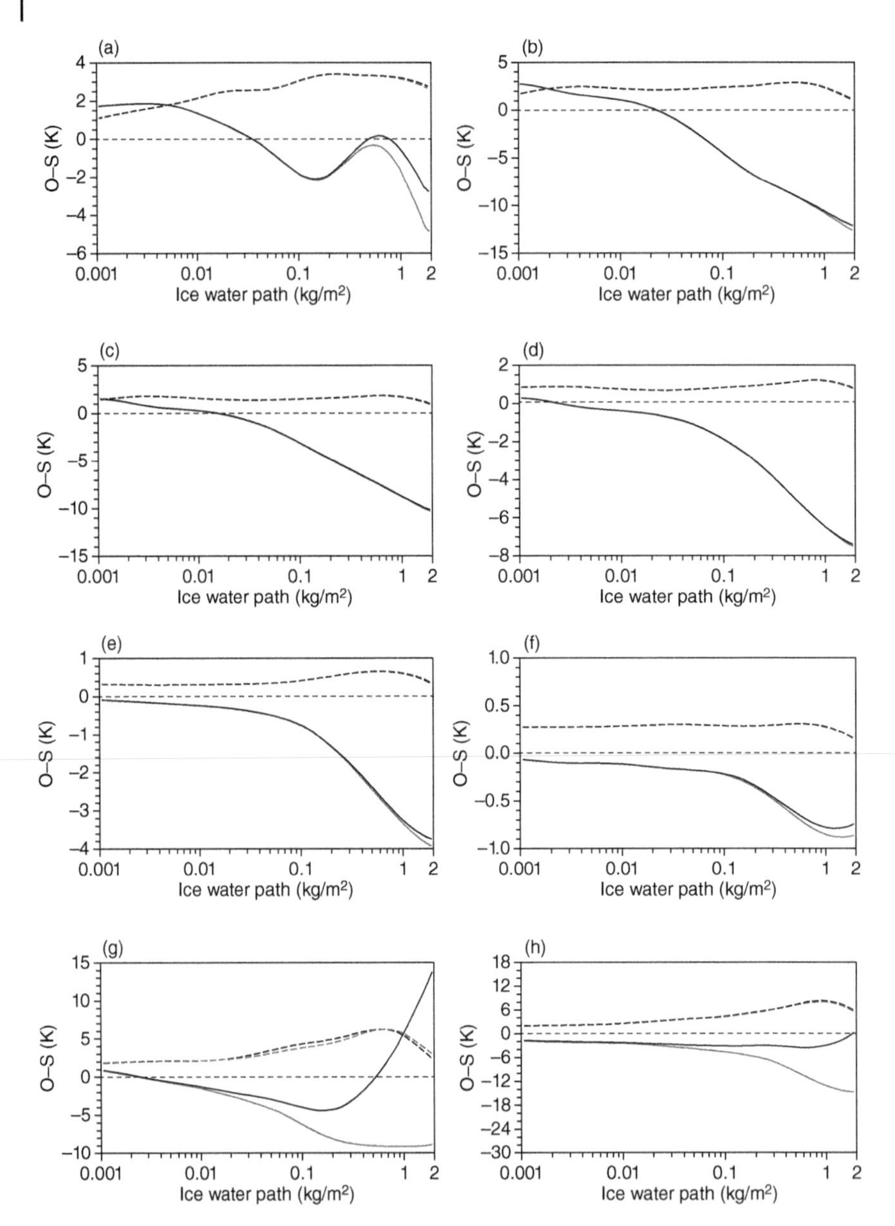

Figure 4.20 Mean (solid curves) and standard deviation (dashed curves) of O–S versus ice water path at ATMS channels (a) 31.4 GHz, (b) 50.3 GHz, (c) 51.76 GHz, (d) 52.8 GHz, (e) 53.6 GHz, (f) 54.4 GHz, (g) 88.2 GHz, (h) 165.5 GHz, (i) 183.31 ± 7 GHz, (j) 183.31 ± 4.5 GHz, (k) 183.31 ± 3GHz, and (l) 183.31 ± 1.8 GHz. Simulations (S) are made through uses of Mie (gray) and DDA (black) scattering table.

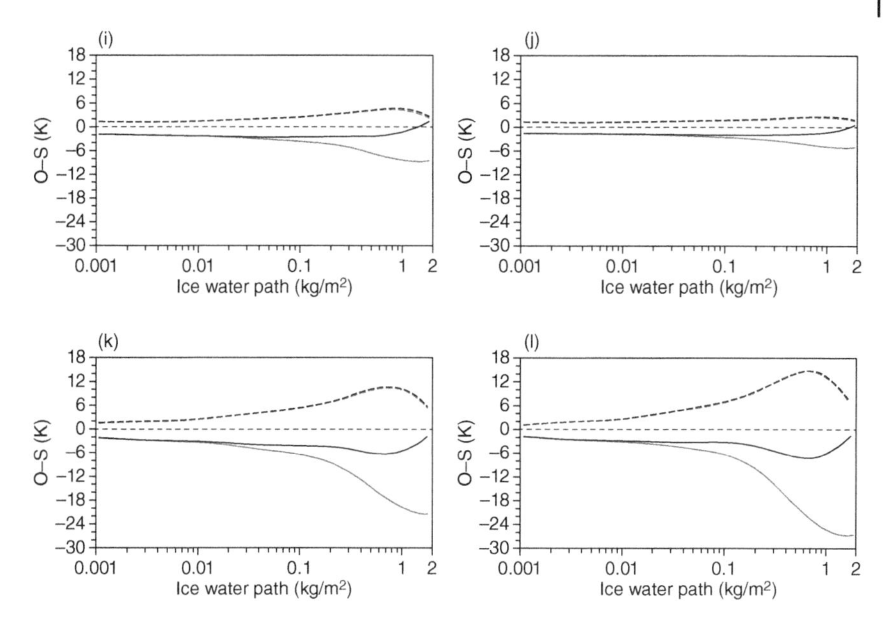

Figure 4.20 *(Continued)*

obvious in channel 17 than in other channels. This is because the water vapor absorption in 183 GHz channels can compensate some of the scattering effect.

Spatial distribution of observed and simulated brightness temperatures for ATMS channels 2, 5, 17, and 22 are shown in Figure 4.21. Consistent with the results shown, for high-frequency channels, simulation using DDA can correct the insufficient scattering in Mie simulation and match with the observation more closely. Moreover, for channels 2, 5, and 22, the differences between Mie and DDA simulations are not obvious at channel 17.

4.7 Advanced Radiative Transfer Simulations

Microwave instruments measuring the full Stokes parameters can also be simulated through some advanced radiative transfer models. Here, we use the mesoscale model version 5 to generate three-dimensional fields of microphysical parameters such as for Hurricane Bonnie in 1998 [157]. The hydrometeors and wind fields are for 48 h simulation with the finest grid size of 4 km. The surface wind speed is between 30 and 60 m/s. Figures 4.22 and 4.23 show that the third and fourth Stokes components at 10.7 and 37 GHz reveal the hurricane vortex very well. Microwave polarimetric measurements correlate well with the wind direction and speed in hurricane environments. Note that the magnitudes of

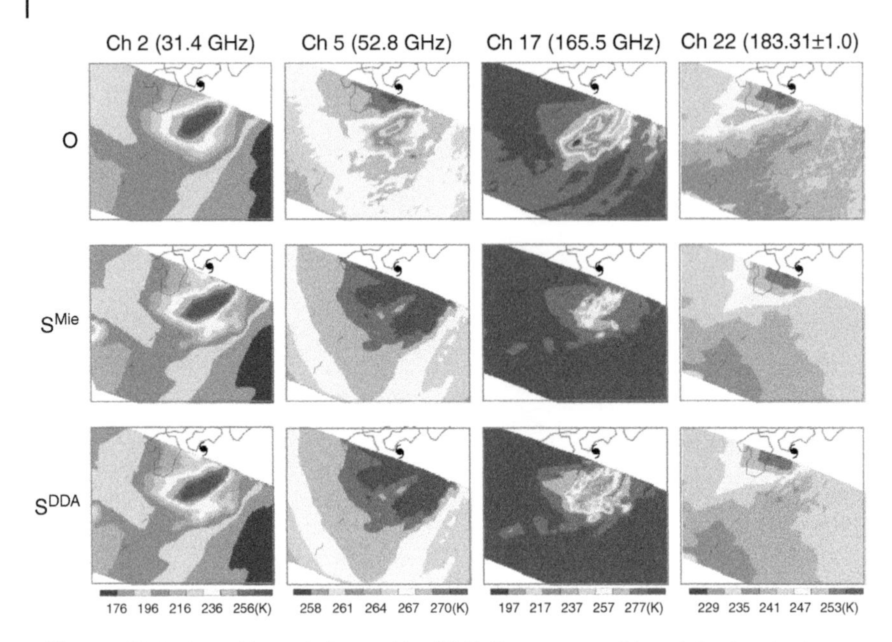

Figure 4.21 Typhoon Neoguri observed by ATMS (O, upper panels), and simulated using Mie (S^{Mie}, middle panels) and DDA (S^{DDA}, lower panels) at ATMS channels (a) 31.4 GHz, (b) 52.8 GHz, (c) 165.5 GHz, and (d) 183.31 ± 1 GHz.

the Stokes components increase as frequency and wind speed increase. The amplitude at 37 GHz is much larger than that at 10.7 GHz.

The spatial heterogeneous properties also modify the Stokes components of microwave measurements. Here, we applied Monte Carlo radiative transfer model [50] to simulate the Stokes signals from Hurricane Bonnie after landfall. As shown in Figure 4.22, the inhomogeneous clouds may indeed produce comparable signatures at 37 GHz to the third and fourth Stokes components because in our simulations, surface emissivity is constant for the first two Stokes components and zero for the third and fourth components. Notice that the third component responds well to the precipitation structures as indicated in the lower right panel.

Microwave observations from WindSat are consistent with the theoretical interpretations and display the third Stokes component near the cloud edge. In addition, the fourth component is strong and has not been fully understood (see Figure 4.23). Notice that over Antarctic and central Greenland, the WindSat fourth component can be as large as 10–20 K [44] at its 10 GHz channel. Of course, vicarious calibration of the WindSat polarimetric channels can reduce to its third and fourth components by 0.5 K over global regions [44]. Physical interpretation of the surface polarimetric signals requires more theoretical understanding (Figures 4.24 and 4.25).

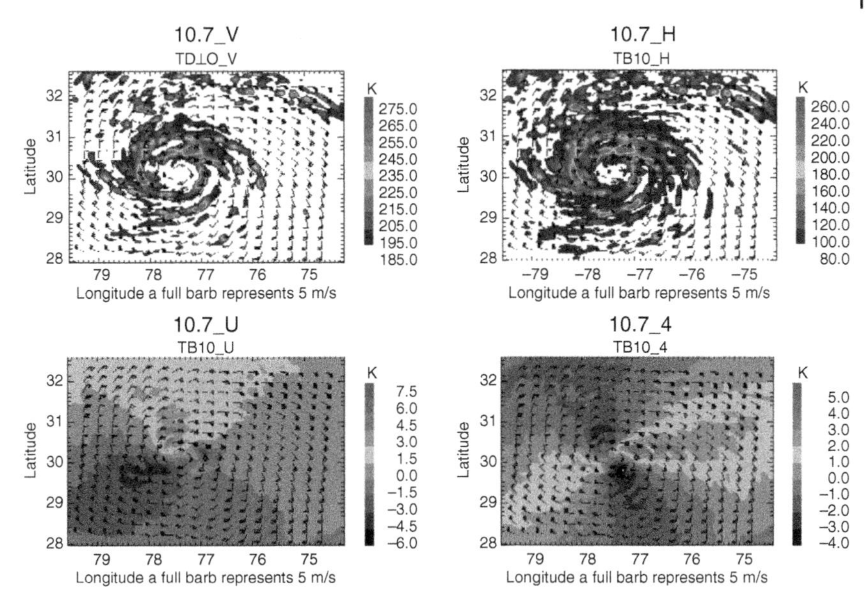

Figure 4.22 Simulated Stokes components at 10.7 GHz using MM5 model hydrometeors and atmospheric parameters of Hurricane Bonnie, August 26, 1998. Wind speed with a full bar represents 5 m/s. (Weng 2002 [157]. Reproduced with permission of Taylor and Francis.)

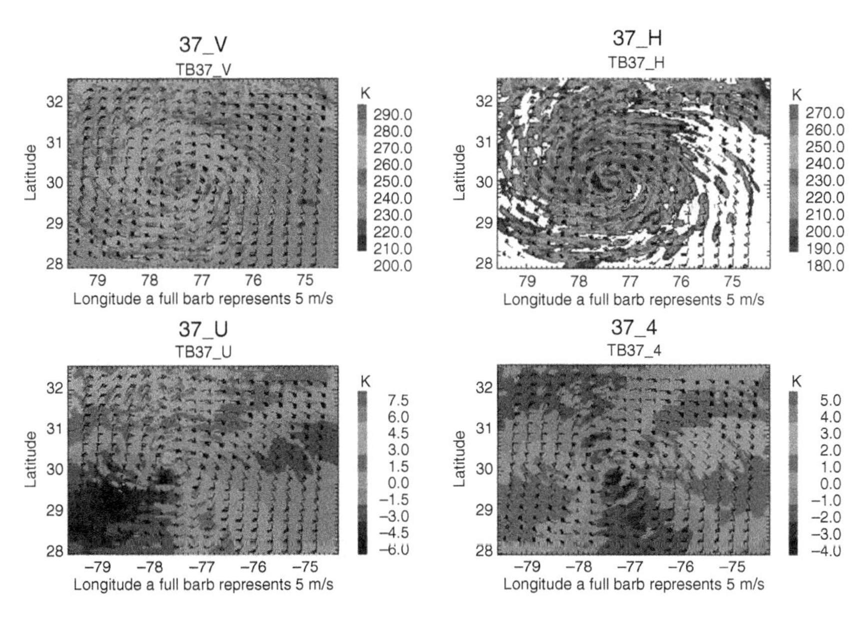

Figure 4.23 Simulated Stokes components at 37 GHz using MM5 model hydrometeors and atmospheric parameters of Hurricane Bonnie, August 26, 1998. Wind speed with a full bar represents 5 m/s. (Weng 2002 [157]. Reproduced with permission of Taylor and Francis.)

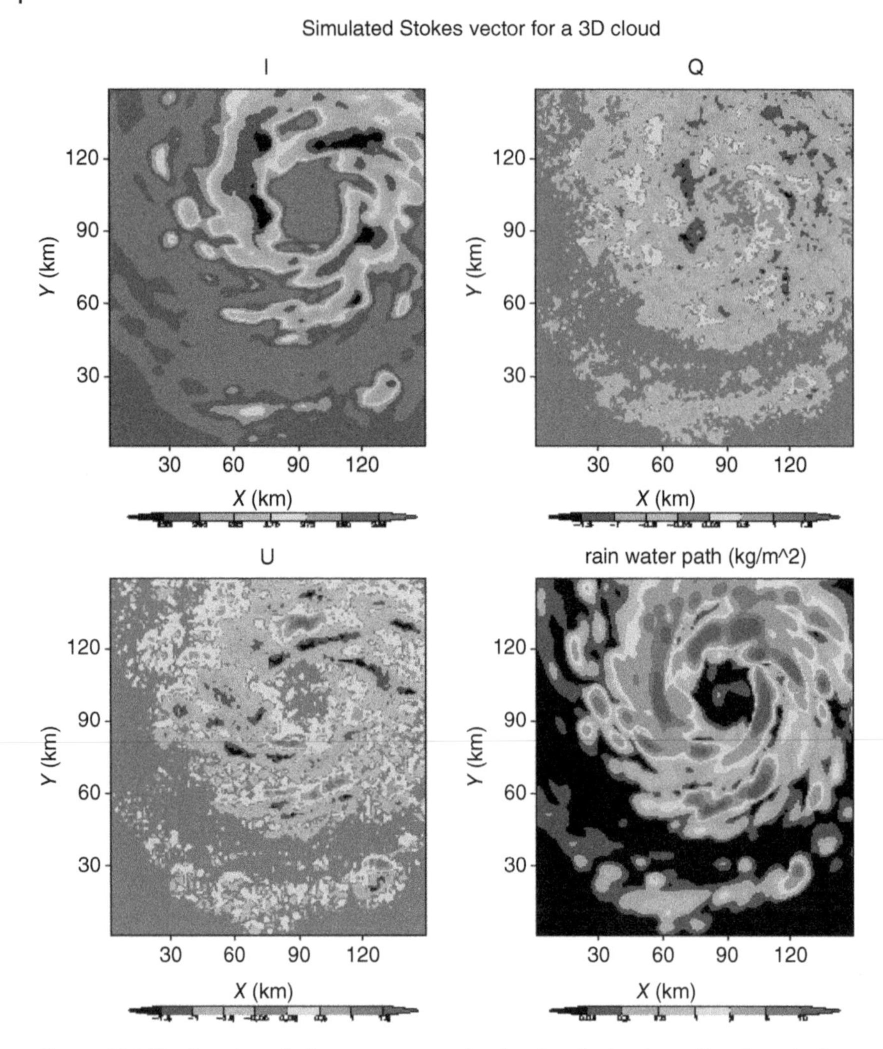

Figure 4.24 The first three Stokes components simulated under hurricane Bonnie and rain water path when it landfalls. The third Stokes component is purely generated from the 3D cloud effects and precipitation inhomogeneity.

4.8 Summary and Conclusions

Understanding the uncertainty in the forward model in simulation of microwave radiance is important for many applications. The uncertainty in simulations can come from the inputs to forward models as well as the accuracy of the radiative transfer scheme. In simulating the microwave sounding channels, we use the atmospheric and surface parameters from NWP models, GPS RO sounding profiles, and TRMM hydrometeor profiles as inputs to CRTM. The current radiative

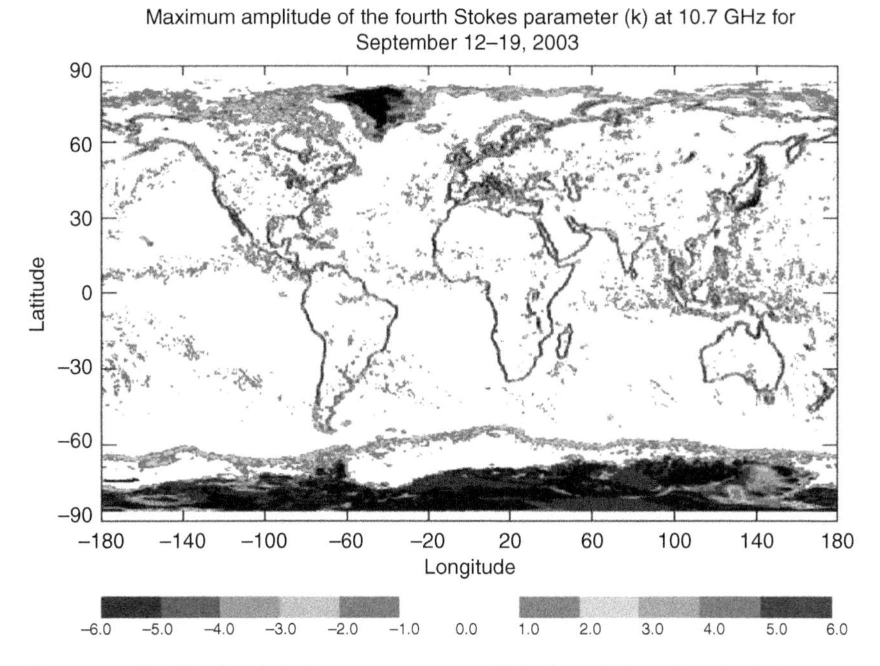

Figure 4.25 WindSat fourth Stokes component at 10 GHz channel. Note that clouds, ice edge, sea ice, clearly delineates water/land, and water/ice boundaries.

transfer solver used in CRTM is a plane-parallel doubling and adding scheme and does not handle polarization. In a strong precipitation condition, the plane-parallel approach may not be able to resolve the radiative effects of clouds in three dimensions. Finally, the instrument calibration may also contribute to the bias between observations and simulation.

5
Calibration of Microwave Sounding Instruments

5.1
Introduction

Over the past two decades, microwave observations improved the numerical weather predictions and contributed to the long-term climate monitoring. In the weather arena, the Advanced Microwave Sounding Unit (AMSU) on board the series of National Oceanic and Atmospheric Administration (NOAA) Polar Orbiting Environmental Satellites has been a major factor in significantly increasing the accuracy of global medium-range forecasts to the point where the current 5-day forecast accuracies are about the same as the 3-day forecast accuracies were 10 years ago. For climate studies, the nine microwave sounding units (MSUs) on board the early NOAA satellites have provided a unique 26-year-old time series of the global tropospheric temperature as well as its trend [159–162]. Differences in instrument calibration are accounted for by intercalibrating the MSU instruments using overlapping orbital data (see, e.g., [162, 163]). In order to extend the temperature and trend analysis to longer time periods, it is necessary to continue the MSU time series using AMSU data. On October 28, 2011, the Suomi National Polar-Orbiting Partnership (SNPP) satellite was successfully launched into an orbit which has an inclination angle of 98.7° to the Equator and is 824 km above the Earth. On board the SNPP satellite, the Advanced Technology Microwave Sounder (ATMS) is a cross-track scanning instrument and has 22 channels at frequencies ranging from 23 to 183 GHz for profiling the atmospheric temperature and moisture under clear and cloudy conditions. The ATMS instrument is well calibrated and is now in its third-year mission for both weather and climate applications [147].

In this chapter, we present various aspects of operational calibration of ATMS instruments. In Section 5.2, a general concept on microwave calibration is introduced. In Section 5.3, ATMS instrument description is presented. In Section 5.4, ATMS calibration algorithm is presented. In Sections 5.5 and 5.6, impacts of ATMS antenna emission on two-point calibration and retrieval of reflector emissivity using ATMS pitch-over data are given, respectively. In Section 5.7, we investigate the uses of Allan deviation for ATMS noise characterization and compare the result with the traditional standard deviation approach.

Passive Microwave Remote Sensing of the Earth: For Meteorological Applications, First Edition. Fuzhong Weng.

In Section 5.8, conversion from antenna to sensor brightness temperature is described. A summary and conclusions are given in Section 5.9.

5.2
Calibration Concept

Calibration refers to the process of quantitatively determining a sensor's response to known, controlled signal inputs. Prelaunch laboratory calibration establishes a sensor's characteristic response function. The prelaunch calibration algorithm can then be carried into the postlaunch era by requiring only straightforward coefficient updates after a satellite is in orbit. However, the need for new postlaunch calibration algorithms becomes evident when unforeseen behavior or an uncharacteristic sensor response is identified through detailed scientific analysis of the on-orbit sensor data. Postlaunch effects leading to abnormal measurements may come from obstructions to the Earth's view and calibration signals, space environment, variation in instrument spectral response, receiver coherent noises, slow deterioration of the electronic system, or even mechanical malfunction following the intense rigors of launch. In such situations, the new postlaunch calibration algorithms must make use of the calibration data from onboard sources, ground truth data, and intersensor comparisons so that the measurements become consistent over time.

During the prelaunch phase, microwave calibration consists of evaluating the sensor in a thermal vacuum (T/V) chamber using three calibration targets: (i) a cold calibration target that is cooled with liquid nitrogen to about 80 K, (ii) a variable temperature target from about 80 to 330 K placed in the scene field of view (FOV), and (iii) the sensor's on-orbit warm calibration load. The radiometer's two-point calibration is determined from the warm and cold targets within the T/V chamber. It is tested at a variety of "scene temperatures" that are simulated by changing the temperature of the variable temperature target.

A combination of model-based and empirical corrections to the two-point instrument calibration is necessary in order to take into account the uncertainties from the antenna spillover energy emanating from the spacecraft, cross-polarization, and antenna side lobes beyond the Earth's horizon, and so on. It is difficult to claim that the on-orbit calibration achieves better than 1 K residual calibration accuracy. Corrections based on prelaunch laboratory data have been found to be in error when applied to the on-orbit sensor calibration. Accordingly, new corrections based on on-orbit measurements can be applied to improve sensor calibration.

5.3
ATMS Instrument Description

ATMS scan angle ranges within 52.725° from the nadir direction. It has 22 channels, with the first 16 channels primarily for temperature soundings from

Table 5.1 Requirements and characteristics of the ATMS 22 channels, including the peak values of the channel weighting-functions of a US standard atmospheric condition.

Channel	Center frequency (GHz)	Total bandpass (GHz)	Polari-zation	Accuracy (K)	NEΔT (K)	EFOV cross-track (°)	EFOV along-track (°)	Peak of weight function (hPa)
1	23.8	0.27	*Qv*	1.00	0.50	6.3	5.2	Window
2	31.4	0.18	*Qv*	1.00	0.60	6.3	5.2	Window
3	50.3	0.18	*Qh*	0.75	0.70	3.3	2.2	Window
4	51.76	0.40	*Qh*	0.75	0.50	3.3	2.2	950
5	52.8	0.40	*Qh*	0.75	0.50	3.3	2.2	850
6	53.596 ± 0.115	0.17	*Qh*	0.75	0.50	3,3	2.2	700
7	54.4	0.40	*Qh*	0.75	0.50	3.3	2.2	400
8	54.94	0.40	*Qh*	0.75	0.50	3,3	2.2	250
9	55.5	0.33	*Qh*	0.75	0.50	3.3	2.2	200
10	57.29	0.33	*Qh*	0.75	0.75	3.3	2.2	100
11	57.29 ± 0.217	0.078	*Qh*	0.75	1.20	3.3	2.2	50
12	57.29 ± 0.322 ± 0.048	0.036	*Qh*	0.75	1.20	3.3	2.2	25
13	57.29 ± 0.322 ± 0.022	0.016	*Qh*	0.75	1.50	3.3	2.2	10
14	57.29 ± 0.322 ± 0.010	0.008	*Qh*	0.75	2.40	3.3	2.2	5
15	57.29 ± 0.322 ± 0.0045	0.003	*Qh*	0.75	3.60	3,3	2.2	2
16	88.2	3	*Qv*	1.00	0.30	2.2	2.2	Window
17	165.5	3	*Qh*	1.00	0.60	2.2	1.1	Window
18	183.31 ± 7.0	2	*Qh*	1.00	0.80	2.2	1.1	800
19	183.31 ± 4.5	2	*Qh*	1.00	0.80	2.2	1.1	700
20	183.31 ± 3.0	1	*Qh*	1.00	0.80	2.2	1.1	500
21	183.31 ± 1.8	1	*Qh*	1.00	0.80	2.2	1.1	400
22	183.31 ± 1.0	0.5	*Qh*	1.00	0.90	2.2	1.1	300

the surface to about 1 hPa (~45 km) and the remaining 6 channels for humidity soundings in the troposphere from the surface to about 200 hPa (~15 km). The ATMS channels 3–16 have a beam width of 2.2°, which is smaller than that of the corresponding AMSU-A channels 3–16. However, the beam width of the ATMS channels 1–2 is 5.2°, which is much larger than that of the corresponding AMSU-A channels 1–2. The ATMS channels 17–22 have a beam width of 1.1°, which is the same as that of the AMSU-B and MHS channels (see Table 5.1).

Figure 5.1 shows a schematic diagram of the assembling position of ATMS on the SNPP platform. The antenna reflectors rotate counterclockwise relative to the spacecraft direction of motion (i.e., the x-direction) to complete three revolutions in 8 s. The scan mechanism is synchronized to the spacecraft clock with a "sync" pulse every 8 s (i.e., at every third revolution). As shown in Figure 5.2, each ATMS

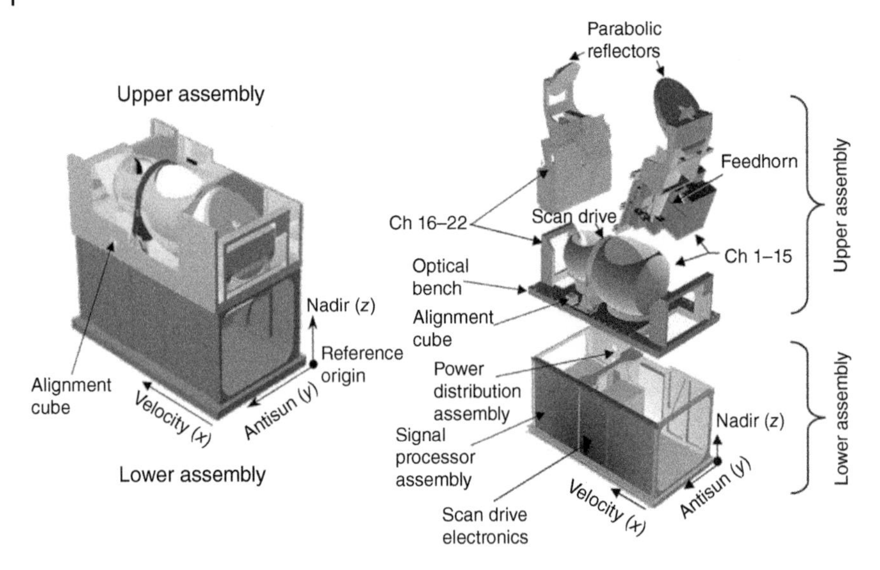

Figure 5.1 Schematic diagram of ATMS instrument layout. (Weng *et al.* 2013 [5]. Reproduced with permission of American Geophysical Union.)

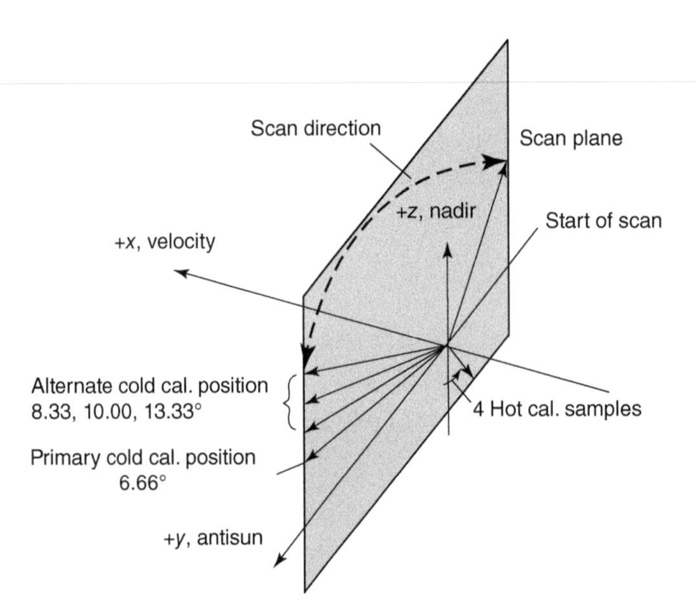

Figure 5.2 ATMS scan cycle during which 96 Earth views, and 4 cold and 4 warm calibrations are made. The angle at each cold calibration position is defined with respect to the *y*-axis of antisun direction. (Weng *et al.* 2013 [5]. Reproduced with permission of American Geophysical Union.)

scan cycle is divided into three segments. In the first segment, the Earth is viewed at 96 different scan angles, which are distributed symmetrically around the nadir direction. Ninety-six such ATMS FOV samples are taken "on the fly," with each FOV sample representing the midpoint of a brief sampling interval of about 18 ms. With a scan rate of 61.6° per second, the angular sampling interval is 1.11°. Therefore, the angular range between the first and last (i.e., 96th) sample centroids is 105.45° (i.e., 52.725° relative to the nadir). As soon as one scan line is completed, the antenna accelerates and moves to a position that points to an unobstructed view of space (i.e., between the Earth's limb and the spacecraft horizon). Then, it resumes the same slow scan speed as it scans across the Earth scenes. In this period, four consecutive cold calibration measurements are performed. Next, the antenna accelerates again to the zenith direction where the blackbody target is located and performs four consecutive warm calibration measurements with the same slow scan speed. Finally, it accelerates back to the starting position and then slows down to the normal scan speed for continuing the next scan cycle. More details of the ATMS scan mechanism can be found in [5, 164].

ATMS has two sets of receiving antenna and reflector. One serves for channels 1–15 with frequencies below 60 GHz, and the other serves for channels 16–22 with frequencies above 60 GHz. Each receiving antenna is paired with a plane reflector mounted on a scan axis at a 45° tilt angle so that the incoming radiation is reflected from a direction perpendicular to the scan axis to a direction along the scan axis (i.e., a 90° reflection) (see Figure 5.3). With the scan axis oriented in the along-track direction, this results in a cross-track scan pattern. The reflected radiation is focused by a stationary parabolic reflector onto a dichroic plate and then either reflected on to or passed through a feedhorn. Each aperture/reflector serves two frequency bands for a total of four bands. With a 45° incident and reflecting

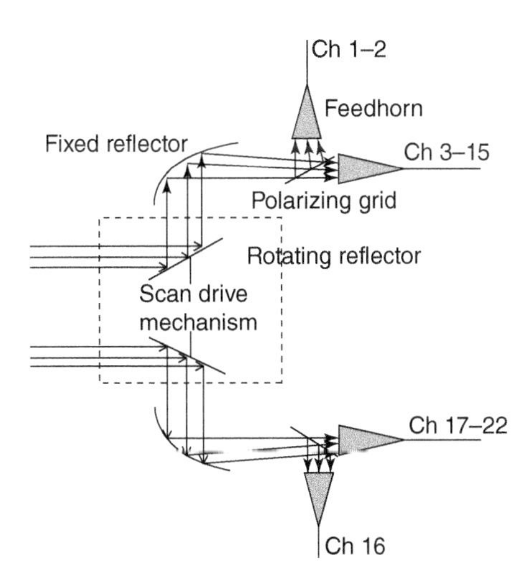

Figure 5.3 Schematic diagram of ATMS antenna subsystem. The top portion shows the antenna subsystem for K/Ka and V bands, whereas the lower portion is for W/G bands. (Weng *et al.* 2013 [5]. Reproduced with permission of American Geophysical Union.)

angle for two reflectors, the polarization state from the scene remains the same, although each individual reflector switches the polarization states.

5.4
ATMS Radiometric Calibration

A radiometric calibration at radiance was derived as follows [165]:

$$R = R_w + w(R_w - R_c)\left(\frac{C_s - \overline{C_w}}{\overline{C_w} - \overline{C_c}}\right) + Q, \tag{5.1}$$

$$Q = \mu(R_w - R_c)^2 \frac{(C_s - \overline{C_w})(C_s - \overline{C_c})}{(\overline{C_w} - \overline{C_c})^2}, \tag{5.2}$$

$$G = \frac{\overline{C_w} - \overline{C_c}}{R_w - R_c}, \tag{5.3}$$

$$\overline{C_x} = \sum_{i=-Ns}^{Ns} W_i C_x, \quad x = w \text{ or } c. \tag{5.4}$$

Here, all variables in the equation should be channel specific. For simplicity, the channel subscript is omitted in all the following deviations. In the history of NOAA operational calibration, Eq. (5.1) is expressed in a quadratic form

$$R = a_0 + a_1 C_s + a_2 C_s^2. \tag{5.5}$$

So, the calibration coefficients, a_0, a_1, and a_2 can be expressed as follows:

$$a_0 = R_w - \frac{\overline{C_w}}{G} + \mu \frac{\overline{C_w}\,\overline{C_c}}{G^2}, \tag{5.6}$$

$$a_1 = \frac{1}{G} - \mu \frac{\overline{C_w} + \overline{C_c}}{G^2}, \tag{5.7}$$

$$a_2 = \frac{\mu}{G^2}. \tag{5.8}$$

In the given radiometric calibration equations, the Earth-scene counts are typically converted to the radiance. In general, the radiance describes the amount of electromagnetic energy radiated from an Earth scene in a specified direction, a solid angle, and a frequency interval. From Eq. (1.5), the radiance can be computed by its kinetic temperature (T) and wave number (υ) as follows:

$$R_\upsilon(T) = \frac{2hc^2\upsilon^3}{\exp\left(\frac{hc\upsilon}{kT}\right) - 1} \equiv \frac{C_1\upsilon^3}{\exp\left(\frac{C_2\upsilon}{T}\right) - 1}, \tag{5.9}$$

where k is the Boltzmann constant, h is the Planck constant, c is the speed of light (in m), $C_1 = 2hc^2 = 1.1909 \times 10^{-8}$ W/m^2/sr/cm cm^3, and $C_2 = \frac{hc}{k} = 1.438786$ cm/K. Equation (5.9) is known as the Planck radiation law.

Assuming $\frac{C_2 v}{T} \ll 1$, the exponential function in Planck function can be expressed in Taylor series:

$$\exp\left(\frac{C_2 \upsilon}{T}\right) = 1 + \frac{C_2 \upsilon}{T} + \frac{1}{2}\left(\frac{C_2 \upsilon}{T}\right)^2 + \cdots + \frac{1}{n!}\left(\frac{C_2 \upsilon}{T}\right)^n + \cdots. \quad (5.10)$$

Substituting the first-order approximation of the given Taylor expansion into Eq. (5.9) results in the following linear relationship between the blackbody temperature (T) and radiance (R_v):

$$R_\upsilon^{RJ}(T) = \frac{C_1 \upsilon^2}{C_2} T. \quad (5.11)$$

Using the Rayleigh–Jeans (RJ) approximation to Planck's function from Eq. (1.13), $C_2\upsilon$ is generally less than 10 K for a range of 23.8 GHz $\leq f \leq$ 190.3 GHz. Thus, the temperature in Eq. (5.10) must be above 100 K. Substituting Eq. (5.11) into Eq. (5.1) results in

$$T_b = T_w + w(T_w - T_c)\left(\frac{C_s - \overline{C_w}}{\overline{C_w} - \overline{C_c}}\right) + Q_b, \quad (5.12)$$

$$Q_b = \mu(T_w - T_c)^2 \frac{(C_s - \overline{C_w})(C_s - \overline{C_c})}{(\overline{C_w} - \overline{C_c})^2}. \quad (5.13)$$

The accuracy of the radiance calculated from RJ approximation varies with frequency and temperature. The radiometric calibration is processed through the use of Eq. (5.12). As a result, the two-point calibration is derived in the form of brightness temperature as

$$T_b = T_w + G_b^{-1}(C_s - \overline{C_w}) + Q_b = T_{b,l} + Q_b, \quad (5.14)$$

where the linear and nonlinear terms are expressed as

$$T_{b,l} = T_w + G_b^{-1}(C_s - \overline{C_w}), \quad (5.15)$$

$$Q_b = \mu G_b^{-2}(C_s - \overline{C_w})(C_s - \overline{C_c}) = \mu(T_w - T_c)^2 x(1 - x), \quad (5.16)$$

$$G_b = \frac{\overline{C_w} - \overline{C_c}}{T_w - T_c}, \quad (5.17)$$

respectively, where

$$x = \frac{T_{b,l} - T_c}{T_w - T_c}.$$

The maximum nonlinearity value can be derived by performing the derivative with respect to x, which is $f'(x) = 1 - 2x$. Using Taylor's expansion for $f(x) = x(1 - x)$ at $x_0 = 0.5$, which is equal to $C_s = 0.5(\overline{C}_w + \overline{C}_c)$, then

$$Q_b = \frac{1}{4}\mu(T_w - T_c)^2[1 - 4(x - 0.5)^2] = Q^{\max}[1 - 4(x - 0.5)^2]. \tag{5.18}$$

If the first two terms in Eq. (5.18) are maintained as the nonlinearity terms,

$$Q^{\max} = \frac{1}{4}\mu(T_w - T_c)^2, \tag{5.19}$$

The ATMS antenna/receiver system measures the radiation from two calibration sources during every scan cycle. The first source is the well-known cosmic background radiation. This source (often called cold space) is viewed immediately after the Earth has been scanned. The second source is an internal blackbody calibration target (often called warm load), whose physical temperature is the same as the instrument internal ambient temperature. This warm source is viewed immediately after the space calibration view. Every scan cycle (8/3 s) contains the aforementioned three consecutive views: Earth scene, cold space, and warm calibration measurements. Such a thorough radiometer calibration procedure allows the most impacts from ATMS system gain variations to be automatically eliminated since the two calibration measurements used for computing the gain involve the same optical and electrical signal paths as those of the Earth scene measurements. Thus, ATMS has an advantage over those calibration systems using switched internal calibration sources, which yields calibration measurements having slightly different signal paths compared to the Earth scene measurements.

The two calibration measurements are used to accurately determine the so-called radiometer transfer function, which converts the measured digitized output (i.e., counts) to a radiometric brightness temperature [5]. The current ATMS antenna brightness temperature at each channel is obtained through the following equation:

$$\begin{aligned} T_{b,ch} &= T^w_{b,ch} + \frac{C^s_{ch} - \overline{C^w_{ch}}}{\overline{C^w_{ch}} - \overline{C^c_{ch}}}(T^w_{b,ch} - T^c_{b,ch}) + Q^{\max}_{ch}[1 - 4(x - 0.5)^2] \\ &\equiv T^w_{b,ch} + (\overline{G_{ch}})^{-1}(C^s_{ch} - \overline{C^w_{ch}}) + Q_{ch}, \end{aligned} \tag{5.20}$$

where Q_{ch} is the channel-based (subscript, ch) quadratic correction term, $\overline{G_{ch}}(i)$ is the averaged gain function at the ith scan line, defined via

$$\overline{C^w_{ch}}(i) = \sum_{k=i-N_s}^{i+N_s} \sum_{j=1}^{4} W_{k-i} C^w_{ch}(k, j), \tag{5.21a}$$

$$\overline{C^c_{ch}}(i) = \sum_{k=i-N_s}^{i+N_s} \sum_{j=1}^{4} W_{k-i} C^c_{ch}(k, j), \tag{5.21b}$$

$$\overline{G_{ch}}(i) = \frac{\overline{C^{w}_{ch}}(i) - \overline{C^{c}_{ch}}(i)}{\overline{T^{w}_{b,ch}}(i) - T^{c}_{b,ch}}. \tag{5.21c}$$

The other terms used in Eq. (5.20) are as follows: $T^{w}_{b,ch}$, the warm-load brightness temperature; C^{s}_{ch}, the scene count; $C^{w}_{ch}(i,j)$, the warm-load count at the ith scan line of the jth sample; $C^{c}_{ch}(i,j)$, the cold-space count at the ith scan line of the jth sample; W_i, the weighting coefficient obtained by using either a triangular or a boxcar function for averaging the warm and cold counts of $(2N_s + 1)$ consecutive scan lines; and $G_{ch}(i)$, the calibration gain at the ith scan line. The overbar on each variable represents an average over a number of scan lines.

Calculations of the averaged warm-load and cold-space counts, $\overline{C^{w}_{ch}}(i)$ and $\overline{C^{c}_{ch}}(i)$, depend on the weighting coefficients as well as the number of scan lines involved in the averaging (i.e., $2N_s + 1$). For example, with the application of a triangular function, the weighting coefficients of all ATMS channels for the kth scan line are averaged as follows:

$$W_k = \frac{1}{N_s + 1}\left(1 - \frac{|k|}{N_s + 1}\right) \tag{5.22}$$

The nonlinearity parameter Q_{ch} in Eq. (5.20) is estimated using the prelaunch thermal vacuum (TVAC) data sets that were measured at different scene temperatures. For the SNPP ATMS TVAC test, the scene temperature is typically measured between 93 and 330 K. However, for ATMS on-orbit calibration, the cold calibration temperature is approximately 3 K. Thus, the nonlinearity value must be estimated by extrapolating the TVAC data down to 3 K. In general, the nonlinear parameter Q_{ch} can be expressed as a quadratic function of the scene temperature ($T_{b,ch}$) as

$$Q_{ch} = b_{0,ch} + b_{1,ch}T_{b,ch} + b_{2,ch}(T_{b,ch})^2, \tag{5.23}$$

where $b_{i,ch}$ $(i = 0, 1, 2;\ ch = 1, \ldots, 22)$ are unknown coefficients. These values can be obtained by applying a least-squares fit to the TVAC data measured within the temperature range of 93–330 K. Then, by applying these coefficients back into Eq. (5.23), one can compute Q_{ch} for all scene temperatures within and beyond the range of 93–330 K. One example of $b_{i,ch}$ can be found in Weng [5]. Figure 5.4 presents an example of the Q_{ch} of ATMS channel 1, obtained by applying the TVAC measured scene temperatures for cold plate (CP) at 5 °C from redundancy configuration 1 (RC1). The peak nonlinearity is located near 170 K, which is about in the middle between 3 and 330 K. The ATMS maximum nonlinearity, $Q^{\max}_{ch}$, can be derived from the TVAC data for a scene temperature range between 3 and 276 K for four redundant configurations [5].

In the current ATMS antenna brightness temperatures (TDR) processing algorithm, the blackbody brightness temperature, $T^{w}_{b,ch}$, is directly determined from its physical (or kinetic) temperature measured by the embedded platinum

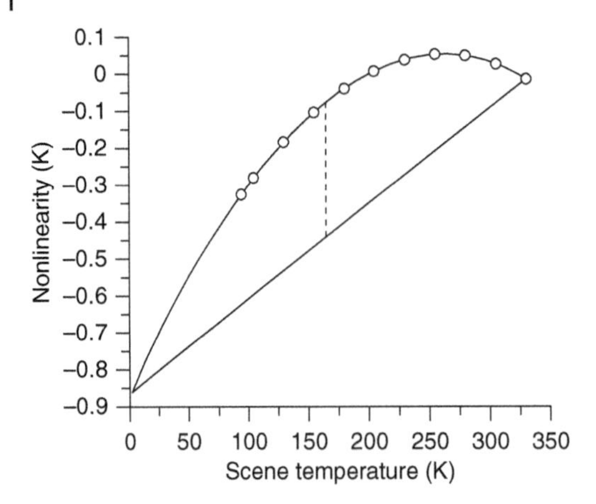

Figure 5.4 Nonlinearity of ATMS channel 1, calculated for cold plate (CP) at 5 °C for redundancy configuration 1 (RC1). Dots represent the measured scene temperatures. Black solid curve represents the regression curve. Dashed line represents the peak nonlinearity. (Weng *et al.* 2013 [5]. Reproduced with permission of American Geophysical Union.)

resistance thermometers (PRTs). This approach may be problematic if the ATMS antenna reflector has its own thermal emission. The details are discussed in the next section. The blackbody and cold-space calibration counts, C_w and C_c, are averaged over several calibration cycles before being used in Eq. (5.21c) to obtain the calibrated gain.

The ith channel cold-space brightness temperature, $T^c_{b,ch}$ $(ch = 1, \ldots, 22)$, is estimated by adding two correction terms to the cold-space temperature, T^c. The first correction term, $\Delta T^{c,SL}_{b,i}$, takes into account the Earth's radiation into the antenna side lobes, and the second correction term, $\Delta T^{c,RJ}_{b,ch}$, corrects the error introduced by the RJ approximation. Specifically, $T^c_{b,ch}$ is written as

$$T^c_{b,ch} = T^c + \Delta T^{c,SL}_{b,ch} + \Delta T^{c,RJ}_{b,ch}. \tag{5.24}$$

Details on these two correction terms can be found in Weng *et al.* [2, 147].

The averaged warm-load temperature for the ith scan is determined from the multiple PRT temperatures $T^w(k,j)$ $(k = i - N_s, \ldots, i + N_s;\ j = 1, \ldots, N_p)$. Depending on the user's need, a temperature-dependent bias correction from either ATMS telemetry file (i.e., $\Delta T^{shelf}_{w,ch}$) or user-defined values can be applied via Eq. (5.25).

$$\overline{T^w_{ch}}(i) = \frac{\sum_{k=i-N_s}^{i+N_s} \sum_{j=1}^{N_p} W_{k,j} T^w(k,j)}{\sum_{k=i-N_s}^{i+N_s} \sum_{j=1}^{N_p} W_{k,j}} + \Delta T^{shelf}_{w,ch}, \tag{5.25}$$

where $T^{w}(k,j)$ is the jth PRT temperature for the kth scan, $\Delta T_{w,ch}^{shelf}$ is the energy contributed by the channel-dependent receiving shelf temperature; and $W_{i,j}$ is the weighting coefficient. If a PRT is deemed to be bad by the user, it is then excluded from the calibration process and the corresponding weighting coefficient is set to zero in the parameter file. Theoretically, the warm-load temperature also needs to be adjusted for the error introduced by the RJ approximation, that is,

$$T_{b,i}^{w} = T_{ch}^{w} + \Delta T_{b,ch}^{w,RJ}. \tag{5.26}$$

Since the normally operating warm-load temperature is above 280 K, the errors are typically negligible at lower ATMS frequencies [2].

5.5 Impacts of ATMS Antenna Emission on Two-Point Calibration

As shown in Figure 5.3, ATMS has a plane reflector mounted on a scan axis at a 45° tilt angle. It was found that this reflector may have its own thermal emission. Thus, the impacts of the reflector emission on calibration need to be investigated further. In Eqs. (5.1) and (5.12), the radiances or brightness temperatures of calibration targets are directly used since the calibration targets are assumed to be blackbodies and unpolarized. While some terms are added to correct the warm-load temperature as shown in Eq. (5.25), there remain additional radiative sources such as the thermal emission from ATMS reflector. For simplicity, we only consider the linear term in Eq. (5.12) for demonstrating the effects of antenna emission on the two-point calibration. For ATMS quasi-vertical and quasi-horizontal polarization channels, we have the two-point calibration equations as follows:

$$T_{b}^{Qv} = T_{b,w}^{Qv} + (T_{b,w}^{Qv} - T_{b,c}^{Qv})\left(\frac{C_{s}^{Qv} - \overline{C_{w}^{Qv}}}{\overline{C_{w}^{Qv}} - \overline{C_{c}^{Qv}}}\right), \tag{5.27a}$$

$$T_{b}^{Qh} = T_{b,w}^{Qh} + (T_{b,w}^{Qh} - T_{b,c}^{Qh})\left(\frac{C_{s}^{Qh} - \overline{C_{w}^{Qh}}}{\overline{C_{w}^{Qh}} - \overline{C_{c}^{Qh}}}\right), \tag{5.27b}$$

where T_{b}^{Qv} and T_{b}^{Qh} are the brightness temperatures at quasi-vertical and quasi-horizontal polarization states, respectively; $T_{b,w}^{Qv}$ and $T_{b,w}^{Qh}$ are the warm-load brightness temperatures at two-quasi polarization states, respectively; $T_{b,c}^{Qv}$ and $T_{b,c}^{Qh}$ are the cold-space brightness temperatures at two polarization states, respectively. In Eq. (5.27), all the counts are also labeled with polarization in consistence with the calibration target definition.

For an emitting antenna, brightness temperatures in the aforementioned two-point calibration should be corrected with the emitted energy. Since the polarization vector is rotated by 90° after reflection, brightness temperatures in both vertical and horizontal polarizations including the emission from ATMS plane reflector are derived as

$$T_{b,r}^{v} = (1 - \varepsilon_h)T_b^v + \varepsilon_h T_r, \tag{5.28a}$$

$$T_{b,r}^{h} = (1 - \varepsilon_v)T_b^h + \varepsilon_v T_r, \tag{5.28b}$$

where T_r is the plane reflector physical temperature, and $\boldsymbol{\varepsilon} = (\varepsilon_v, \varepsilon_h, \varepsilon_3, \varepsilon_4)$ is the reflector emissivity vector. For ATMS plane reflector, the incident angle of the radiation at horizontal and vertical polarization states is 45° to the normal, the reflective waves after the plane reflector have a 90° rotation in its Stokes vector. From Eq. (1.18) neglecting the third Stokes component,

$$T_{b,r}^{Qv} = T_{b,r}^{h}\cos^2\theta + T_{b,r}^{v}\sin^2\theta, \tag{5.29a}$$

$$T_{b,r}^{Qh} = T_{b,r}^{h}\sin^2\theta + T_{b,r}^{v}\cos^2\theta, \tag{5.29b}$$

Substituting Eq. (5.28) into Eq. (5.29) results in

$$T_{b,r}^{Qv} = T_b^{Qv} + \varepsilon_h(T_r - T_b^v) + [\varepsilon_v(T_r - T_b^h) - \varepsilon_h(T_r - T_b^v)]\sin^2\theta, \tag{5.30a}$$

$$T_{b,r}^{Qh} = T_b^{Qh} + \varepsilon_h(T_r - T_b^v) + [\varepsilon_v(T_r - T_b^h) - \varepsilon_h(T_r - T_b^v)]\cos^2\theta. \tag{5.30b}$$

Thus, the total antenna brightness temperatures at quasi-vertical and horizontal polarization states are contributed by additional terms related to the antenna emissivity and reflector physical temperatures. Applying Eq. (5.30) to cold-space and warm-load calibration targets yields

$$T_{b,w}^{Qv} = T_w + \varepsilon_h(T_r - T_w) + [\varepsilon_v(T_r - T_w) - \varepsilon_h(T_r - T_w)]\sin^2\theta_w, \tag{5.31a}$$

$$T_{b,w}^{Qh} = T_w + \varepsilon_h(T_r - T_w) + [\varepsilon_v(T_r - T_w) - \varepsilon_h(T_r - T_w)]\cos^2\theta_w, \tag{5.31b}$$

and

$$T_{b,c}^{Qv} = T_c + \varepsilon_h(T_r - T_c) + [\varepsilon_v(T_r - T_c) - \varepsilon_h(T_r - T_c)]\sin^2\theta_c, \tag{5.32a}$$

$$T_{b,c}^{Qh} = T_c + \varepsilon_h(T_r - T_c) + [\varepsilon_v(T_r - T_c) - \varepsilon_h(T_r - T_c)]\cos^2\theta_c. \tag{5.32b}$$

In the ATMS scan cycle, $\theta_c = 83.3°$ and $\theta_w = 195°$. Equations (5.31) and (5.32) can be used to improve the calibration accuracy if the reflector emissivity is known, and the corrected warm-load and cold-space temperatures can be used in Eqs. (5.20) and (5.28).

The effects of the plane reflector emission on simulated brightness temperatures at six ATMS channels are simulated using Eq. (5.30). In the past, the radiative transfer simulation for ATMS channels did not include the reflector emission, which is associated with the second and third terms. We define the total of the second and third terms as the biases. The biases are simulated and shown in Figure 5.5. When ATMS scans the Earth, the bias of the brightness temperature for quasi-vertical polarization channels varies with scan position and is the smallest at the nadir, increases to a maximum at a local zenith angle of 45° off the nadir, and decreases with the angle after 45°. The most significant angular-dependent feature is shown at 31.4 GHz, where the atmosphere is less opaque. For the quasi-horizontal polarization, the maximum bias occurs at the nadir and decreases with the scan angle. Comparing the two O2-sounding channels at 50.3 and 54.4 GHz, there are less number of scan-angle-related biases for the upper-air sounding channels.

When ATMS scans the cold space, the bias at the quasi-vertical polarization is also the smallest at the nadir and increases with the scan angle, and there are no fluctuations in the trend of increase. This is referred to as a "smile pattern." For the quasi-horizontal polarization, the biases are the maximum at the nadir and decrease with the scan angle. This is a "frown" pattern.

5.6 Retrieval of Reflector Emissivity Using ATMS Pitch-Over Data

Fresnel's equations simplify at normal incidence so that the reflection coefficients in both vertical and horizontal polarization are the same. At the Brewster angle, the reflected beam is completely polarized in the s direction or perpendicular to the incident plane (in microwave remote sensing, s direction is the same as horizontal polarization direction). However, there is another angle where Fresnel's equations also simplify. In general, the perpendicular and parallel reflection amplitudes (R_s and R_p) are

$$R_s = \frac{n_1 \cos\theta_i - n_2 \cos\theta_r}{n_1 \cos\theta_i + n_2 \cos\theta_r} E_s, \tag{5.33a}$$

$$R_p = \frac{n_2 \cos\theta_i - n_1 \cos\theta_r}{n_2 \cos\theta_i + n_1 \cos\theta_r} E_p, \tag{5.33b}$$

where n_1 and n_2 are the refractive indices of media 1 and 2, respectively. Using the Snell law ($n_1 \sin\theta_i = n_2 \sin\theta_r$),

$$R_s = -\frac{\sin(\theta_i - \theta_r)}{\sin(\theta_i + \theta_r)} E_s, \tag{5.34a}$$

$$R_p = \frac{\tan(\theta_i - \theta_r)}{\tan(\theta_i + \theta_r)} E_p. \tag{5.34b}$$

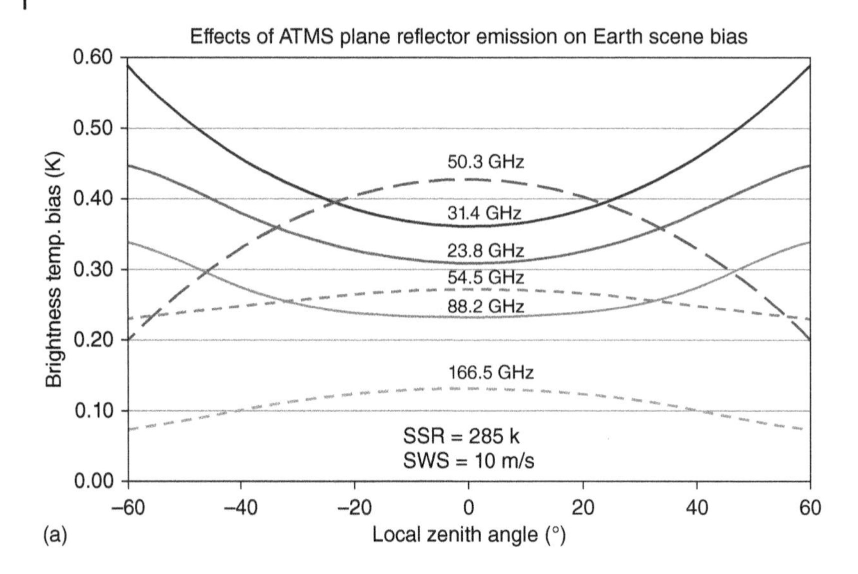

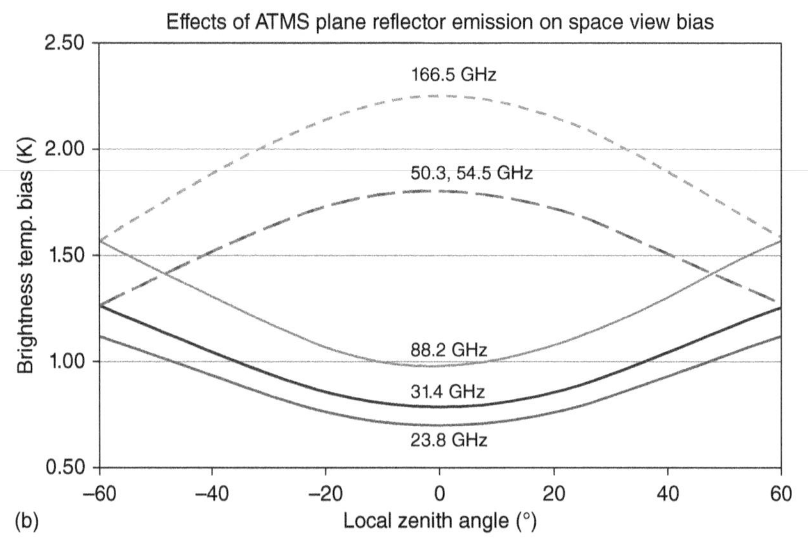

Figure 5.5 (a) Brightness temperature biases simulated for an ocean condition with a mid-latitude atmospheric profile, ocean wind speed of 10 m/s and surface temperature of 285 K. (b) Brightness temperature biases simulated for cold space view where the microwave radiation is uniform across the scan angle. Notice that the biases at 50.3 and 54.4 GHz are the same and overlay each other. Reflector temperature is assumed as 283 K and its emissivity varies from 0.0025 to 0.0065.

At an incident angle of $\theta_i = 45°$,

$$R_s = \frac{\cos\theta_r - \sin\theta_r}{\cos\theta_r + \sin\theta_r} E_s, \tag{5.35a}$$

$$R_p = \left[\frac{\cos\theta_r - \sin\theta_r}{\cos\theta_r + \sin\theta_r}\right]^2 E_p. \tag{5.35b}$$

Let $r_s = R_s/E_s$, and $r_p = R_p/E_p$. Thus,

$$r_v = r_h^2. \tag{5.36}$$

Remarkably, the resulting simplification in Fresnel's equations (5.36) appears to have been first noticed by Humphreys-Owen only around 1960. The emissivity between two polarization states can also be related to each other:

$$\varepsilon_v = 1 - r_v \equiv 2\varepsilon_h - \varepsilon_h^2. \tag{5.37}$$

When ATMS scans over the cold space, the scene brightness temperatures from Eq. (5.30) become

$$T_{c,r}^{Qv} = T_c + \varepsilon_h(T_r - T_c) + [\varepsilon_v(T_r - T_c) - \varepsilon_h(T_r - T_c)]\sin^2\theta_s, \tag{5.38a}$$

$$T_{c,r}^{Qh} = T_c + \varepsilon_h(T_r - T_c) + [\varepsilon_v(T_r - T_c) - \varepsilon_h(T_r - T_c)]\cos^2\theta_s. \tag{5.38b}$$

On February 20, 2012, the Suomi NPP satellite made its pitch-over maneuver. The spacecraft is pitched completely off the Earth to enable ATMS to acquire full scans of deep space. During about 18-min pitch-over maneuver, ATMS continually scans over cold space across its 96 FOVs. These data sets are collected from homogenous and unpolarized cold space. To reduce the impacts of the Earth's contamination in deep space view, only ± 25 scan lines at the maximum pitch angle were selected for retrievals of antenna emissivity.

Substituting Eqs. (5.30), (5.31), (5.32), (5.37), (5.38) into Eq. (5.27) results in the emissivity equation for the horizontal polarization. For the quasi-vertical polarization channels,

$$\varepsilon_h = \frac{\delta^{Qv}(T_w - T_r)}{\delta^{Qv}[(T_w - T_r)sin^2\theta_w - (T_c - T_r)sin^2\theta_c] - (T_c - T_r)(sin^2\theta_s - sin^2\theta_c)}. \tag{5.39a}$$

For the quasi-horizontal polarization channels,

$$\varepsilon_h = \frac{\delta^{Qh}(T_w - T_r)}{\delta^{Qh}[(T_w - T_r)\cos^2\theta_w - (T_c - T_r)\cos^2\theta_c] - (T_c - T_r)(\cos^2\theta_s - \cos^2\theta_c)}, \tag{5.39b}$$

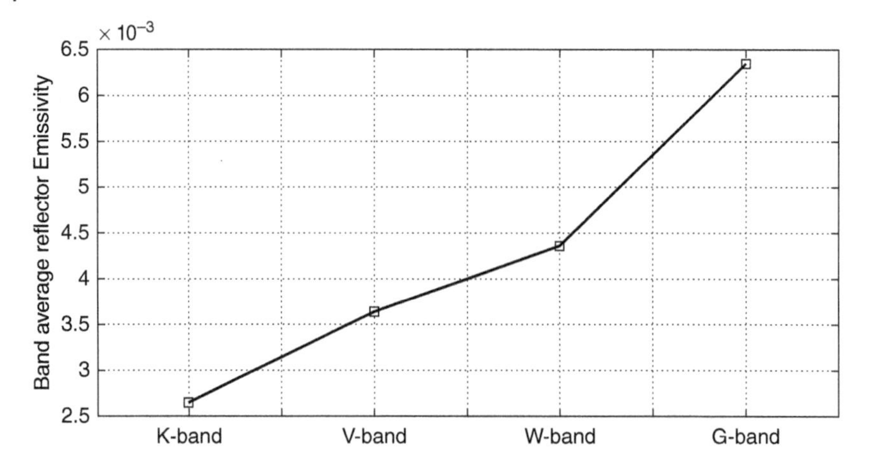

Figure 5.6 SNPP ATMS K, V, W, and G band reflector emissivity retrieved from the pitch-maneuver data on February 20, 2012 from the orbit number 1637 with $T_r = 283$ and $T_c = 3$.

where

$$\delta^{Qv} = \left(\frac{\overline{C_s^{Qv}} - \overline{C_w^{Qv}}}{\overline{C_w^{Qv}} - \overline{C_c^{Qv}}} \right),$$

$$\delta^{Qh} = \left(\frac{\overline{C_s^{Qh}} - \overline{C_w^{Qh}}}{\overline{C_w^{Qh}} - \overline{C_c^{Qh}}} \right).$$

Figure 5.6 shows the retrieved ATMS emissivity spectrum at the horizontal polarization. The spectral emissivity is in the range of 0.002–0.007 and increases with frequency.

5.7
ATMS Noise-Equivalent Difference Temperature (NEDT)

Satellite data quality is mainly characterized by accuracy, precision (i.e., sensitivity), and stability. From a statistical point of view, the standard deviation can properly represent the precision of measurements that have a stable mean. However, for the data having a nonstationary mean, the standard deviation is dependent on the total sample size [166]. In our earlier studies, it was shown that the mean value of the ATMS calibration measurements (i.e., warm counts) varies along an orbit. As a result, the standard deviation is, in general, inappropriate for ATMS noise characterization if the total sample size used in the calculation is not optimized.

However, currently, the ATMS noise magnitudes at all channels are computed for a nonoptimal sample size, and thus they are inaccurate and should be corrected in various applications.

The overlapping Allan deviation was proposed for noise characterization [17, 166] and was suggested for ATMS noise characterization [5, 167]. It was indicated that the overlapping Allan deviation is much more stable than the standard deviation by using the ATMS warm counts. However, the noises estimated by the overlapping Allan deviation are much smaller than the standard deviation for all the ATMS channels [5]. The causes of those discrepancies were not well known. In this research, an optimal averaging window size in the overlapping Allan deviation is derived through a theoretical analysis.

The overlapping Allan deviation was introduced for assessing the ATMS measurement sensitivity. In that study, for a comparison purpose, the averaging window size used in the overlapping Allan deviation was set to be the same as that used in the current ATMS operational calibration system for all the channels. The noise magnitudes calculated by the two methods, the overlapping Allan deviation and the standard deviation, showed large discrepancies in all ATMS channels. The following three major questions arise before a final decision can be made: (i) how to optimally determine the window size from a trend of convergence, (ii) how to optimally determine the window size when the convergence trend is not clear, and (iii) what do the values calculated under an optimal window size actually represent? Do they really represent the sensitivity or instrument noise? To answer the given questions, the so-called two-sample (i.e., neighborhood) Allan deviation is studied here along with the overlapping Allan deviation.

The overlapping Allan deviation is defined as follows:

$$\sigma^2_{Allan}(M,m) = \frac{1}{2m^2(M-2m+1)} \sum_{j=1}^{M-2m+1} \left[\sum_{i=j}^{j+m-1} \left(y_{i+m} - y_i \right) \right]^2, \tag{5.40}$$

where m is the averaging window size. If m is set to 1, the overlapping Allan deviation is simplified as

$$\sigma^2_{Allan}(M) = \frac{1}{2(M-1)} \sum_{i=1}^{M-1} (y_{i+1} - y_i)^2, \tag{5.41}$$

where M is the total sample size of a data series $\{y_i\}$, and Eq. (5.40) constitutes the two-sample Allan deviation.

For comparison purposes, two data sets are constructed to represent slowly and rapidly varied noise measurements, respectively. The first data set, $\{y_i, i = 1, \ldots, 1000\}$, is constructed by adding a Gaussian noise, $N(0,1)$, to a constant signal, 10. The second data set is constructed by adding the same Gaussian noise to

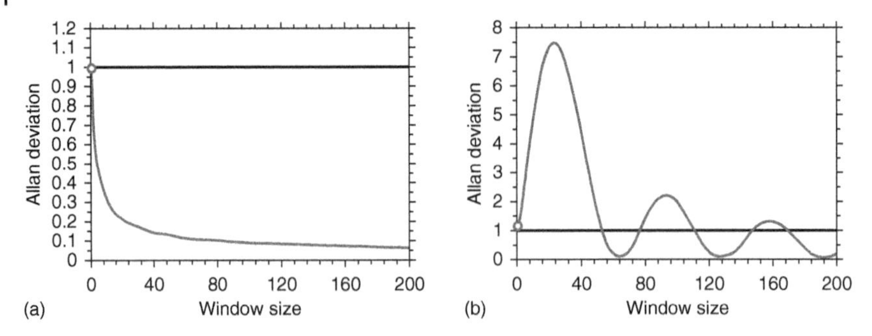

Figure 5.7 Dependence of the overlapping Allan deviation (gray curve) on the window size for the datasets constructed by the addition of (a) Gaussian noise and constant signal; (b) Gaussian noise and sinusoidal signal, respectively. The standard deviation of the added Gaussian noise is denoted by the dark black line and the two-sample Allan deviation is denoted by the gray circle. (Tian *et al.* 2015 [167]. Reproduced with permission of IEEE.)

a periodic signal $y_i^{signal} = \alpha \sin(\omega x)$, where α represents the amplitude and ω is the frequency. The values of the amplitude α vary from 10^{-3} to 10^3, and the frequency ω ranges from 10^0 to 10^{-4}. As shown in Figure 5.7a, for the slowly varied data, the noise magnitude calculated from the overlapping Allan deviation (gray curve) monotonically decreases as the averaging window size increases. On the other hand, the magnitude calculated from the two-sample Allan deviation (gray circle) is very close to the standard deviation of the Gaussian noise (dark black line). For the sinusoidal signal with $\alpha = 10$ and $\omega = 0.1$, the magnitudes of the overlapping Allan deviation oscillate around the noise standard deviation, as shown in Figure 5.7b. However, the two-sample Allan deviation still derives a noise very close to the true value. Hence, it suggests that the two-sample Allan deviation is more appropriate for quantifying the measurement precision compared to the overlapping Allan deviation.

Taking the expectation on both sides of Eq. (5.41) gives

$$
\begin{aligned}
E(\sigma^2_{Allan}(M,1)) &= E\left(\frac{1}{2\,(M-1)}\sum_{i=1}^{M-1}[y_{i+1}-y_i]^2\right) \\
&= \frac{1}{2(M-1)}\left(\sum_{i=1}^{M-1}E\left(y_{i+1}^2\right) - 2\sum_{i=1}^{M-1}E(y_{i+1}y_i) + \sum_{i=1}^{M-1}E(y_i^2)\right).
\end{aligned}
\tag{5.42}
$$

If the measurements are independent of time, then it yields

$$
\sum_{i=1}^{M-1}E(y_{i+1}y_i) = \sum_{i=1}^{M-1}E(y_{i+1})E(y_i). \tag{5.43}
$$

Substituting Eq. (5.43) into Eq. (5.42) yields

$$\begin{aligned}
E(\sigma^2_{Allan}(M,1)) &= \frac{1}{2(M-1)}\left(\sum_{i=1}^{M-1} E\left(y^2_{i+1}\right) - 2\sum_{i=1}^{M-1} E(y_{i+1})E(y_i) + \sum_{i=1}^{M-1} E(y_i^2)\right)\\
&= \frac{1}{2(M-1)}\left\{\sum_{i=1}^{M-1}\left[\mathrm{Var}\left(y_{i+1}\right) + E^2(y_{i+1})\right]\right.\\
&\qquad \left. -2\sum_{i=1}^{M-1} E\left(y_{i+1}\right)E(y_i) + \sum_{i=1}^{M-1}[\mathrm{Var}(y_i) + E^2(y_i)]\right\}\\
&= \frac{1}{2(M-1)}\left\{\sum_{i=1}^{M-1}\left[\sigma^2+\mu^2\right] - 2\sum_{i=1}^{M-1}\mu^2 + \sum_{i=1}^{M-1}[\sigma^2+\mu^2]\right\}\\
&= \frac{1}{2(M-1)}\{2(M-1)\sigma^2\}\\
&= \sigma^2.
\end{aligned} \tag{5.44}$$

Equation (5.44) theoretically proves that the two-sample Allan variance gives an unbiased estimate of the noise true variance.

For the second data set (i.e., nonstationary), two methods, the two-sample Allan deviation and the standard deviation, are compared numerically, and the relative errors are derived against the true noise. As shown in Figure 5.8a and 5.8b, when the amplitude factor α is less than 10, both methods can provide estimates that are very close to the true noise standard deviation within 10% of error (see the light-gray shaded areas). This is because in this situation, the

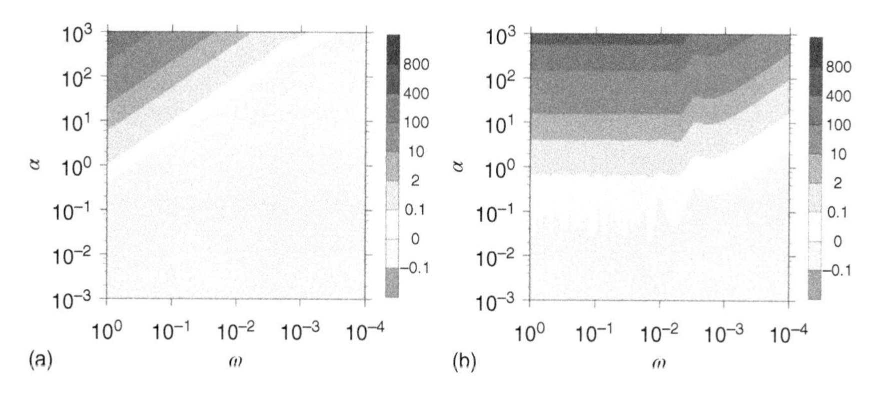

Figure 5.8 Variation of the relative errors defined via (a) Allan deviation: $\frac{(\sigma_{Allan}-\sigma_{noise})}{\sigma_{noise}}$ and (b) standard deviation: $\frac{(\sigma_{Std}-\sigma_{noise})}{\sigma_{noise}}$ with varying the frequency factor, ω and the amplitude factor, α of the signal defined as $f(x) = \alpha\sin(\omega x)$. The added noise follows a Gaussian distribution $N(0,1)$. (Tian *et al.* 2015 [167]. Reproduced with permission of IEEE.)

sinusoidal data series is actually very close to the data set having a stable mean, and, therefore, the two-sample Allan deviation coincides with the standard deviation. As the amplitude factor α increases, the relative errors increase in both methods. However, the area with errors less than 10% from the two-sample Allan deviation is much larger than that from the standard deviation. This is due to the fact that the standard deviation involves the calculation of the distance between each sample and the mean of data, while the two-sample Allan deviation only uses the distance between two consecutive samples. The latter avoids including the variation from any long period trend as the data size increases.

The SNPP with ATMS onboard was launched on October 28, 2011. ATMS is a cross-track scanning instrument and has 22 channels at frequencies ranging from 23 to 183 GHz. It measures the atmospheric temperature and moisture under most weather conditions. Since its on-orbit, the functioning of S-NPP ATMS is stable, and all specifications are well within the requirements. As mentioned before, the ATMS channel sensitivity is currently computed through noise-equivalent differential temperature (NEDT), which is fundamentally derived from the standard deviation via the following formula [5]:

$$NEDT_{ch} = \left[\frac{1}{4M} \sum_{i=1}^{M} \sum_{j=1}^{4} \left(\frac{C_{ch}^{w}(i,j) - \overline{C_{ch}^{w}}(i)}{\overline{G_{ch}}(i)} \right)^2 \right]^{1/2} \quad \text{(Unit: in K)}, \tag{5.45}$$

where C represents the warm counts per channel and G is the averaged calibration gain. For comparison purposes, the two-sample Allan deviation is expressed either as

$$Adev_{ch} = \left[\frac{1}{2(M-1)} \sum_{i=1}^{M-1} \left(\frac{C_{ch}^{w}(i+1,j) - C_{ch}^{w}(i,j)}{\overline{G_{ch}}(i)} \right)^2 \right]^{1/2}, \tag{5.46}$$

for the use of the warm counts from one of its four scan positions (i.e., $j = 1, \ldots, 4$) or as the averaged version as

$$Adev_{ch} = \frac{1}{4} \sum_{j=1}^{4} \left[\frac{1}{2(M-1)} \sum_{i=1}^{M-1} \left(\frac{C(i+1,j) - C(i,j)}{\overline{G_{ch}}(i)} \right)^2 \right]^{1/2}. \tag{5.47}$$

The ATMS on-orbit data are utilized in this study to further explore the difference in the NEDT calculations based on the standard deviation and the Allan deviation. Figure 5.9 shows the typical ATMS on-orbit warm counts of four selected channels, using the entire orbit data obtained on December 15, 2014. These warm counts are all from the second warm scan position. As shown in Figure 5.9, in comparison with channels 16 and 22, channels 14 and 15 exhibit smaller orbital variations in their warm counts; thus, both channels have more stable mean values. In this case, the noise magnitudes of the two channels estimated by the two-sample Allan deviation and the standard deviation should be very close. On the

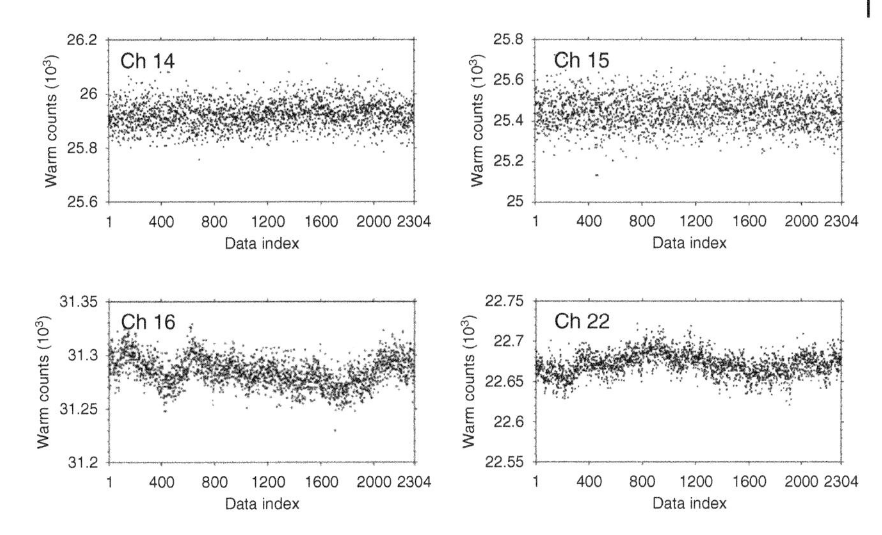

Figure 5.9 On-orbit warm counts of ATMS channels 14, 15, 16, and 22. The warm counts of the second scan position from the orbit 16 299, December 15, 2014, are used. (Tian *et al.* 2015 [167]. Reproduced with permission of IEEE.)

contrary, since channels 16 and 22 exhibit much stronger orbital variations, a larger discrepancy between the estimated noise magnitudes can be expected.

The dependencies of the two-sample Allan deviation on the total sample size for the ATMS channels are shown in Figure 5.10. The Allan deviations based on Eqs. (5.46) and (5.47) are illustrated by different colors. In general, the noise magnitudes are stabilized at $M = 300$ for most channels. Thus, for ATMS, the total sample size, M, should be set to at least 300 in the operational calculation.

Figure 5.11 compares NEDTs calculated by the two-sample Allan deviation and the standard deviation using the first 300 warm counts from the same orbit for all 22 ATMS channels. As expected, for the channels with stable warm counts (i.e., channels 1–15), NEDTs calculated from both the methods are very close to each other. However, it is clear that for the channels with relatively strong orbital variations (i.e., channels 16–22), the discrepancy between the two methods is large, since the current operational NEDT overestimates the noise magnitudes by including the impact of those orbital variations. Thus, the two-sample Allan deviation is recommended for noise characterization of ATMS and other instruments alike.

5.8 Conversion from Antenna to Sensor Brightness Temperature

For a cross-track scanning microwave radiometer, pure vertical (v) or horizontal (h) polarization measurements only occur in the nadir direction. At the other scan angles, the measurements represent a mixed contribution from both

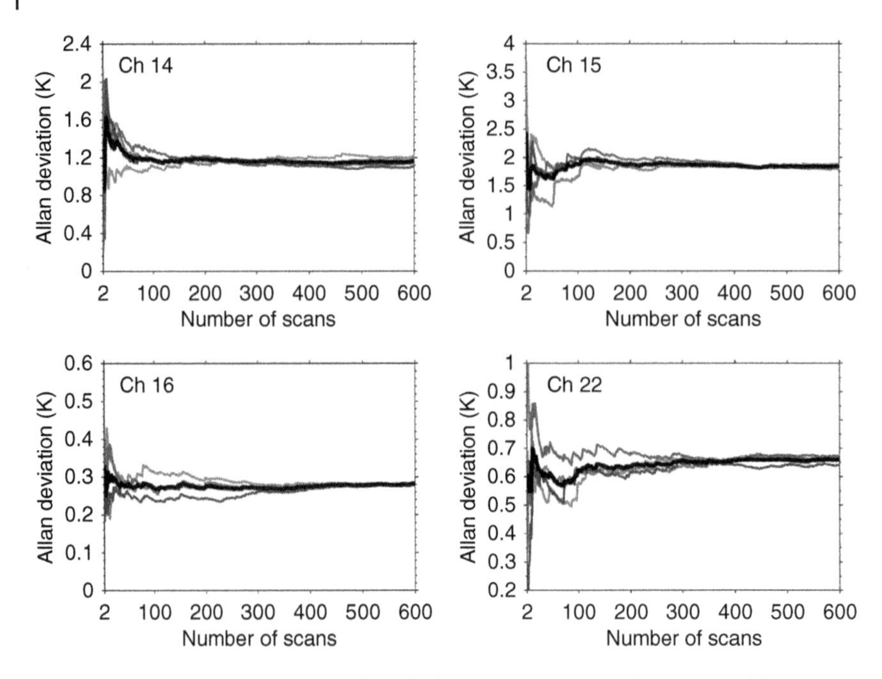

Figure 5.10 The noise magnitudes of ATMS channels 14, 15, 16, and 22 obtained by altering the number of scans used in the two-sample Allan deviation. The Allan deviation is calculated separately using on-orbit warm counts measured at four scan positions, denoted by four gray lines, respectively. The mean of these four is denoted by the thick black line. (Tian *et al.* 2015 [167]. Reproduced with permission of IEEE.)

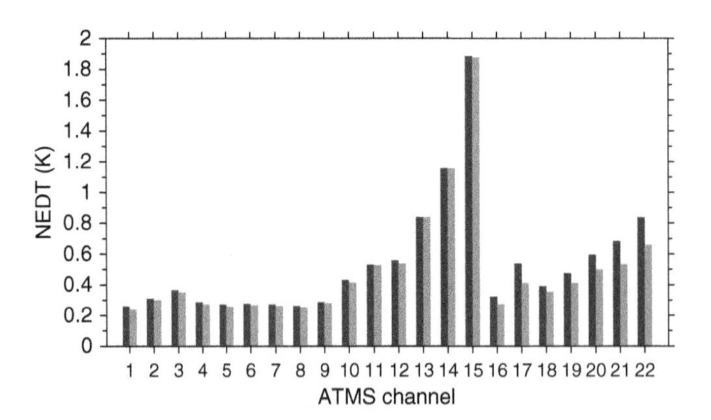

Figure 5.11 The NEDTs estimated from the two-sample Allan deviation (gray) and the current way of using the standard deviation (black) for all ATMS 22 channels by using the warm counts of the first 300 scan lines from the orbit 16 229 on December 15, 2014. (Tian *et al.* 2015 [167]. Reproduced with permission of IEEE.)

v and h polarizations. Thus, it is necessary to define the quasi-vertical and quasi-horizontal TDR, T_a^{Qv} and T_a^{Qh}, via [9]

$$T_a^{Qv} = \eta_{me}^{vv} T_b^{Qv} + \eta_{me}^{hv} T_b^{Qh} + \eta_{se}^{vv} T_{b,se}^{Qv} + \eta_{se}^{hv} T_{b,se}^{Qh} + (\eta_{sc}^{vv} + \eta_{sc}^{hv}) T_{c,RJ} + S_a^{Qv}, \tag{5.48a}$$

$$T_a^{Qh} = \eta_{me}^{hh} T_b^{Qh} + \eta_{me}^{vh} T_b^{Qv} + \eta_{se}^{hh} T_{b,se}^{Qh} + \eta_{se}^{vh} T_{b,se}^{Qv} + (\eta_{sc}^{hh} + \eta_{sc}^{vh}) T_{c,RJ} + S_a^{Qh}, \tag{5.48b}$$

where $(\eta_{me}^{vv}, \eta_{me}^{hh})$ are the co-polarized antenna main beam efficiencies; $(\eta_{me}^{vh}, \eta_{me}^{hv})$ are the cross-polarized antenna main beam efficiencies; $(\eta_{se}^{vv}, \eta_{se}^{hh})$ and $(\eta_{se}^{vh}, \eta_{se}^{hv})$ are the co-polarized and cross-polarized antenna side-lobe beam efficiencies, respectively; $(\eta_{sc}^{vv}, \eta_{sc}^{hh})$ and $(\eta_{sc}^{vh}, \eta_{sc}^{hv})$ are the cold-space co-polarized and cross-polarized side-lobe beam efficiencies. It is worth pointing out that each ATMS frequency channel measures only one polarization, that is, either horizontal or vertical polarization (see Table 5.1). Therefore, in Eq. (5.48), there are, correspondingly, only one co-polarization antenna beam efficiency and one cross-polarization antenna beam efficiency in pair for each of the antenna main beam, antenna side lobe, and cold-space side lobe.

The quasi-vertical and quasi-horizontal sensor brightness temperatures, T_b^{Qv} and T_b^{Qh}, are related to the pure vertically and horizontally polarized brightness temperatures, T_b^v and T_b^h, through the following relationships, neglecting the third Stokes components:

$$T_b^{Qv} = T_b^v \cos^2\theta + T_b^h \sin^2\theta, \tag{5.49a}$$

$$T_b^{Qh} = T_b^v \sin^2\theta + T_b^h \cos^2\theta, \tag{5.49b}$$

where θ is the scan angle. From Eq. (5.49), it can easily be seen that both T_b^{Qv} and T_b^{Qh} vary with the scan angle and are the same at the nadir and 45° scan angle.

The last terms in Eqs. (5.48a) and (5.48b), S_a^{Qv} and S_a^{Qh}, are considered as the radiation contributions from the antenna near-field side lobe or other effects. As discussed in the previous section, emitted radiation from ATMS flat reflector is mostly removed in TDR data. Here, we can neglect these two terms for simplicity.

Assuming $T_{b,se}^{Qv} \approx T_b^{Qv}$ and $T_{b,se}^{Qh} \approx T_b^{Qh}$, Eqs. (5.48a) and (5.48b) can be rewritten as

$$T_a^{Qv} = (\eta_{me}^{vv} + \eta_{se}^{vv}) T_b^{Qv} + (\eta_{me}^{hv} + \eta_{se}^{hv}) T_b^{Qh} + (\eta_{sc}^{vv} + \eta_{sc}^{hv}) T_{c,RJ}, \tag{5.50a}$$

$$T_a^{Qh} = (\eta_{me}^{hh} + \eta_{se}^{hh}) T_b^{Qh} + (\eta_{me}^{vh} + \eta_{se}^{vh}) T_b^{Qv} + (\eta_{sc}^{hh} + \eta_{sc}^{vh}) T_{c,RJ}. \tag{5.50b}$$

For a fixed scan angle and under a given surface condition, T_b^{Qv} and T_b^{Qh} are related to each other via the following empirical models:

$$T_b^{Qh} = A^h(\theta) T_b^{Qv},$$

$$T_b^{Qv} = A^v(\theta) T_b^{Qh},$$

where $A^h(\theta)$ and $A^v(\theta)$ are functions dependent on the scan angle. At scan angles of 0° and 45°, and for the channels that are not impacted by the surface polarization, $A^h(\theta) = 1$ and $A^v(\theta) = 1$. Thus, Eqs. (5.50a) and (5.50b) can be further written as

$$T_a^{Qv} = [(\eta_{me}^{vv} + \eta_{se}^{vv}) + A^v(\eta_{me}^{hv} + \eta_{se}^{hv})]T_b^{Qv} + (\eta_{sc}^{vv} + \eta_{sc}^{hv})T_{c,RJ}, \tag{5.51a}$$

$$T_a^{Qh} = [(\eta_{me}^{hh} + \eta_{se}^{hh}) + A^h(\eta_{me}^{vh} + \eta_{se}^{vh})]T_b^{Qh} + (\eta_{sc}^{hh} + \eta_{sc}^{vh})T_{c,RJ}. \tag{5.51b}$$

Tables 5.2 and 5.3 list all the coefficients needed for the aforementioned conversion between antenna temperatures (TDR) and brightness temperatures (SDR). Note that ATMS channels 1–3 and 16 have 1% to 4% polarization spillover radiation. Thus, a correction must be made to account for the contribution of polarization spillover effect between TDR and SDR conversions. Moreover, for ATMS W- and G-bands (channels 16–22), the beam efficiencies listed in the table remain highly uncertain. The ATMS vendor, NGES, has provided the W- and G-band beam efficiencies; however, the results have not been verified by others. Further investigation on this issue is required. Thus, for ATMS W- and G-bands, the

Table 5.2 ATMS antenna main beam efficiencies analyzed from co- and cross-polarization antenna gain distribution functions.

Channel	η_{me}^{pp} (%)			η_{me}^{pq} (%)		
	B1	B48	B96	B01	B48	B96
1	95.5	95.3	95.9	0.84	0.73	0.81
2	97.0	96.4	96.8	0.64	0.65	0.64
3	96.2	95.6	96.3	1.01	1.05	0.90
4	96.2	95.7	96.6	0.95	0.94	0.70
5	96.2	95.8	96.1	0.87	0.91	0.98
6	96.3	95.9	96.2	0.88	0.94	1.04
7	96.5	96.1	96.6	0.87	0.86	0.82
8	96.6	96.1	96.2	0.90	0.90	1.13
9	96.7	96.2	96.6	0.90	0.88	0.86
10	97.3	97.1	97.2	0.92	0.91	0.93
11	97.3	97.1	97.2	0.92	0.91	0.93
12	97.3	97.1	97.2	0.92	0.91	0.93
13	97.3	97.1	97.2	0.92	0.91	0.93
14	97.3	97.1	97.2	0.92	0.91	0.93
15	97.3	97.1	97.2	0.92	0.91	0.93
16	90.9	91.3	91.7	4.71	4.65	4.54
17	86.2	83.9	86.6	3.71	3.40	5.18
18	86.5	85.2	85.2	3.31	3.46	5.12
19	86.0	87.4	89.3	4.03	2.25	1.85
20	86.0	87.4	89.3	4.03	2.25	1.85
21	86.0	87.4	89.3	4.03	2.25	1.85
22	86.0	87.4	89.3	4.03	2.25	1.85

Table 5.3 ATMS antenna side-lobe Earth beam efficiencies analyzed from co- and cross-polarization antenna gain distribution functions.

Channel	η_{se}^{pp} (%)			η_{se}^{pq} (%)			$\eta_{sc}^{pp} + \eta_{sc}^{pq}$ (%)		
	B1	B48	B96	B01	B48	B96	B1	B48	B96
1	2.30	3.10	2.01	0.56	0.54	0.35	0.78	0.29	0.95
2	1.55	2.25	1.53	0.35	0.37	0.22	0.49	0.36	0.76
3	1.71	2.46	1.74	0.45	0.51	0.44	0.60	0.38	0.58
4	1.93	2.49	1.83	0.43	0.45	0.33	0.52	0.42	0.57
5	1.86	2.40	1.80	0.47	0.50	0.39	0.56	0.44	0.72
6	1.75	2.32	1.72	0.44	0.51	0.44	0.60	0.35	0.63
7	1.66	2.21	1.66	0.44	0.43	0.32	0.52	0.41	0.61
8	1.62	2.11	1.54	0.37	0.46	0.45	0.53	0.40	0.66
9	1.63	2.13	1.61	0.33	0.41	0.34	0.46	0.34	0.55
10	1.18	1.47	1.12	0.27	0.28	0.26	0.35	0.22	0.48
11	1.18	1.47	1.12	0.27	0.28	0.26	0.35	0.22	0.48
12	1.18	1.47	1.12	0.27	0.28	0.26	0.35	0.22	0.48
13	1.18	1.47	1.12	0.27	0.28	0.26	0.35	0.22	0.48
14	1.18	1.47	1.12	0.27	0.28	0.26	0.35	0.22	0.48
15	1.18	1.47	1.12	0.27	0.28	0.26	0.35	0.22	0.48
16	1.34	2.12	1.45	1.33	1.36	0.87	1.70	0.53	1.40
17	3.83	5.68	3.49	1.73	1.83	1.66	4.53	5.23	3.08
18	5.10	5.30	4.80	1.41	1.59	1.51	3.69	4.42	3.36
19	5.10	5.30	4.80	1.41	1.59	1.51	3.69	4.42	3.36
20	5.17	5.37	5.01	1.44	1.37	0.95	3.41	3.59	2.89
21	5.17	5.37	5.01	1.44	1.37	0.95	3.41	3.59	2.89
22	5.17	5.37	5.01	1.44	1.37	0.95	3.41	3.59	2.89

current TDR data are only corrected for the near-field side-lobe contributions for the SDR data by setting the antenna main beam efficiency to 1.

For ATMS channels 1, 2, and 16, TDRs are converted to SDRs using the following equation:

$$T_b^{Qv} = [T_a^{Qv} - (\eta_{sc}^{vv} + \eta_{sc}^{hv})T_{c,RJ}]/[\eta_{me}^{vv} + \eta_{se}^{vv} + A^v(\eta_{me}^{hv} + \eta_{se}^{hv})]. \tag{5.52a}$$

For the other ATMS channels,

$$T_b^{Qh} = [T_a^{Qh} - (\eta_{sc}^{hh} + \eta_{sc}^{vh})T_{c,RJ}]/[\eta_{me}^{hh} + \eta_{se}^{hh} + A^h(\eta_{me}^{vh} + \eta_{se}^{vh})]. \tag{5.52b}$$

5.9 Summary and Conclusion

In this chapter, the ATMS Earth scene counts are calibrated to antenna brightness temperatures (TDR) through a two-point calibration algorithm with a quadratic nonlinear correction. The nonlinearity term is derived from the prelaunch TVAC data with a maximum value less than 0.5 K. After applying the nonlinear correction, the absolute accuracy of TDR for all ATMS channels is generally about 0.2–0.5 K, which meets the specification. Unlike AMSU-A/MHS calibration operating at radiance, ATMS calibration to TDR is directly carried out at brightness temperature based on the RJ approximation. Thus, the cold-space temperatures are corrected to the apparent brightness temperatures prior to use in the two-point calibration. In future, we plan to modify the current ground processing system into a full radiance space.

The current algorithm for quantifying the precision of the ATMS radiometric measurements, *NEDT*, is described. The *NEDT* values for all the channels are well within the instrument specification. However, through the sensitivity study of *NEDT* and the Allan deviation, we found that the Allan deviation may be a better metric for precision.

Currently, noise magnitudes of the operational satellite instruments are mostly quantified by computing the standard deviation of the measurements from their calibration targets. The standard deviation is valid for describing the spread of a statistical distribution of the measured values around its mean that is stable. However, the measurements of a warm calibration target such as ATMS blackbody can exhibit a considerable variation in each orbit. In this study, we propose to use Allan deviation to characterize the ATMS noise. It is found that in the overlapping Allan deviation formula, the averaging window size has to be set to 1 in order to accurately assess the noises for both stationary and nonstationary time series. From the ATMS on-orbit data, the noise magnitudes at several channels show a large discrepancy between the Allan deviation and the current operational NEDT. Thus, the Allan deviation method is recommended for the noise characterization of all the ATMS channels.

The two-sample Allan deviation is introduced here as an alternative approach for characterizing the ATMS measurement precision, NEDT. Firstly, it is mathematically proved that the two-sample Allan variance provides an unbiased estimate of a quasi-stationary data. In addition, this chapter shows that the two-sample Allan deviation always provides an estimate closer to the true noise standard deviation compared with the traditional standard deviation. For characterizing the ATMS on-orbit channel precision, this chapter shows that for the channels with stable warm counts, both the two-sample Allan deviation and the current operational NEDT provide very close estimates. However, for the channels with unstable warm counts, the current NEDT that is based on the standard deviation may overestimate the noise magnitudes by including the impact of the variations. Moreover, use of the two-sample Allan deviation avoids the determination of the optimal sample size, which is critical, but is vague and *ad hoc* in the calculation of the current NEDT. The aforementioned merits render the two-sample Allan deviation superior to both the overlapping Allan deviation and the standard deviation. It is thus suggested that the two-sample Allan deviation is a better choice for noise characterization of ATMS and other satellite microwave sensors.

The conversion from TDR to SDR is very important since SDR products can be directly used in the Numerical Weather Prediction (NWP) models for satellite data assimilation. One thing that needs to be pointed out is that the convertibility is not always unique if ATMS antenna subsystem has a significant polarization spillover effect and/or a side-lobe contribution from the nearby scene cells. While ATMS antenna gain distribution functions were measured during the prelaunch period, there remain some uncertainties in the characterization of side lobe and cross-polarization at high frequencies. Under the conditions where ATMS brightness temperatures at quasi-vertical and quasi-horizontal polarization states are the same, the conversion from TDR to SDR becomes unique, assuming that all the side-lobe contributions are estimated. At 45° scan angle, ATMS SDR can be uniquely derived from its TDR and therefore directly compared with the simulations. It is shown that the biases of ATMS SDRs with respect to GPS RO and GFS simulation are similar in magnitude. The largest biases are found for surface-sensitive channels at both low and high frequencies. Further investigation is planned to assess the errors in the forward modeling associated with surface emissivity and surface parameters obtained from GFS.

6
Detection of Interference Signals at Microwave Frequencies

6.1
Introduction

Microwave instruments on board satellites observe the Earth's surface, and atmospheres have some channels operating at L-, C-, and X-bands. The measurements of the natural Earth's thermal emission from satellites could be mixed with the signals from lower frequency active transmitters including radar, air traffic control, cell phone, garage door remote control, GPS signal on highway, defense tracking, vehicle speed detection for law enforcement, and so on. Such a phenomenon of satellite-measured passive microwave thermal emission being mixed with the signals from the active sensors is referred to as radio-frequency interference (RFI). The operating frequencies by some ground active sensors are shown in Figure 6.1 and are within the bandwidth of microwave radiometers. For example, satellite television broadcasts in Europe are transmitting the signals using 10 – 13 GHz and thus interfere with the microwave radiometer measurements in the X-band while the direct TV in the United States operates in 18 – 24 GHz and contaminates the microwave K-band data. Some ground-based sources such as garage door openers and radar transmit the signals at lower frequencies near microwave radiometer L-band frequencies. With the expanding demand for fixed-satellite service (FSS) technology, increasing amounts of Television-frequency interference (TFI) are now affecting the oceanic measurements from satellite passive microwave instruments. The RFI signatures in brightness temperature measurements, if not identified and removed, would introduce errors in microwave products. It is therefore important to identify RFI-contaminated data before carrying out the product retrieval and data assimilation.

Existing methods for identifying RFI signatures at C-, X-, and K-band channels of various sensors include the spectral difference method [168], the mean and standard deviation method [169], principal component analysis (PCA) method [7], and the normalized PCA method [6]. For global land and ocean RFI detection, we present several methods for detecting the RFI from variable sources in all seasons. In this chapter, the double principal component analysis (DPCA) method is developed to detect the RFI signals in advanced microwave scanning radiometer – EOS (AMSR-E) data over the global oceans.

Passive Microwave Remote Sensing of the Earth: For Meteorological Applications, First Edition. Fuzhong Weng.

Active remote sensing usually uses low-frequency channels.
(C-band: 4-8 GHz, X-band: 8-12 GHz, K-band: 18-26.5 GHz)

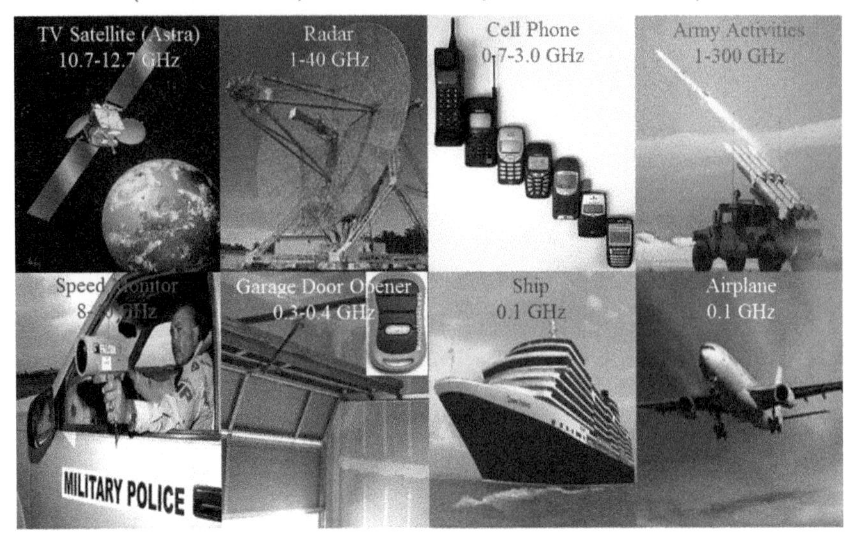

Figure 6.1 Examples of radio-frequency interference sources from human activities including communication satellite, radar, cell phone, vehicle speed monitor, and garage door opener.

Table 6.1 Satellite missions carrying microwave radiometers on board.

Sensor	WindSat	AMSR-E	AMSR2
Spacecraft	Coriolis	Aqua	GCOM-W1
Agency	NRL	NASA	JAXA
Launch date	January 6, 2003	May 4, 2002	May 17, 2012
Altitude (km)	840	705	700
EIA (°)	49.9–55.3	53.1	55.0
Swath (km)	1025/350 (Fore/Aft)	1445	1450
ECT	6:00 am	1:30 pm	1:30 pm

6.2 Microwave Imaging Radiometers and Data Sets

Microwave imaging radiometers onboard satellite platforms are selected to demonstrate the RFI signals over different regions. They include WindSat, AMSR-E and AMSR2, and SSM/I as listed in Table 6.1. The data sets from these radiometers for retrieval of environmental products will be further used in later chapters.

WindSat radiometer is carried on board the Coriolis satellite. From an altitude of 840 km, WindSat scans the Earth in a forward-looking swath of 1000 km and an aft-looking swath of 400 km. The antenna beams view the Earth at different incidence angles, varying from 50° to 55°, for different frequency channels. The four low-frequency channels of WindSat at 6.8, 10.7, 18.7, and 23.8 GHz measure the radiation for both vertical and horizontal polarization. At 10.7, 18.7, and 37.0 GHz, the third and fourth components of Stokes vector are also observed, and these frequencies are referred to as full polarimetric channels. The WindSat raw data counts are converted into antenna temperatures at its footprint resolution after applying calibration, geolocation, and quality control. The antenna temperature data at this stage is called temperature data record (TDR). The footprint resolution of antenna temperatures in TDR varies with frequency. The antenna temperatures in TDR are further processed into the so-called sensor data record (SDR), in which the antenna temperatures at different frequencies are resampled to the same common resolution and beam width. There are two WindSat SDRs with three different resolutions: the coarse resolution has the same footprint resolution as 6.8 GHz, the middle resolution SDR has the same footprint resolution as 10.7 GHz, and the high resolution SDR has the same footprint resolution as 18.7 GHz. The coarse resolution swath WindSat SDR data is finally binned into a latitudinal and longitudinal grid with a 1/3 degree resolution over the globe. The gridded data set in February 2011, are used in this study.

AMSR-E is one of the six instruments on board the NASA EOS Aqua satellite. For the Aqua satellite, its equator crossing time (ECT) at its ascending node is 1:30 pm. From an altitude of 705 km, AMSR-E can scan the Earth with a swath width of 1445 km at an incidence angle of 55°. AMSR-E is a 12-channel, 6-frequency, total power passive-microwave conical-scanning radiometer system. It measures vertically and horizontally polarized brightness temperatures at 6.925, 10.65, 18.7, 23.8, 36.5, and 89.0 GHz. The across-track and along-track spatial resolutions of the individual ground instantaneous field-of-view (IFOV) measurements are 75×43 km at 6.925 GHz, 51×29 km at 10.65 GHz, 27×16 km at 18.7 GHz, 32×18 km at 23.8 GHz, 14×8 km at 36.5 GHz, and 6×4 km at 89.0 GHz. The sampling interval is 10 km for 6–36 GHz channels and 5 km for the 89 GHz channel. The AMSR-E TDR data at its original resolution is used for RFI study.

It is worth mentioning that the successor of AMSR-E is AMSR-2, which was on board the Global Change Observation Mission 1st–Water (GCOM-W1) satellite launched on May 18, 2012 [170]. AMSR-2 is a conical-scanning microwave imager with 14 channels located at the following 7 frequencies: 6.925, 7.3, 10.65, 18.7, 23.8, 36.5, and 89.0 GHz. It has a local incident angle of 55° from an orbit of 700 km above the Earth's surface. The AMSR-2 antenna reflector size is 2 m, which is larger than AMSR-E and therefore provides better spatial resolution. Specifically, the across-track and along-track spatial resolutions of the individual ground IFOV measurements are 62×35 km at both 6.925 and 7.3 GHz frequencies, 42×24 km at 10.65 GHz, 22×14 km at 18.7 GHz, 26×15 km at 23.8 GHz, 12×7 km at 36.5 GHz, and 5×3 km at 89.0 GHz frequencies. The sampling intervals between two neighboring field of views are 5 km for the 89 GHz channels and 10 km for the remaining channels. The two additional channels at 7.3 GHz channels allow for mitigating the RFI effectively at C-bands [169, 171, 172].

6.3 Radio-Frequency Interference Signals in Microwave Data

Under clear-sky conditions, brightness temperature spectrum at microwave frequency is driven by surface emissivity. Over land free from surface scattering medium, the brightness temperature displays a typical emitting spectrum as shown in Figure 6.2 (solid lines). In horizontal polarization, the brightness temperature increases as the frequency increases. However, when a channel is RFI-contaminated, the brightness temperature spectrum then departs its typical emission or scattering spectrum. Over land, the typical RFI contamination occurs in the C-, X-, and K-bands and the RFI signals can significantly increase the brightness temperatures near these bands as shown in Figure 6.2 (dashed line). Note that natural phenomena such as flooding and wet surface at lower frequencies tend to decrease the brightness temperatures at this channel. Human-made radiation from active microwave transmitters is different from natural radiation in terms of intensity, spatial variability, spectral characteristics, and channel correlations. RFI signals typically originate from a wide variety of coherent point target sources, that is, radiating devices and antennas, and are often directional, isolated, narrow-banded. On the other hand, as distributed targets, the Earth's surfaces often produce smooth, ultra-wideband, and incoherent microwave radiation. At 30 GHz and below, scattering effects from natural targets are relatively weaker than the emission signals. RFI can significantly increase the brightness temperatures at a particular frequency and generate a negative spectral gradient [171].

The spectral difference in brightness temperatures at lower frequencies is used to quantify the RFI magnitude and extent. For example, the land RFI at

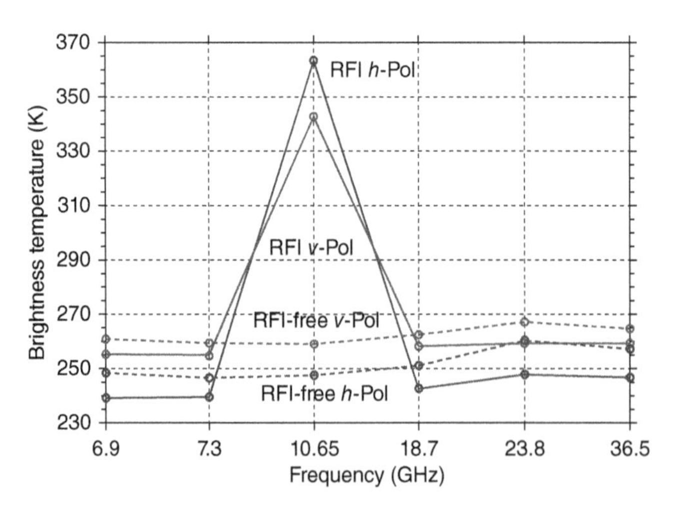

Figure 6.2 Observed brightness temperatures for horizontal and vertical polarization at all frequencies observed by AMSR2 on January 28, 2016. The locations of the selected observation pixels are at [51.5N 0E] (London) and [51.5N, 7.5E].

AMSR-E 6.925 GHz horizontal polarization channel is defined, if it exceeds 5 K, as follows:

$$RFI = TB_{6h} - TB_{10h}$$
$$= \begin{pmatrix} 5-10\ \text{K} & \text{weak} \\ 10-20\ \text{K} & \text{moderate} \\ >20\ \text{K} & \text{strong} \end{pmatrix}. \quad (6.1)$$

The exact threshold used for each microwave radiometer depends on the radiometer sensitivity, channel frequency and polarization, surface type, and intensities of active sources. A generalized detection scheme is discussed in the following sections.

6.4
Detection of RFI over Land

6.4.1
Double Principal Component Analysis (DPCA)

The DPCA method developed for RFI detection consists of two PCA steps. In the first PCA step, a vector of 10-component brightness temperature is defined at each grid (i.e., 1/3 degrees latitudinal and longitudinal resolution) as

$$\vec{V}_i = (TB_{6h,i}, TB_{6v,i}, TB_{10h,i}, TB_{10v,i}, TB_{18h,i}, TB_{18v,i}, TB_{23h,i}, TB_{23v,i}, TB_{37h,i}, TB_{37v,i})^T, \quad i = 1, 2, \ldots, N, \quad (6.2)$$

where the superscript T represents the transpose to a matrix and N is the total number of data points over the domain of interest.

A data matrix is first defined for PCA of $\vec{V}_i$ as follows:

$$\mathbf{A}_{10\times N} = \begin{pmatrix} TB_{6h,1} & TB_{6h,2} & \cdots\cdots & TB_{6h,N} \\ TB_{6v,1} & TB_{6v,2} & \cdots\cdots & TB_{6v,N} \\ \vdots & \vdots & \ddots & \vdots \\ \vdots & \vdots & \ddots & \vdots \\ TB_{37v,1} & TB_{37v,2} & \cdots\cdots & TB_{37v,N} \end{pmatrix}. \quad (6.3)$$

A 10×10 covariance matrix $\mathbf{R}_{10\times10}$ is then constructed from $\mathbf{A}_{10\times N}$:$\mathbf{R}_{10\times10} = \mathbf{A}\mathbf{A}^{\mathrm{T}}$, for an eigenvalue/eigenvector analysis. Specifically, the eigenvalues λ_i ($i = 1, 2, \ldots, 10$) and eigenvectors $\vec{e}_i = [e_{1,i}, e_{2,i}, \ldots, e_{10,i}]^{\mathrm{T}}$ are found by solving the following equation:

$$\mathbf{R}\vec{e}_i = \lambda_i \vec{e}_i, \quad i = 1, 2, \ldots, 10, \quad (6.4)$$

where i indicates the ith PC mode ($i = 1, 2, \ldots, 10$) and $\vec{e}_i$ is called the ith principal component (PC) mode. The ith eigenvalue λ_i quantifies the variance contribution of the ith PC mode to the total variance of data. The first PC mode $\vec{e}_1$ spans in the direction of the maximum variance in the data, the second PC mode $\vec{e}_2$ spans in the direction of the largest variance not accounted for by the first vector, and so on.

By expressing the eigenvalues and eigenvectors in matrix form,

$$\Lambda = \begin{pmatrix} \lambda_1 & \cdots & 0 \\ \vdots & \ddots & \vdots \\ 0 & \cdots & \lambda_{10} \end{pmatrix}, \quad \mathbf{E} = [\vec{e}_1, \vec{e}_2, \ldots, \vec{e}_{10}]. \tag{6.5}$$

Equation (6.4), can, be equivalently written as

$$\mathbf{RE} = \mathbf{E}\Lambda \text{ or } \mathbf{R} = \mathbf{E}\Lambda\mathbf{E}^{\mathrm{T}}, \tag{6.6}$$

where $\mathbf{E}$ is an orthogonal matrix; thus, $\mathbf{E}^{-1} = \mathbf{E}^{\mathrm{T}}$.

By projecting the data matrix $\mathbf{A}$ onto an orthonormal space spanned by the eigenvectors $\vec{e}_1, \vec{e}_2, \ldots, \vec{e}_{10}$, we obtain the so-called PC coefficients:

$$\mathbf{U}_{10\times N} = \mathbf{E}^{\mathrm{T}}\mathbf{A} = \begin{pmatrix} \vec{u}_1 \\ \vec{u}_2 \\ \vdots \\ \vec{u}_{10} \end{pmatrix}, \tag{6.7}$$

where $\vec{u}_i = [u_{i,1}, u_{i,2}, \ldots, u_{i,N}]$ is the PC coefficient for the ith PC mode.

The data matrix $\mathbf{A}$ in Eq. (6.3) can finally be reconstructed using the PC coefficients and the PC modes:

$$\mathbf{A} = \sum_{i=1}^{10} \vec{e}_i \vec{u}_i, \tag{6.8}$$

which could be decomposed into the following two parts: $\mathbf{A} \equiv \mathbf{A}_1 + \mathbf{A}_2$, where

$$\mathbf{A}_1 = \sum_{i=1}^{\alpha} \vec{e}_i \vec{u}_i, \quad \mathbf{A}_2 = \sum_{i=\alpha+1}^{10} \vec{e}_i \vec{u}_i, \tag{6.9}$$

where α is an integer parameter to be determined later. The matrix $\mathbf{A}_2$, which is the sum of the PC modes from brightness temperatures from the $(\alpha + 1)$th to 10th, is used in the normalized PCA analysis. The first α PC modes ($\mathbf{A}_1$) reflect the strong channel-by-channel correlations of scattering medium. The residual matrix $\mathbf{A}_2$ contains the RFI signals in snow surfaces.

The second step of the DPCA method is a normalized PCA applied to the matrix $\mathbf{A}_2$. A normalized RFI index vector is defined as

$$\vec{R}^{A_2}_{indices} = \begin{pmatrix} \dfrac{TB_{10h} - TB_{18h} - \mu}{\sigma} \\ \dfrac{TB_{18v} - TB_{23v} - \mu}{\sigma} \\ \dfrac{TB_{18h} - TB_{23h} - \mu}{\sigma} \\ \dfrac{TB_{23v} - TB_{37v} - \mu}{\sigma} \\ \dfrac{TB_{23h} - TB_{37h} - \mu}{\sigma} \end{pmatrix}_{A_2}, \tag{6.10}$$

where μ and σ are the mean and standard deviation of the corresponding five RFI indices, and is used for another PCA analysis. The higher values of the PC coefficient for the first PC mode of $\mathbf{A}_2$ would suggest larger probabilities of RFI. The multichannel correlations of radiometer data from natural land and ice surface radiations are much higher than those RFI-induced signatures. A PCA of multichannel brightness temperatures is thus used to separate the correlated components ($\mathbf{A}_1$) from less correlated components ($\mathbf{A}_2$).

Figure 6.3 presents WindSat-measured brightness temperatures at 6.8, 10.7, and 18.7 GHz for horizontal polarization over Greenland during February 1–10, 2011. The brightness temperature varies by more than 100 K, from the coastal land area

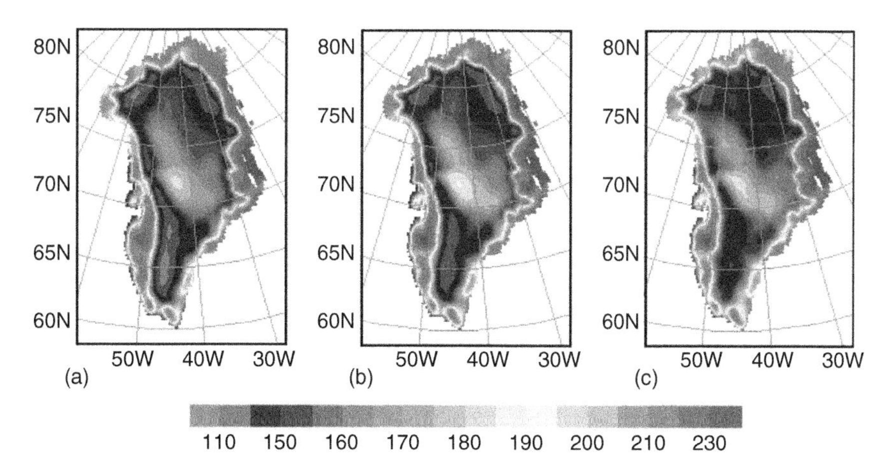

Figure 6.3 Brightness temperatures at (a) 6.8, (b) 10.7, and (c) 18.7 GHz for horizontal polarization averaged over the period of February 1–10, 2011. (Zhao *et al.* 2013 [7]. Reproduced with permission of IEEE.)

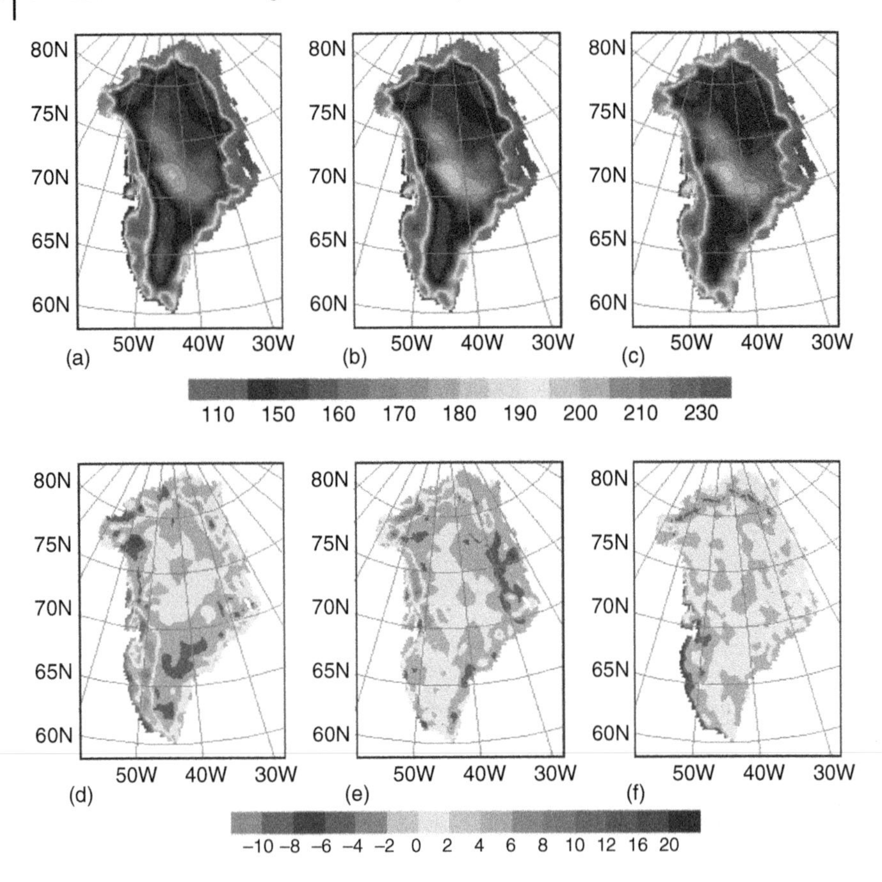

Figure 6.4 Brightness temperatures of 6.8 GHz (left panels), 10.7 GHz (middle panels), and 18.7 GHz (right panels) from the 10-channel average for horizontal polarization reconstructed by (a)–(c) the first to the fourth and (d)–(f) the fifth to the tenth PC modes. (Zhao *et al.* 2013 [7]. Reproduced with permission of IEEE.)

to frozen ice sheet away from the coast (Figure 6.3b), with a significant jump in the brightness temperature values occurring at the edge of frozen ice.

The spatial distributions of the two components of brightness temperatures $\mathbf{A}_1$ and $\mathbf{A}_2$ at 6.8, 10.7, and 18.7 GHz are shown in Figure 6.4, when α is set to 4. Large variations in the brightness temperature from coastal land area (~210–230 K) to frozen ice ground (~110–150 K) as seen in Figure 6.3 are captured by $\mathbf{A}_1$ (Figure 6.4a–c). The sharp gradient found at the edge of frozen ice is not seen in the $\mathbf{A}_2$ fields (Figure 6.4d–f). The first PC mode explains more than 99.67% of the total data variances. The first four PCs capture the majority of data variance in Greenland (>99.99%).

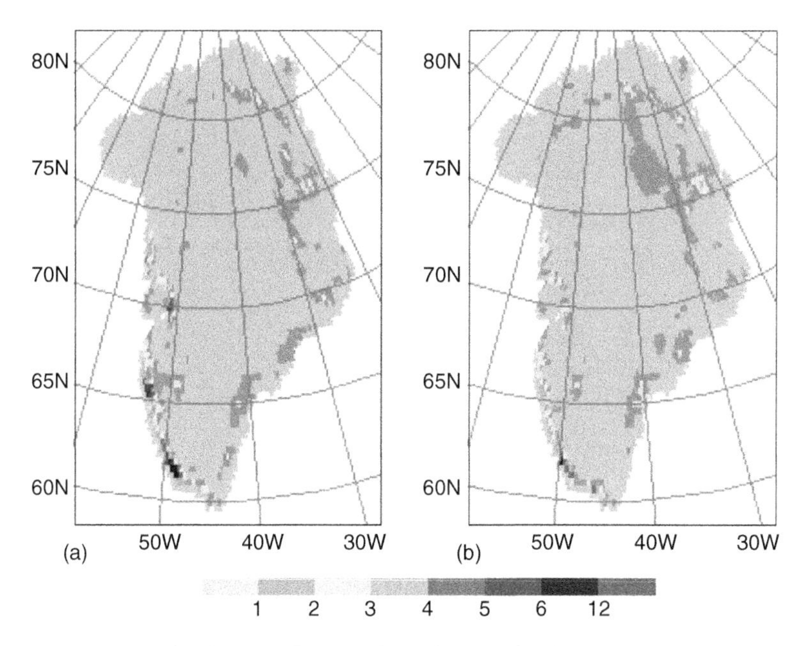

Figure 6.5 RFI distributions of 6.8 GHz for (a) horizontal polarization and (b) vertical polarization over Greenland using the DPCA method. (Zhao *et al.* 2013 [7]. Reproduced with permission of IEEE.)

By applying a normalized PCA to $\mathbf{A}_2$, we may obtain RFI distributions. Sensitivities of the DPCA method to parameter α for RFI detection could be assessed by examining the variation of the first PC coefficients in the second step of DPCA for different values of α. For WindSat 6.8 GHz channel for both polarization, we set $\alpha = 4$. Figure 6.5 shows RFI distributions over Greenland for both horizontal and vertical polarization at 6.8 GHz. The RFI signals are found mostly near the southwest coastal areas where research stations are more populated. The RFI for horizontal polarization is slightly stronger than that for the vertical polarization. The same DPCA method is also applied to WindSat data over Antarctic, and it is shown that RFI-contaminated WindSat data correlates well with the research stations located within two longitude zones of 75W – 55W and 75E – 90E [7].

Having demonstrated the performance of the DPCA method over Greenland and Antarctic, it is important to test if the same method also works in other areas of the globe. Figure 6.6 displays the RFI distributions of 6.8 GHz for horizontal polarization identified by the PCA method [172] and the DPCA proposed in this study over the United States. The PCA method cannot detect RFI in WindSat winter data. The anomaly associated with the snow in high latitudes dominates, making the RFI signals undetectable. The DPCA, however, successfully detects winter RFI signals over the United States.

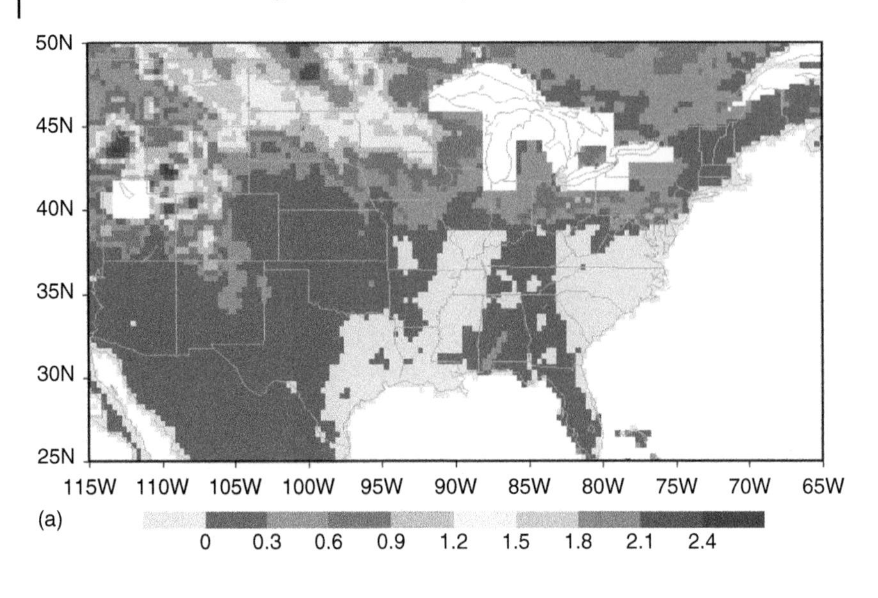

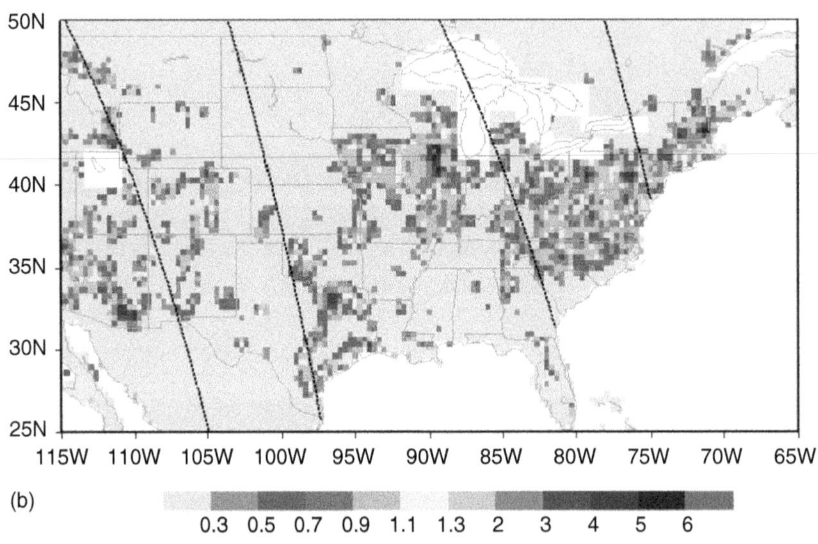

Figure 6.6 RFI distributions of 6.8 GHz for horizontal polarization identified by (a) PCA and (b) DPCA methods over the United States during February 1–10, 2011.

6.4.2 Spectral Difference Method

In general, land surface emissivity increases with frequency, resulting in higher brightness temperatures at 10.65 GHz (channels 3–4) compared to those at

6.925 GHz, that is, $TB_{6v} < TB_{10v}$ or $TB_{6h} < TB_{10h}$. Natural phenomena such as flooding and wet surfaces further decrease the brightness temperatures, especially at lower microwave frequencies such as 6.925 GHz. The measured brightness temperatures at low frequencies can thus be used for retrieving the soil moisture content. The presence of RFI at 6.925 GHz, however, increases the brightness temperature at this frequency, resulting in a reversed spectral gradient, that is, $TB_{6v} > TB_{10v}$ or $TB_{6h} < TB_{10h}$ [168]. By examining the spatial distributions of the inequality about RFI-sensitive spectral difference indices $TB_{6v} - TB_{10v}$ and/or $TB_{6h} - TB_{10h}$ (e.g., differences between brightness temperatures at two different frequencies for a given polarization), RFI-contaminated data can be identified. Since RFI signals typically originate from a wide variety of coherent point target sources and are often directional and narrow-banded, they are often isolated in space and persistent in time.

Figure 6.7 presents the spatial distributions of the spectral differences $TB_{6h} - TB_{10h}$ (Figure 6.7a), $TB_{7h} - TB_{10h}$ (Figure 6.7b), $TB_{6v} - TB_{10v}$ (Figure 6.7c) and $TB_{7v} - TB_{10v}$ (Figure 6.7d) for AMSR-2 data from descending nodes over North America on December 11, 2012. Since the presence of RFI at 6.925 GHz would increase the brightness temperature at this frequency, RFI-contaminated data at the 6.925 GHz horizontal polarization state could be identified by their excessively positive values of spectral differences, that is, $TB_{6k} - TB_{10k} \gg 0$. Similarly, RFI-contaminated data at the 7.3 GHz horizontal polarization state could be identified

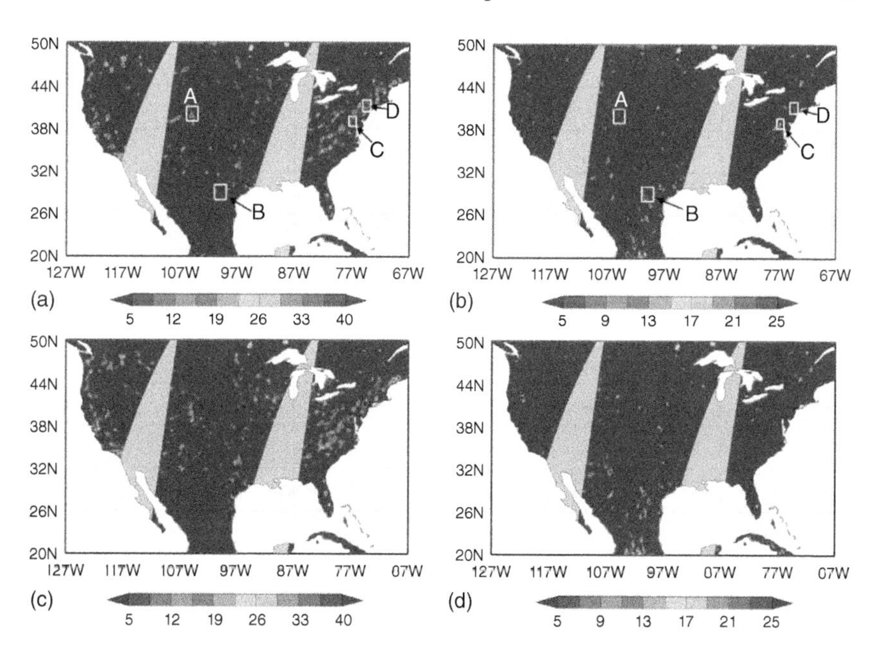

Figure 6.7 Spatial distributions of AMSR-2 RFI signals in descending nodes at (a) 6.925 GHz (left panels) and 7.3 GHz (right panels) for (a)–(b) horizontal and (c)–(d) vertical polarization using the spectral difference approach over North America on December 2, 2012.

by their excessively positive values of spectral differences, $TB_{7k} - TB_{10k} \gg 0$. Figure 6.7 shows that isolated RFI signals characterized by large positive spectral differences in brightness temperatures at 6.925 GHz are found in many places over the United States (Figure 6.7a,c), while RFI signals at 7.3 GHz seem to occur only in Mexico, Washington DC, and New York (Figure 6.7b,d).

6.5
RFI Detection over Oceans

At microwave frequencies, the ocean surface has higher reflectance compared to land surface due to the high permittivity of seawater. Thus, the transmitted signals from the communication geostationary satellites can be reflected by the ocean surfaces, and the reflected signals can also be received by microwave radiometer and mixed with the thermal emission from the oceanic environments. The geostationary satellites such as Astra, Hot Bird, Atlantic Bird 4A, and DirecTV-10/11 transmit the signals at frequencies near X- and K-bands and cause interference in microwave data. Figure 6.8 schematically illustrates the RFI to the microwave radiometer from satellite TV broadcast system [173].

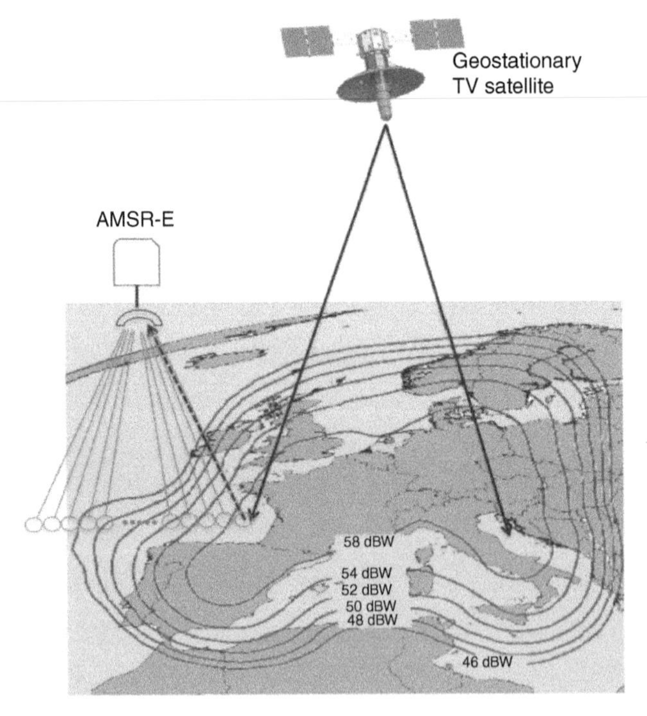

Figure 6.8 A schematic illustration of RFI of AMSR-E Earth views (light gray) with TV signals reflected off from ocean surfaces (dark black dashed). Satellite downlink beam coverage is shown in moderate gray curves. Numbers on the contours indicate the strength of the TV signal expressed in the decibel watt (dBW). (Zou *et al.* 2014 [173]. Reproduced with permission of American Meteorological Society.)

Microwave emissivity over oceans is much lower than that over land. Cloud and precipitation can also increase the thermal emission and therefore significantly increase the brightness temperature at lower frequencies. Such an increase in the measured brightness temperatures could be of a similar magnitude to the increase introduced by RFI from the ocean surface. It is important to develop a robust technique for detecting the oceanic RFI signals so that the weather signals do not appear as "false" RFI signals. There are two methods that were employed for identifying RFI signatures over oceans: a chi-square probability method [174] and a regression method [172]. In the chi-square probability algorithm, an RFI detection method is developed using a time-averaged statistical quantity based on the fact that the source of the oceanic RFI, TV signals, is fixed in location and time, while weather signals associated with cloud and precipitation are transient. The goodness of fit (i.e., chi-square probability) is used for RFI detection. The lower the goodness of fit, the higher the probability of the presence of RFI is expected. A regression model is first established to predict RFI-free brightness temperatures at X- and K-bands from the other WindSat channels. The difference between microwave observations and regression-model-predicted brightness temperatures is then used for oceanic RFI detection. The larger the difference, the stronger the RFI intensity is likely to be. The DPCA was applied for WindSat data and shown to work at any geographical location including snow and sea ice conditions. It is also applied for RFI detection over oceans. Since AMSR-E data are used for RFI studies, a vector of normalized brightness temperatures including a high frequency at 89 GHz is first constructed as follows:

$$\vec{V}_i = \begin{pmatrix} \dfrac{TB_{6h,i} - \mu_{6h}}{\sigma_{6h}} \\ \dfrac{TB_{6v,i} - \mu_{6v}}{\sigma_{6v}} \\ \cdots\cdot \\ \dfrac{TB_{89h,i} - \mu_{89h}}{\sigma_{89h}} \\ \dfrac{TB_{89v,i} - \mu_{89v}}{\sigma_{89v}} \end{pmatrix}, \quad i = 1, 2, \dots, N, \tag{6.11}$$

where N is the total number of data points over an AMSR-E swath (about 2000 scan lines) excluding data over land and sea ice, μ_λ and σ_λ are the mean and standard deviation of brightness temperatures at the frequency λ at all the data points.

A data matrix is first defined for PCA of $\vec{V}_i$ as follows:

$$\mathbf{A}_{12\times N} = \begin{pmatrix} TB_{6h,1} & TB_{6h,2} & \cdots\cdots & TB_{6h,N} \\ TB_{6v,1} & TB_{6v,2} & \cdots\cdots & TB_{6v,N} \\ \vdots & \vdots & \ddots & \vdots \\ \vdots & \vdots & \ddots & \vdots \\ TB_{89v,1} & TB_{89v,2} & \cdots\cdots & TB_{89v,N} \end{pmatrix}. \tag{6.12}$$

A 12×12 covariance matrix $\mathbf{R}_{12\times12} = \mathbf{A}\mathbf{A}^\mathrm{T}$ is then constructed from $\mathbf{A}_{12\times N}$. The eigenvalues and eigenvectors of the covariance matrix $\mathbf{R}$ are found by solving the following equation:

$$\mathbf{R}\vec{e}_i = \lambda_i \vec{e}_i, \quad (i = 1, 2, \dots, 12), \tag{6.13}$$

where λ_i is the ith eigenvalue, and $\vec{e}_i = [e_{1,i}, e_{2,i}, \dots e_{12,i}]^\mathrm{T}$ is called the ith PC mode, $i = 1, 2, \dots, 12$. The ith eigenvalue λ_i $(i = 1, 2, \dots, 12)$ quantifies the variance contribution of the ith PC mode to the total variance of data. The eigenvalues/eigenvectors are sorted by the magnitude of the eigenvalue in decreasing order.

The eigenvalues and eigenvectors can be expressed in a matrix form:

$$\Lambda = \begin{pmatrix} \lambda_1 & \cdots & 0 \\ \vdots & \ddots & \vdots \\ 0 & \cdots & \lambda_{12} \end{pmatrix}, \quad \mathbf{E} = [\vec{e}_1, \vec{e}_2, \dots, \vec{e}_{12}]. \tag{6.14}$$

Equation (6.13), can, be equivalently written as

$$\mathbf{RE} = \mathbf{E}\Lambda \ \text{ or } \ \mathbf{R} = \mathbf{E}\Lambda\mathbf{E}^\mathrm{T} \tag{6.15}$$

Since $\mathbf{E}$ is an orthogonal matrix, $\mathbf{E}^{-1} = \mathbf{E}^\mathrm{T}$.

By projecting the data matrix $\mathbf{A}$ onto an orthogonal space spanned by the eigenvectors $\vec{e}_1, \vec{e}_2, \dots, \vec{e}_{12}$, we obtain the so-called PC coefficients:

$$\begin{pmatrix} \vec{u}_1 \\ \vec{u}_2 \\ \vdots \\ \vec{u}_{12} \end{pmatrix} = \mathbf{E}^\mathrm{T}\mathbf{A}, \tag{6.16}$$

where $\vec{u}_i = (u_{i,1}, u_{i,2}, \dots, u_{i,N})$ is the PC coefficient for the ith PC mode.

The data matrix, $\mathbf{A}$, in Eq. (6.12) can finally be reconstructed using PC coefficients and PC modes:

$$\mathbf{A} = \sum_{i=1}^{12} \vec{e}_i \vec{u}_i, \tag{6.17}$$

which could be further separated into the following two parts: $\mathbf{A} \equiv \mathbf{A}_1 + \mathbf{A}_2$, where

$$\mathbf{A}_1 = \sum_{i=1}^{\alpha} \vec{e}_i \vec{u}_i, \quad \mathbf{A}_2 = \sum_{i=\alpha+1}^{12} \vec{e}_i \vec{u}_i, \tag{6.18}$$

where α is the integer parameter to be determined later. The matrix $\mathbf{A}_2$, which is the sum of the PC modes from the $(\alpha + 1)$th to the 12th, is called a residual data matrix. The large-scale variability in natural emission associated with weather systems is mostly captured by the first α PC modes ($\mathbf{A}_1$) due to strong channel-by-channel correlations. Small-scale weather features and RFI signals are contained in the residual matrix $\mathbf{A}_2$.

The data after taking out the large-scale part (i.e., $\mathbf{A}_1$) are then used to create the spectral difference vectors for detecting RFI signals. Since the multichannel correlations of microwave data are often high for natural radiations and are low for RFI signatures, the spectral difference vectors (also called *RFI indices*) are analyzed using the PCA, technique, which linearly transforms a set of correlated RFI indices into a smaller set of uncorrelated variables so that RFI signals can be effectively separated from natural radiations. Specifically, the following RFI index vectors are defined after $\mathbf{A}_1$ is removed from the original data. For RFI detection at 10 GHz horizontal polarization state, the following spectral difference vector is constructed:

$$\vec{R}_{10h}^{\mathbf{A}_2} = \begin{pmatrix} TB_{10h}^{A_2} - TB_{6h}^{A2} \\ TB_{10h}^{A2} - TB_{18h}^{A2} \\ TB_{18v}^{A2} - TB_{23v}^{A2} \\ TB_{18h}^{A2} - TB_{23h}^{A2} \\ TB_{23v}^{A2} - TB_{36v}^{A2} \\ TB_{23h}^{A2} - TB_{36h}^{A2} \end{pmatrix}, \tag{6.19a}$$

which consists of the spectral differences of the 10 GHz horizontal polarization channels from its neighboring channels (6 and 18 GHz) at the same horizontal polarization state, (i.e., the first and second components in $\vec{R}_{10h}^{\mathbf{A}_2}$), as well as the spectral differences between two neighboring channels with the same polarization states for the remaining frequencies (i.e., the fourth to sixth components in $\vec{R}_{10h}^{\mathbf{A}_2}$). Similarly, RFI detections for the 10 GHz vertical polarization channel and 18 GHz horizontal and vertical polarization channels are constructed as follows:

$$\vec{R}_{10v}^{\mathbf{A}_2} = \begin{pmatrix} TB_{10v}^{A2} - TB_{6v}^{A2} \\ TB_{10v}^{A2} - TB_{18v}^{A2} \\ TB_{18v}^{A2} - TB_{23v}^{A2} \\ TB_{18h}^{A2} - TB_{23h}^{A2} \\ TB_{23v}^{A2} - TB_{36v}^{A2} \\ TB_{23h}^{A2} - TB_{36h}^{A2} \end{pmatrix}, \quad \vec{R}_{18h}^{\mathbf{A}2} = \begin{pmatrix} TB_{10h}^{A2} - TB_{6h}^{A2} \\ TB_{18h}^{A2} - TB_{10h}^{A2} \\ TB_{18h}^{A2} - TB_{23h}^{A2} \\ TB_{23v}^{A2} - TB_{36v}^{A2} \\ TB_{23h}^{A2} - TB_{36h}^{A2} \end{pmatrix},$$

$$\vec{R}_{18v}^{\mathbf{A}2} = \begin{pmatrix} TB_{10v}^{A2} - TB_{6v}^{A2} \\ TB_{18v}^{A2} - TB_{10v}^{A2} \\ TB_{18v}^{A2} - TB_{23v}^{A2} \\ TB_{23v}^{A2} - TB_{36v}^{A2} \\ TB_{23h}^{A2} - TB_{36h}^{A2} \end{pmatrix}, \tag{6.19b}$$

A data matrix $\mathbf{B}_{6\times N}$ (or $\mathbf{B}_{5\times N}$) is constructed from each of the vectors $\vec{R}^{\mathbf{A}_2}_{10h}$, $\vec{R}^{\mathbf{A}_2}_{10v}$, $\vec{R}^{\mathbf{A}_2}_{18h}$, and $\vec{R}^{\mathbf{A}_2}_{18v}$, where N is the total number of data points over an AMSR-E swath. The eigenvectors of the covariance matrix $\mathbf{S}_{6\times 6} = \mathbf{BB}^{\mathrm{T}}$ (or $\mathbf{S}_{5\times 5} = \mathbf{BB}^{\mathrm{T}}$) are then calculated, which is denoted as $\vec{e}^{A_2}_1, \vec{e}^{A_2}_2, \ldots, \vec{e}^{A_2}_5$ and called PC modes. The data matrix $\mathbf{B}$ can be reconstructed from the total five PC modes:

$$\mathbf{B} = \mathbf{EU} = \sum_{i-1}^{5} \vec{e}^{A_2}_i \vec{u}^{A_2}_i \tag{6.20}$$

where $\vec{u}^{A_2}_i = [u^{A_2}_{i,1}, u^{A_2}_{i,2}, \ldots, u^{A_2}_{i,N}]$ is the PC coefficient for the ith PC mode. The high values (greater than 0.4) of the PC coefficient for the first PC mode, $\vec{u}_1$, indicate the presence of RFI.

As the large-scale weather systems have been mostly captured in $\mathbf{A}_1$, the remaining small-scale weather variability associated with wind, cloud, and precipitation and RFI signals remains in $\mathbf{A}_2$. In order to effectively distinguish between false alarms associated with small-scale weather features and RFI signals, an additional condition on satellite glint angle is added. RFI occurs only when the RFI glint angle is small enough so that the TV broadcast signals can be reflected into AMSR-E IFOV. The satellite glint angle being smaller than 50° is used as an additional criterion for the identification of RFI signals.

Figure 6.9a provides a spatial distribution of horizontally polarized brightness temperatures at 10.65 GHz on February 16, 2011 over ocean around Europe. Brightness temperatures at this channel vary from 80 to 170 K. Warmer temperatures are found in the coastal areas in the Mediterranean Sea and B area in Figure 6.9a as well as over the Atlantic Ocean. The latter shows some spatial patterns closely related to variations in ocean surface wind, cloud, and rains. In order to see the temperature differences among various AMSR-E channels in different regions, we select two pairs of data points: points A and B located at the same latitude 45.4N and points C and D located at the same latitude 39.2N. In general, brightness temperatures increase with frequency, and the horizontally polarized brightness temperatures are lower than those at vertical polarization (Figure 6.9b,c). Brightness temperatures at all frequencies at a clear-sky data point A located at (45.4N, 33.0W) in the middle of Atlantic ocean are consistently warmer than those at point B except for 10.65 GHz channels (Figure 6.9b), indicating a potential RFI at 10.65 GHz at point B located at (45.4N, 2.1W). Brightness temperatures at all AMSR-E frequencies at a clear-sky data point C located at (39.2N, 34.3W) in the middle of the Atlantic Ocean are consistently colder than those at point D located at (39.2N, 12.0W) except for 89 GHz channels (Figure 6.9c), indicating the presence of cloud and precipitation at point D. The cloud scattering has the largest impact on the highest AMSR-E frequency channels at 89.0 GHz. The scattering effects of cloud particles reduce the 89 GHz brightness temperatures at both vertical and horizontal polarization channels. Impact of cloud emission dominates the remaining AMSR-E channels. Due to low ocean emissivity and higher cloud-emitted radiation, the presence of

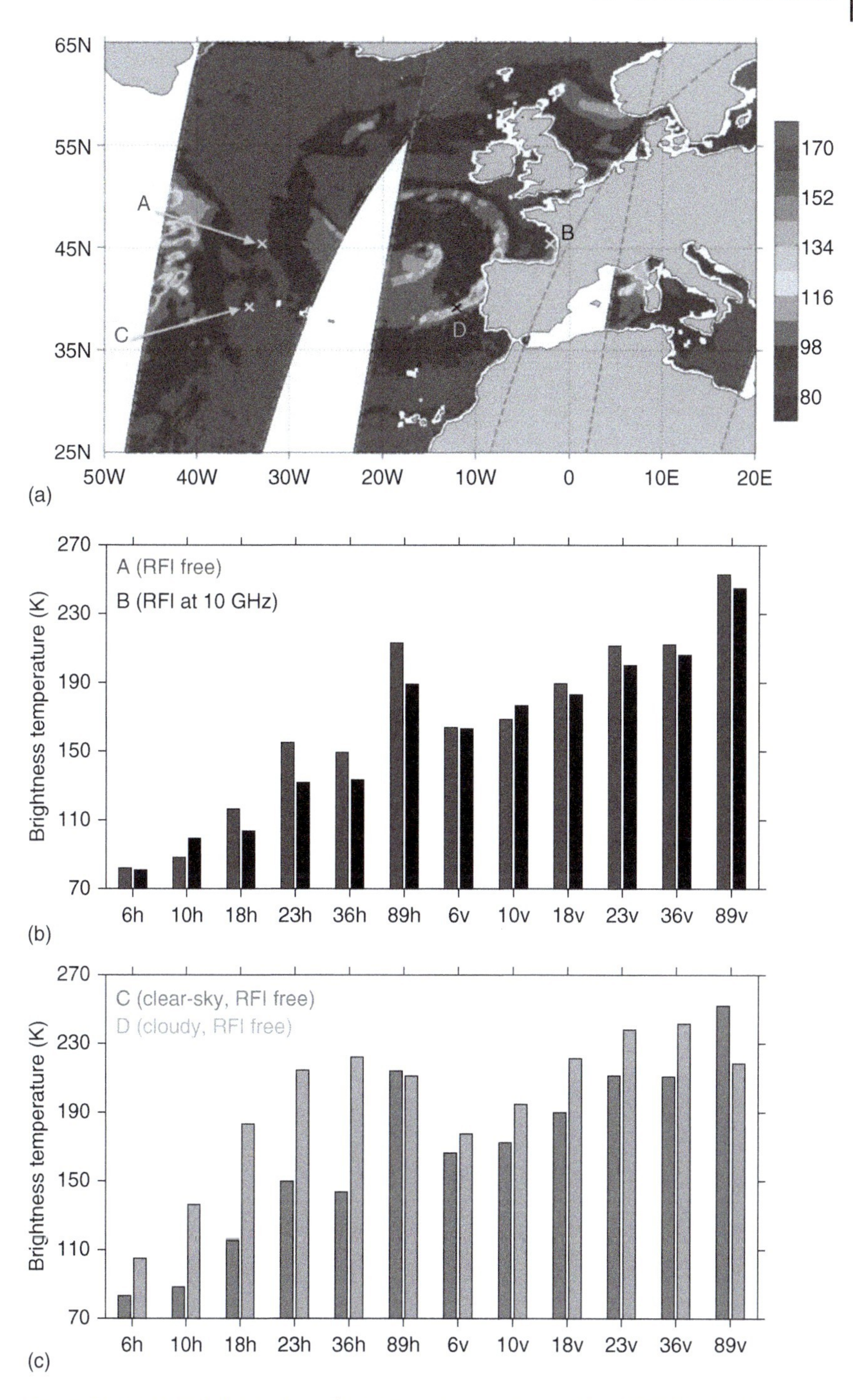

Figure 6.9 (a) Spatial distribution of brightness temperatures of the 10.65 GHz horizontally polarized channel on February 16, 2011, over oceans around Europe. Brightness temperatures of all AMSR-E channels at four arbitrarily chosen data points (b) A (left bars) and B (right bars), and (c) C (left bars) and D (right bars) on February 16, 2011. The geographic locations of points A–D are indicated in (a). (Zou *et al.* 2014 [173]. Reproduced with permission of American Meteorological Society.)

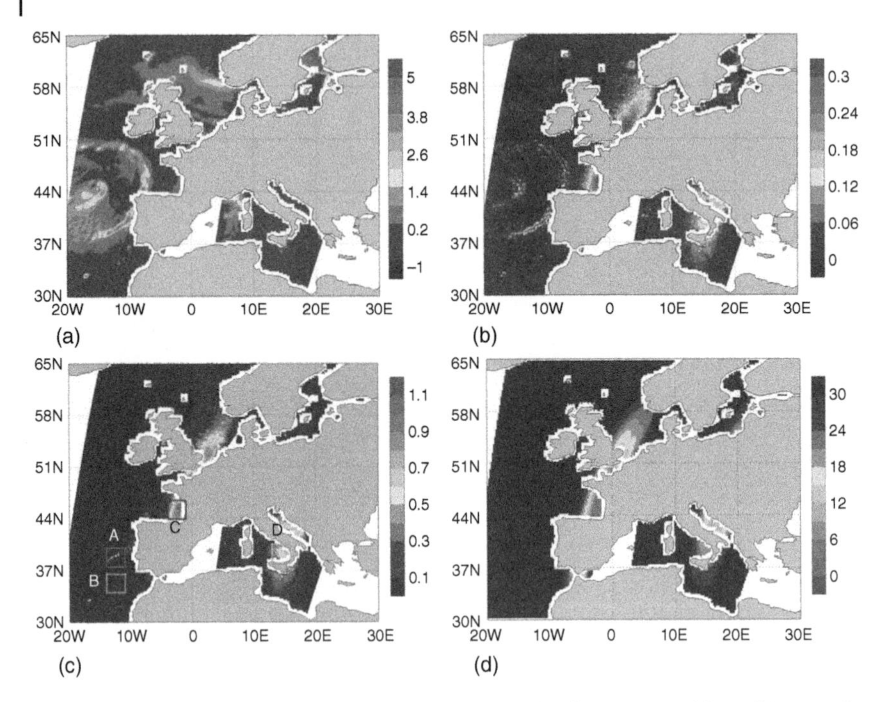

Figure 6.10 Brightness temperatures corresponding to the data matrices (a) $\mathbf{A}_1$, (b) $\mathbf{A}_2$, and (c) RFI signal intensity found by the DPCA method when $\alpha = 7$ for the 10.65 GHz horizontal polarization channel observations on February 16, 2011. (d) Satellite glint angle.

cloud increases the brightness temperatures at lower frequencies from 6.925 to 36.5 GHz.

Figure 6.10 shows the spatial distributions of brightness temperatures corresponding to the data matrices $\mathbf{A}_1$ (Figure 6.10a) and $\mathbf{A}_2$ (Figure 6.10b) and the RFI intensity map (Figure 6.10c) when the parameter α is set to 7 for the 10.65 GHz horizontal polarization channels, which can be compared to the spatial distribution of satellite glint angle (Figure 6.10d), in two descending passes on February 16, 2011, near Europe. It is seen that weather-related features are maintained mostly in the $\mathbf{A}_1$ matrix (Figure 6.10a). The RFI signals found by the DPCA method are consistent with the distributions of small glint angles (Figure 6.10d, which could be used to flag for RFI from the geostationary TV satellites) and are in the $\mathbf{A}_2$ matrix (Figure 6.10b). The DPCA successfully identifies oceanic RFI signals (Figure 6.10d) located mostly in the Mediterranean Sea, Bay of Biscay, and North Sea.

The eigenvalues and eigenvectors of the covariance matrix $\mathbf{AA}^\mathrm{T}$ are also provided in Figure 6.11. The ith eigenvalue (Figure 6.11a) quantifies the variance in the data that is explained by the ith eigenvector ($i = 1, \ldots, 12$) (Figure 6.11b–d). It is seen that the first PC mode explains the largest amount of the total data variances, the second PC mode explains the second largest amount of the total data variances, and so on. The first seven PC modes capture the majority of data variances. When the RFI indices are not normalized, the first PC remains positive, reflecting

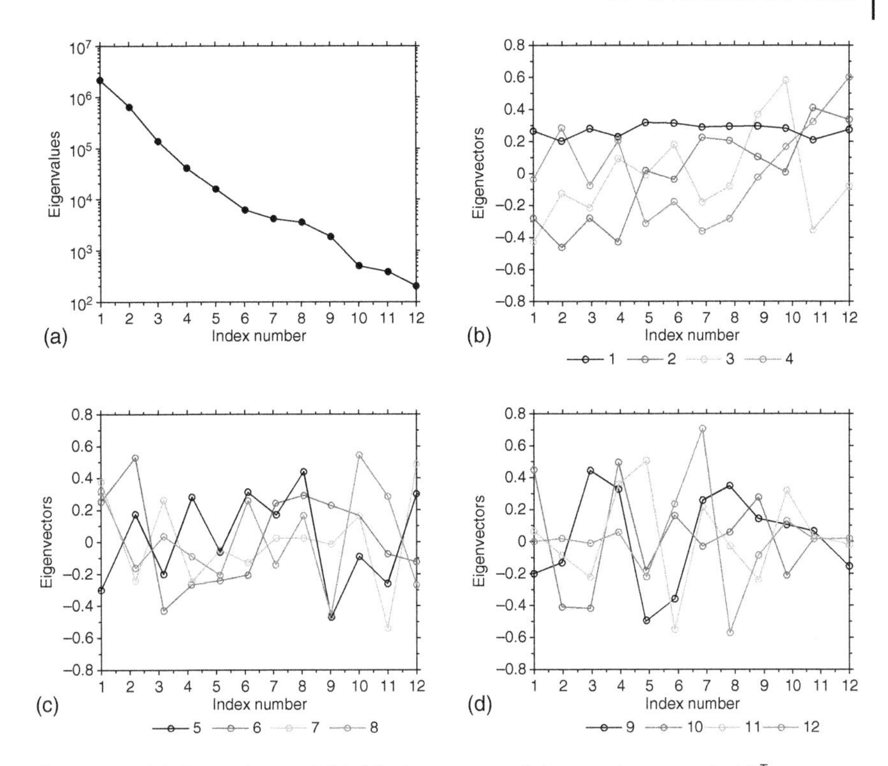

Figure 6.11 (a) Eigenvalues and (b)–(d) eigenvectors of the covariance matrix $\mathbf{AA}^T$ corresponding to data in Figure 6.10.

the overall magnitude of all radiometer channels. The variability of eigenvectors increases with the increase in mode number.

When applying the DPCA for identifying the RFI signals in AMSR-E data over ocean, a key parameter (i.e., α) must be determined for defining the $\mathbf{A}_2$ fields. Sensitivities of the DPCA method to parameter α for RFI detection could be assessed by examining the variation of the first PC coefficients with respect to parameter α in the second step of DPCA. Figure 6.12 presents such a variation for RFI detection of 10.65 GHz horizontally polarized channel in four selected regions indicated in Figure 6.10c. The first PC coefficient in both C and D regions has a rapid increase when α is equal to 6–8, suggesting the appropriateness of choosing $\alpha = 7$ for the RFI detection using the DPCA method, since large values of the first PC coefficient suggest higher probabilities of the presence of RFI signals. The first PC coefficient in box A, which is located in a cloud system, peaks at $\alpha - 3$. By choosing $\alpha - 7$, the weather-related signals are removed for the RFI detection in the second step of the DPCA analysis. A variation of the first PC coefficient with parameter α under a clear-sky condition (box B) is also provided as a benchmark. It is seen that the first PC coefficient decreases nearly monotonically with respect to α under a clear-sky condition, which is very different from the cases when either RFI signals or clouds are present.

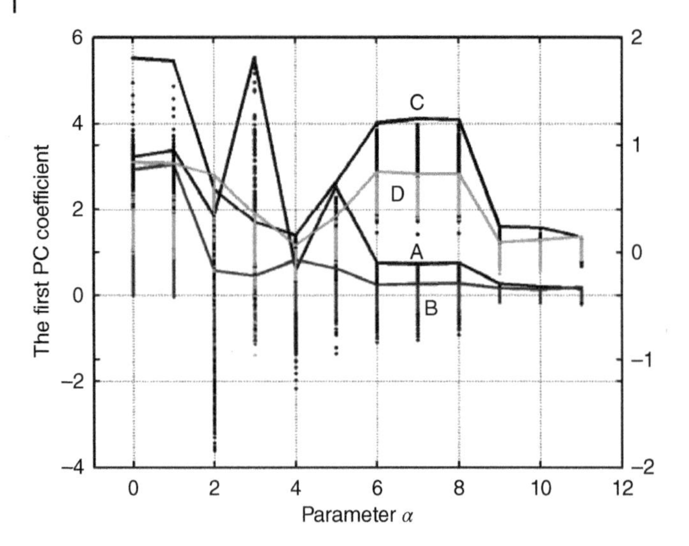

Figure 6.12 Variations of the first PC coefficient with respect to parameter α used in the DPCA method for all the data points averaged in a cloudy box A, a clear-sky box B, as well as two boxes C and D with RFI signals detected by the DPCA using $\alpha = 7$, respectively. Boxes A–D are indicated in Figure 6.10c.

A quantitative relationship between the RFI signal strength in 10.65 GHz horizontal polarization channel and the magnitude of the spectral differences between this channel and its neighboring higher frequency channels at the same polarization on February 16, 2011, is provided in Figure 6.13b. For convenience, spectral differences between 10.65 and 18.7 GHz with horizontal polarization (i.e., $T_{b,\,10.65h} - T_{b,\,18.7h}$) are also provided (Figure 6.13a). Although the spectral difference method could in fact detect much of the ocean RFI signals seen in Figure 6.13a, it also misidentified a cloud weather system west of 8W between 37 and 51N as "RFI" signals. Similarly, Figure 6.14 shows the spatial distributions of RFI signals detected at 18.7 GHz with horizontal polarization by the DPCA (Figure 6.14a), spectral differences between 18.7 and 23.8 GHz with horizontal polarization (i.e., $T_{b,\,18.7h} - T_{b,\,23.8h}$, Figure 6.14b), and a scatter plot between RFI signals and the spectral differences (Figure 6.14c) for data on February 16, 2011. In general, the spectral differences increase with the RFI intensity. However, the spectral differences for those observations without RFI could have the same magnitude as those with strong RFI intensity, suggesting a need to apply the DPCA method for RFI detection.

Figure 6.15 presents daily and accumulative RFI intensity maps at 10.65 GHz horizontal polarization channel for the period of February 5–12, 2011 around west Europe. The exact geographical distribution of AMSR-E swath over west Europe varies daily. The geographical areas of high RFI potential shifting with respect to the location within the AMSR-E scan suggest a fixed, directional source.

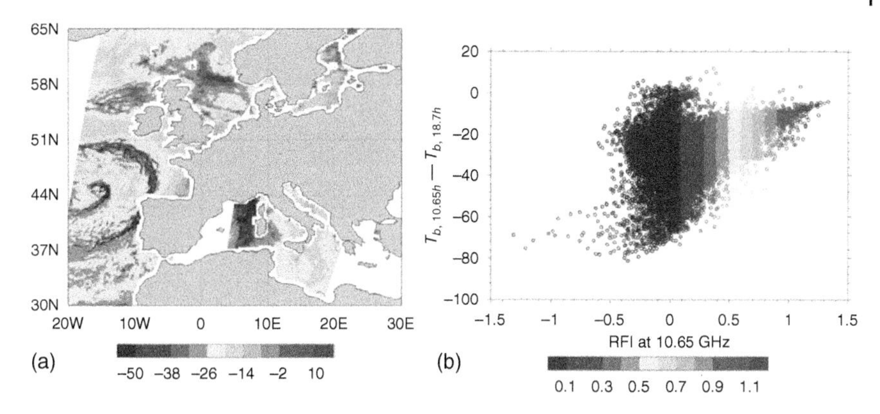

Figure 6.13 (a) Spectral differences between 10.65 and 18.7 GHz with horizontal polarization (i.e., $T_{b,\ 10.65h} - T_{b,\ 18.7h}$) on February 16, 2011 and (b) scatter plot between RFI signals shown in Figure 6.10c, and the spectral differences shown in (a). (Zou *et al.* 2014 [173]. Reproduced with permission of American Meteorological Society.)

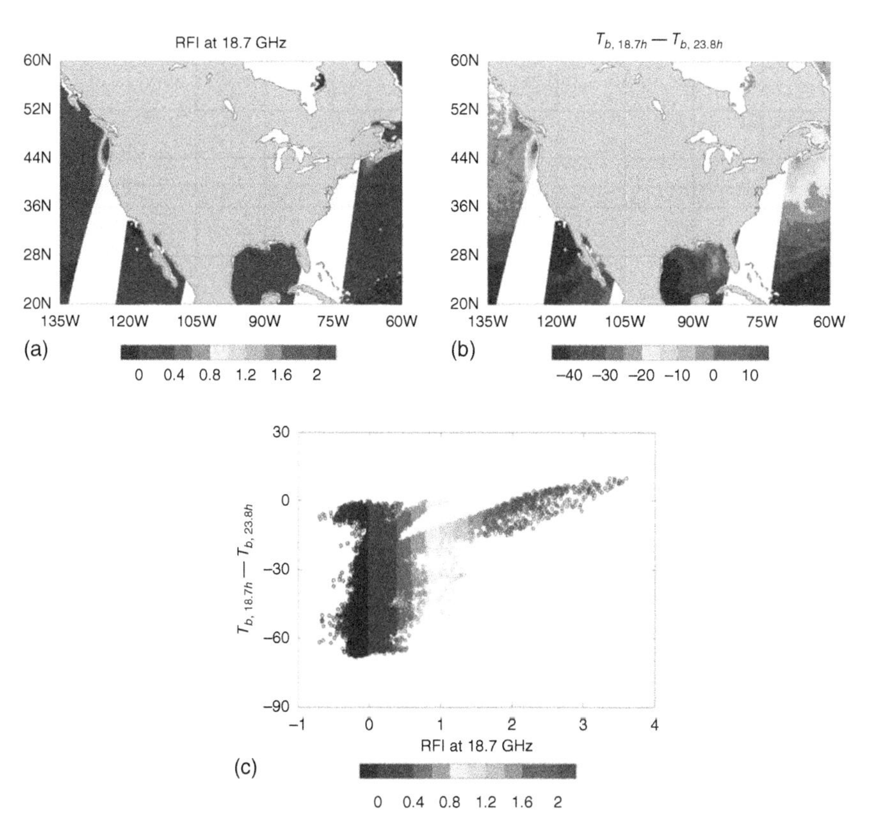

Figure 6.14 (a) RFI signals detected at 18.7 GHz with horizontal polarization by the DPCA, (b) spectral differences between 18.7 and 23.8 GHz with horizontal polarization (i.e., $T_{b,\ 18.7h} - T_{b,\ 23.8h}$), and (c) scatter plot between RFI signals shown in (a), and the spectral differences shown in (b) for data on February 16, 2011.

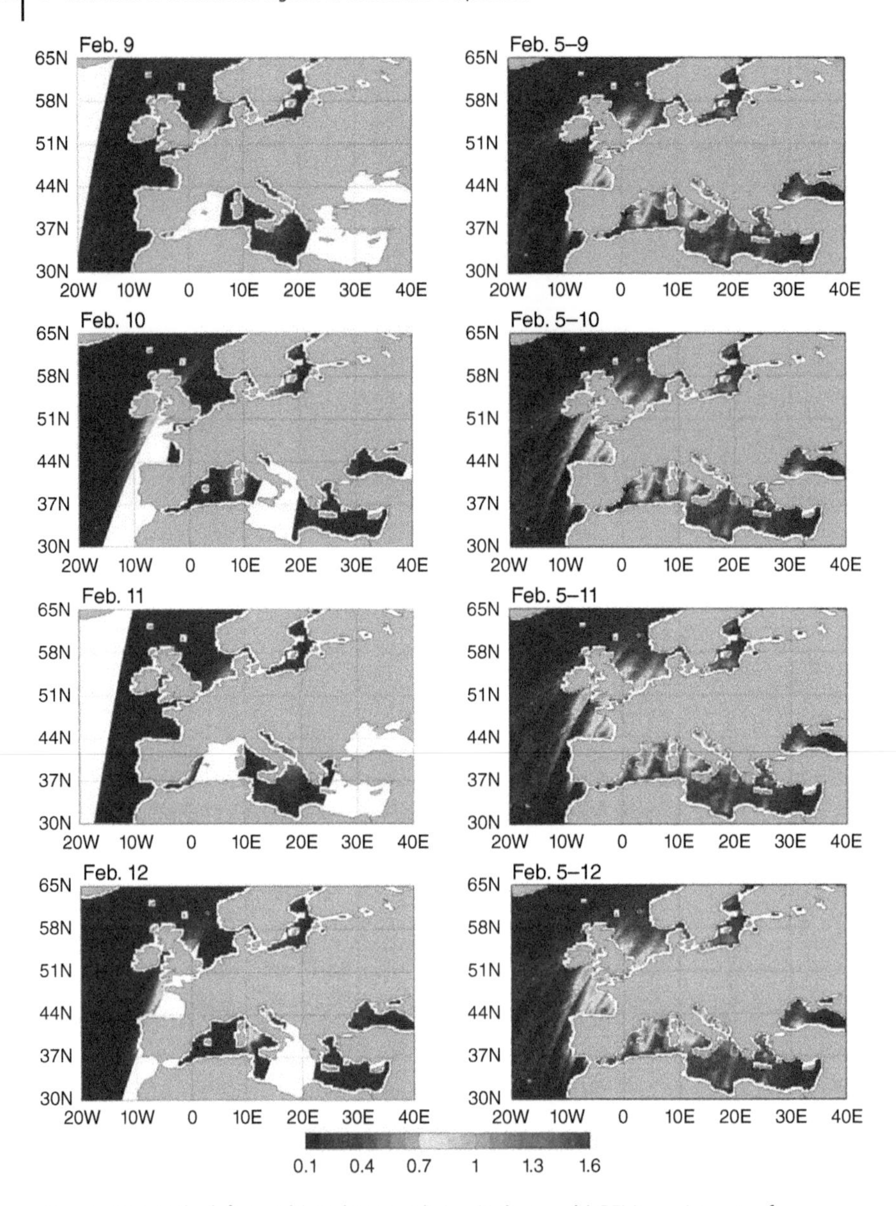

Figure 6.15 Daily (left panels) and accumulative (right panels) RFI intensity maps for 10.65 GHz horizontal polarization during February 5–12, 2011 around Europe. (Zou *et al.* 2014 [173]. Reproduced with permission of American Meteorological Society.)

RFI signals are found in areas with small satellite glint angles [6]. Strong RFI signals are not found when the strength of the TV signal is weak even if the satellite glint angle is small.

The RFI signals are seen in the Bay of Biscay and North Sea on the east side of those AMSR-E swaths west of 13E, the Mediterranean Sea, the Adriatic Sea along the longitude 13E independent of AMSR-E FOVs, or on the west side of the AMSR-E swaths east of 13E. The likely sources of these RFI signals are broadcasting signals from European geostationary television (TV) satellites. The European FSS satellites, such as Hot Bird 6, 7A, or 8 operated by Eutelsat1, transmit signals within the bands from 10.7 to 12.75 GHz, which are very close to AMSR-E channels at 10.65 GHz. The geostationary satellites identified as Hot Bird 6, 7A, or 8 were positioned at 13E longitude, which are reflected off the ocean surfaces and received together with the Earth's passive microwave radiation by AMSR-E. The RFI locations and strength are only found in the descending pass of AMSR-E because the TV satellite signals are in the forward-looking direction of AMSR-E. RFI signals are quite persistent for every AMSR-E overpass with similar observation geometry, eliminating any other possibility such as mobile-source- or weather-system-induced false alarms over the ocean.

The RFI distribution in the US coast is shown in AMSR-E 18.7 GHz (Figure 6.16). There are two DirecTV satellites, DirecTV 10 and DirecTV 11, which transmit to the United States in AMSR-E's K-band (18.7 GHz) channel. DirecTV 10 and DirecTV 11 are positioned at 103W [175] and 99W [165], respectively, above the equator. Both satellites use a nationwide beam for general broadcasting and multiple spot beams for local high-definition channels. The nationwide beam for general broadcasting operates from 18.3 to 18.6 GHz, and multiple spot beams for local high-definition channels operate from 18.6 to 18.8 GHz. It is anticipated that the AMSR-E channels at 18.7 GHz are interfered with DirecTV signals reflected off the ocean surfaces around the US coastal areas. Figure 6.16 shows daily and accumulative RFI intensity maps at 18.7 GHz horizontal polarization during February 5–12, 2011 around the United States. The RFI-contaminated AMSR-E data are found near the east edge of the AMSR-E swath west of 99W or 103W and on the west side of the AMSR-E swath east of 99W or 103W around the coastal areas. Such a characteristic of RFI contamination of AMSR-E data is determined by the geometric relationship between Aqua and DirecTV satellites. Therefore, there are many RFI-free AMSR-E data over coastal areas, depending on their scan positions.

When examining the monthly average global RFI intensity maps for AMSR-E 18.7 GHz channels for all descending portions of AMSR-E orbits during February 1–18, 2011, RFI signals are found only in the east coast, west coast, and Gulf of Mexico coast. It is therefore important to separate weather signals from RFI signals in these coastal regions before AMSR-E data is used for geophysical parameter retrieval and data assimilation.

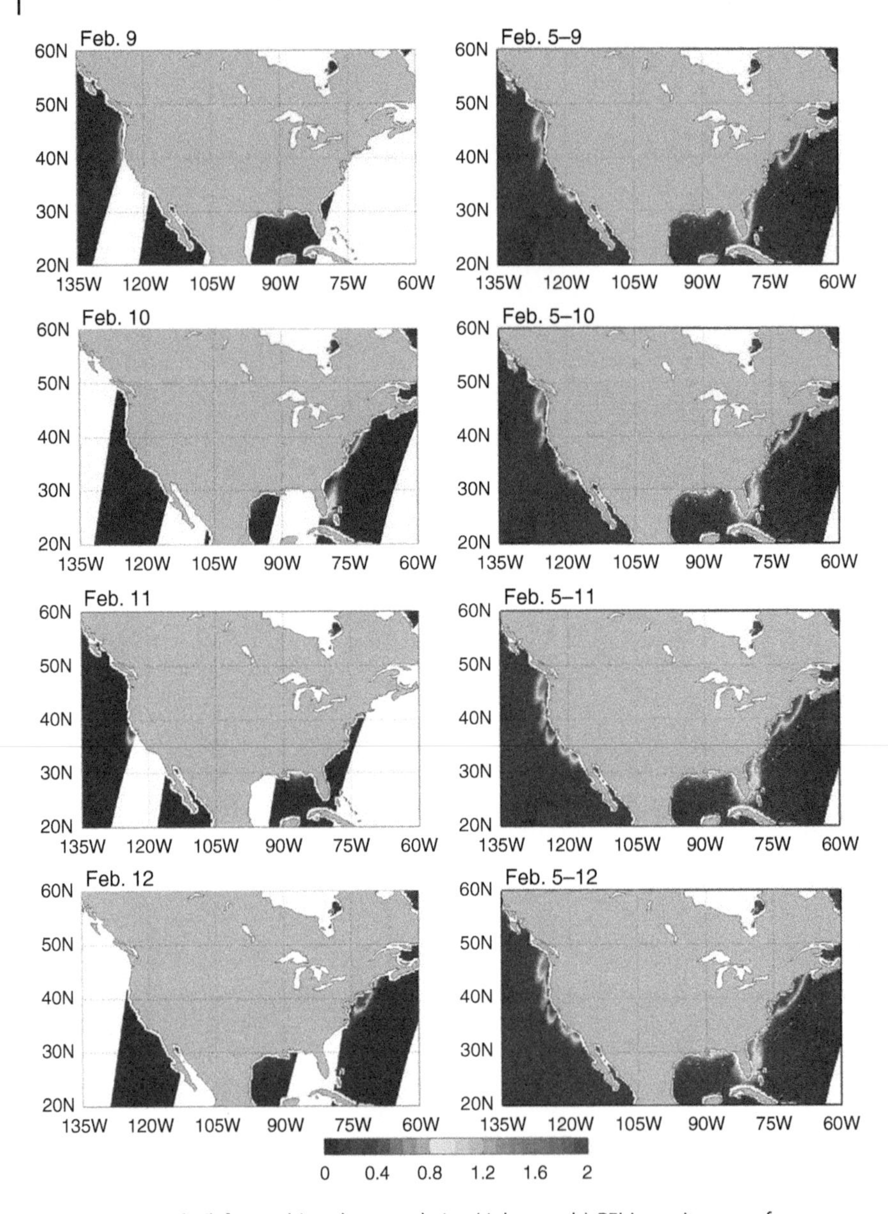

Figure 6.16 Daily (left panels) and accumulative (right panels) RFI intensity maps for 18.7 GHz horizontal polarization during February 5–12, 2011 around United States. (Zou *et al.* 2014 [173]. Reproduced with permission of American Meteorological Society.)

6.6 Summary and Conclusions

RFI detection for satellite low-frequency microwave imager radiances over land is extremely important before these data could be used for either geophysical retrieval or data assimilation in numerical weather prediction models. In addition to using the traditional spectral difference method, a generalized DPCA method is utilized for detecting RFI signals in WindSat, AMSR-E, and other microwave imagers over global areas including Greenland, Antarctic, and the United States.

A strong RFI is seen at C-band channels for both horizontal and vertical polarization over Greenland. Over Antarctic, strong RFI is also found at both C- and X-bands. The RFI signals are more populated over areas where research stations exist over Greenland and Antarctic. Strong RFI signals are populated over the cities of United States at the 6.8 GHz horizontal polarization channel.

The DPCA works at any geographical location over the globe and on both non-scattering and scattering surfaces. It is also tested that the double PCA can be applied at the granule data level, offering a real-time RFI detection method before C- and X-band data are delivered to the users.

In order to mitigate the RFI in C-band channels, two new C-band channels centered at 7.3 GHz are added to the AMSR-2. These C-band channels can be used for validating RFI signals derived from the spectral difference channels over the United States and Central American Continents. For the case studied, a strong RFI signal is detected at the AMSR-2 C-band channels at 6.925 GHz for both horizontal and vertical polarization over North America. The RFI signals are populated near the metropolitan areas of the United States. However, the newly added C-band channels at 7.3 GHz are mostly RFI-free, except in Mexico, Washington DC, and New York. There are no RFI over Mexico at 6.925 GHz for both polarization states. The only places where RFI occurs at both C-bands of AMSR-2 are Washington DC and New York, for the horizontal polarization state. Thus, it can be concluded that a successful mitigation of RFI is achieved in AMSR-2 observations over the United States and Central American Continents.

RFI detection for satellite low-frequency microwave imager radiances over ocean is challenging. The DPCA method is proposed for detecting RFI signals in AMSR-E data over oceans. A strong RFI is visible for X- and K-band channels for both horizontal and vertical polarization over ocean near coastal regions. Consistent with the cause of the ocean RFI signals, measurements of the natural thermal emission from AMSR-E satellite over ocean are interfered by the geostationary satellite signals reflected off the ocean surfaces. Strong RFI signals are populated along the east, south, and west coastal areas of the United States in the AMSR-E K-band data in descending nodes. There are also very strong oceanic RFI signals in the AMSR-E X-band data in the Mediterranean Sea, the Adriatic Sea north of Italy, and around Sicily. The RFI locations are also quite persistent for every AMSR-E descending swath passing over these regions with similar observation geometry. It is believed that the likely sources of the oceanic RFI occurrences come from the broadcasting signals from European geostationary satellites above the equator.

7
Microwave Remote Sensing of Surface Parameters

7.1
Introduction

Sea surface temperature (SST) and sea surface wind (SSW) are critical parameters affecting the formation and development of tropical cyclones. Warm oceans having temperatures greater than 26.5 °C are essential for the genesis of a tropical cyclone [176, 177]. This is because only such warm water can provide sufficient sensible and latent heats. In addition, SSW provides an essential link of heat transfer from the surface into marine boundary layer. The stronger the surface winds, the more effective the heat transfer from the ocean [178]. For tropical systems, the maximum SSW also indicates their intensity: a tropical depression has a maximum sustained surface wind of less than 17 m/s, a tropical storm has at least 17 m/s, and a hurricane has at least 33 m/s.

SST and SSW can be observed from *in situ* sensors on ships, buoys, and remotely from satellites and airborne radiometers. With satellite infrared measurements, SSTs are retrieved with high accuracy (e.g., the root-mean-square errors better than 0.5 K) under mostly clear-sky conditions. However, the infrared retrievals become unreliable under cloudy and high aerosol conditions [179–181]. By using satellite microwave measurements, the retrievals are extended to cloudy conditions because microwave radiation emitted from oceanic surface can penetrate through cloudy atmospheres. With the special sensor microwave imager (SSM/I) measurements, SSW speed is derived under clear-to-cloudy conditions [76]. Since the SSM/I frequency ranges from 19 to 85 GHz and the measurements tend to be saturated by high winds, the retrievals for wind speeds above 15 m/s are rather difficult [76, 182]. Although the SSM/I SSW algorithm is calibrated and validated using globally comprehensive buoy data sets [151], the retrievals are applicable in rain-free atmospheres. Using the tropical rainfall measuring mission (TRMM) tropical microwave imager (TMI) observations, simultaneous retrievals of SST and SSW are improved primarily under nonprecipitating conditions [183–185]. In this chapter, the retrievals of SST and SSW in the presence of precipitating clouds are explored by using the measurements from Earth Observing System (EOS) aqua satellite advanced microwave scanning radiometer – EOS (AMSR-E) and Navy Coriolis WindSat.

Passive Microwave Remote Sensing of the Earth: For Meteorological Applications, First Edition. Fuzhong Weng.

Remotely sensed radiation at various wavelengths has been widely used to derive the surface temperature over land. The infrared split-window technique for obtaining the land surface temperature is based on the fact that transmission through atmospheres at one wavelength is closely correlated with that of a nearby wavelength (e.g., [186, 187]). Since the surface emissivity is essentially the same, the difference between two measurements in nearby channels is due to differential absorption by atmospheric water vapor. Following the removal of atmospheric effects, the land surface temperature is obtained using a priori estimates of surface emissivity. However, the accuracy of the land surface temperature is affected by the surface emissivity used in the algorithm. The studies also show that the infrared emissivity may vary significantly under certain conditions. The largest variation occurs in the spectral range of 8–10 μm. The land surface temperature based on infrared split-window techniques is also limited to clear-sky conditions. Microwave measurements are also used to derive land surface properties such as soil moisture [188–190], canopy cover [191], and surface temperature [192–194]. In comparison with the infrared methods, the microwave measurements can provide useful information on the land surface properties under nearly all weather conditions. One of the unique microwave sensors used for land surface studies is the SSM/I. Here, two SSM/I Ku-band channels are utilized for physical retrieval of land surface temperature.

In addition to the retrievals of land surface temperature, land surface emissivity can be derived from microwave measurements [105, 107, 108, 195]. In this chapter, an analytic scheme is used to demonstrate the process for deriving the land surface emissivity from microwave measurements.

7.2 Remote Sensing of Ocean Surface Parameters

7.2.1 Retrievals of Surface Wind Vector

For an atmosphere in the absence of scattering and having an emission azimuthally independent, radiative transfer equations can be simplified as [196]

$$T_v = \varepsilon_v T_s \Upsilon + T_u + (1 - \varepsilon_v) T_d, \tag{7.1}$$

$$T_h = \varepsilon_h T_s \Upsilon + T_u + (1 - \varepsilon_h) T_d, \tag{7.2}$$

$$T_3 = \varepsilon_3 T_s \Upsilon, \tag{7.3}$$

$$T_4 = \varepsilon_4 T_s \Upsilon, \tag{7.4}$$

where T_s is the surface temperature, T_u and T_d are the upwelling and downwelling radiation, $\Upsilon = \exp(-\tau_s/\mu)$ is the atmospheric transmittance, $\varepsilon_v, \varepsilon_h, \varepsilon_3$, and ε_4 are the components of the surface emissivity vector. Note that the specular surface is assumed for approximating the reflection in Eqs. (7.1–7.4). The third and fourth

Stokes components of brightness temperatures are only contributed by the surface, and they are attenuated by the atmosphere.

Eqs. (7.1–7.4) have six unknown parameters. SST is assumed to be obtained from the operational SST product [197]; therefore, SST is treated as a known variable. The upward and downward atmospheric radiations at frequencies lower than 37 GHz are almost identical [196] with the correlation coefficient of 0.99 and a maximum difference of 0.2 K, and thus, they are denoted as variable T_a hereafter. This high correlation is due to the fact that most of the water vapor locates near the surface, thereby reducing the difference between the SST and the effective atmospheric temperature. In addition, atmospheric radiation is not sensitive to the vertical distribution of the atmospheric temperature because the absorption coefficient decreases slightly with temperature. Thus, the six unknown parameters are reduced to four unknown parameters, which lead to a closure in the physical inversion.

Using Eqs. (7.1–7.4), the inversion equations can be also derived as

$$\begin{pmatrix} \Delta w \\ \Delta\phi \\ \Delta\Upsilon \\ \Delta T_a \end{pmatrix} = (\mathbf{A}^t\mathbf{A} + \mathbf{E})^{-1}\mathbf{A}^t \begin{pmatrix} \Delta T_v \\ \Delta T_h \\ \Delta T_3 \\ \Delta T_4 \end{pmatrix}, \tag{7.5}$$

where

$$\mathbf{A} = \begin{pmatrix} \Upsilon\left(T_s - T_a\right)\dfrac{\partial\varepsilon_r}{\partial w} & \Upsilon(T_s - T_a)\dfrac{\partial\varepsilon_r}{\partial\phi} & T_a(1-\varepsilon_r) + T_s\varepsilon_r & 1+\Upsilon(1-\varepsilon_r) \\ \Upsilon(T_s - T_a)\dfrac{\partial\varepsilon_l}{\partial w} & \Upsilon(T_s - T_a)\dfrac{\partial\varepsilon_l}{\partial\phi} & T_a(1-\varepsilon_l) + T_s\varepsilon_l & 1+\Upsilon(1-\varepsilon_l) \\ \Upsilon T_s\dfrac{\partial\varepsilon_3}{\partial w} & \Upsilon T_s\dfrac{\partial\varepsilon_3}{\partial\phi} & T_s\varepsilon_3 & 0 \\ \Upsilon T_s\dfrac{\partial\varepsilon_4}{\partial w} & \Upsilon T_s\dfrac{\partial\varepsilon_4}{\partial\phi} & T_s\varepsilon_4 & 0 \end{pmatrix}, \tag{7.6}$$

where the matrix **E** represents an error matrix, depending on the measurement and radiative transfer model errors.

The inversion equation can be solved using an iterative method. For a set of microwave frequencies, an initial set of atmospheric and surface parameters (e.g., $w_0, \phi_0, T_{a0}, \Upsilon_0$) is used to calculate the brightness temperatures at the four Stokes components. Then, the increment vector between the observed brightness temperatures and simulated values are derived. The matrix **A** in Eq. (7.6) can also be calculated from the initial parameters. The emissivity gradients are derived from the theoretical emissivity model. The increment vector is then used to determine the change in the atmospheric and surface parameters. The iteration continues until the retrieval converges.

The variation of the Stokes vector to the wind direction is generally less than 3 K. The amplitudes of the variation for the vertically and horizontally polarized components are about 2 K. The amplitudes of the variation are about 3 and 0.5 K for the third and the fourth components of Stokes vector, respectively. For the retrieval of the wind direction, the algorithm must be designed carefully for convergence with accurate knowledge of instrument error and radiative transfer model errors through the matrix **E**.

Since the retrieval is a nonlinear process and requires an iterative process, careful selections of initial values can result in fast convergence. An initial value of the wind speed is derived from the regression equation developed by Goodberlet [76]. The total precipitable water vapor is based on an algorithm by Petty [182]. For the wind direction, it is found that the ratio of the third to the fourth Stokes component can be used (see Figure 7.1). Using the two-scale emissivity model, the wind direction is in the range of $0°-90°$ or $270°-360°$ when the ratio at 37 GHz is less than -4. Wind direction is within $90°-270°$ as the ratio is greater than -2. The phase of the third and fourth Stokes component allows for further separation of the wind direction. The third component is typically negative in the first quadrant ($0°-90°$) and positive in the fourth quadrant ($270°-360°$). The fourth Stokes component is negative in the second quadrant ($90°-180°$) and positive in the third quadrant ($180°-270°$). Thus, the ratio of the third and the fourth components and

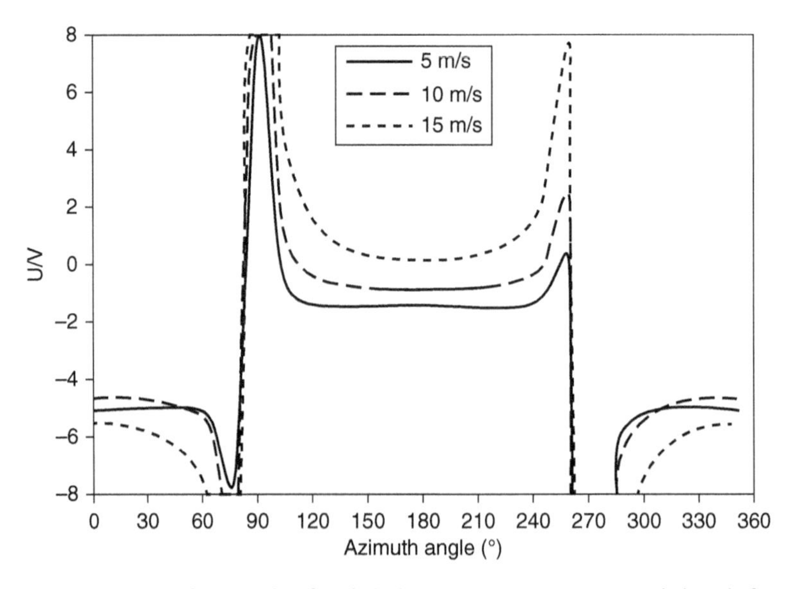

Figure 7.1 Ratio of the third to fourth Stokes component versus azimuthal angle for a local zenith angle of 53°. Solid, dashed–dotted, and dotted lines denote the wind speed of 5, 10, and 15 m/s, respectively. (Liu and Weng 2003 [196]. Reproduced with permission of Wiley.)

themselves provides the first guess of the wind direction with a sufficient accuracy. If the third and fourth components are approximated as

$$T_3 = U_1 \sin\phi + U_2 \sin 2\phi, \tag{7.7}$$

$$T_4 = V_1 \sin\phi + V_2 \sin 2\phi, \tag{7.8}$$

then,

$$\phi = \arccos\left[\frac{T_3 V_1 - T_4 U_1}{2\left(T_4 U_2 - T_3 V_2\right)}\right]. \tag{7.9}$$

It is important to know that this algorithm is based on radiative transfer equation without atmospheric scattering. In the presence of large raindrops or ice particles, brightness temperatures may be affected by particle scattering. To quantify the raindrop-induced scattering effects, the difference between the brightness temperatures computed from the emission-based RTM and those from the complete radiative transfer process is calculated. Here, the vector RTM (scattering radiative transfer model (RTM)) developed by Weng and Liu [46, 47] is used to simulate the brightness temperature in the scattering atmosphere. The vector RTM is for a vertically stratified scattering and emitting atmosphere, where the optical parameters for cloud drops, raindrops, and ice particles are calculated using the Mie theory [30] as discussed in Chapter 3. In our calculations, cloud and rain water contents are vertically distributed between 3 and 6 km. The raindrop size is distributed according to the Marshall–Palmer function [129] with an effective diameter of 0.5 mm. The brightness temperatures from the emission-based RTM can be readily computed from the full RTM by simply setting a zero value of single-scattering albedo at each model layer. Figure 7.2 shows the scattering effects at 6.925 and 10.65 GHz, respectively. Note that scattering intensity increases as frequency and liquid water increase: at 6.925 GHz, the scattering intensity of clouds and precipitation is typically small and less than 1 K, whereas the magnitude at 10.65 GHz can be several degrees in Kelvin. It is concluded that under a raining atmospheric condition, the scattering effect at 10.65 GHz remains important and must be taken into account in the retrievals.

The retrieval algorithm is applied to the WindSat datafor ocean wind vector retrieval. WindSat is a microwave polarimetric radiometer developed by the Naval Research Laboratory and designed to demonstrate the capability of polarimetric microwave radiometry to measure the ocean surface wind vector from space. It is the primary payload on the WindSat/Coriolis mission, which is jointly sponsored by the DoD Space Test Program and the U.S. Navy. The satellite was launched on a Titan II rocket from Vandenberg Air Force Base on January 6, 2003. The environmental parameters derived from WindSat include SST, total precipitable water, integrated cloud liquid water, and rain rate over the ocean. Table 7.1 provides the

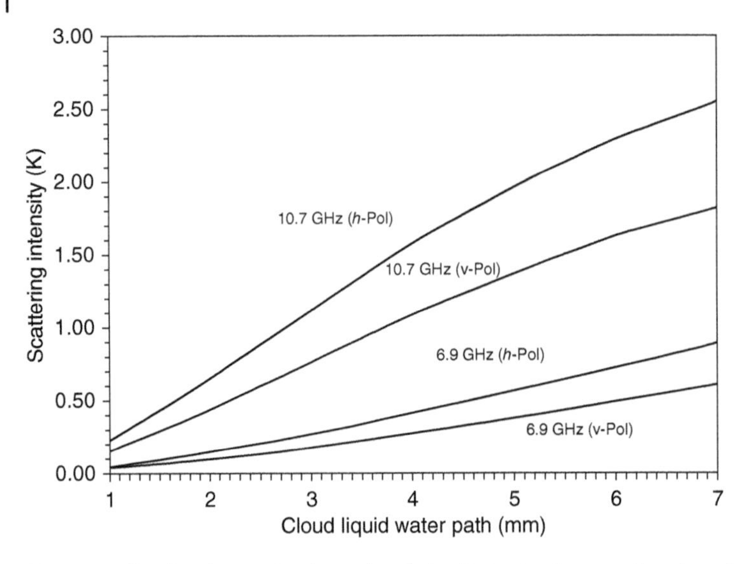

Figure 7.2 Simulated scattering intensity of cloud/rain droplets as a function of cloud liquid water path. (Yan and Weng 2008 [198]. Reproduced with permision of Springer.)

Table 7.1 WindSat instrument characteristics and parameters.

Frequency (GHz)	Channel	BW (MHz)	τ (ms)	NEDT	EIA (°)	IFOV (km)
6.8	v, h	125	5.0	0.48	53.5	40 × 60
10.7	v, h, ±45,lc,rc	300	3.5	0.37	49.9	25 × 38
18.7	v, h, ±45,lc,rc	750	2.00	0.39	55.3	16 × 27
23.8	v, h	500	1.48	0.55	53.0	12 × 20
37.0	v, h, ±45,lc,rc	2000	1.00	0.45	53.0	8 × 13

key instrument performance parameters. The 10.7, 18.7, and 37.0 GHz channels are fully polarimetric. At 6.8 and 23.8 GHz, the measurements under both vertical and horizontal polarization are performed. At the remaining frequencies, the brightness temperatures are measured under vertical and horizontal polarization as well as linear polarization of ±45° and circular polarization (left and right).

WindSat has a 1.8 m offset antenna and 11 feed horns feeding from the antenna. The antenna scans the Earth at various incidence angles ranging from 50 to 55°. The satellite orbits the Earth at an altitude of 840 km in a Sun-synchronous orbit and completes over 14 orbits per day. The orbit and antenna geometry result in a forward-looking swath of approximately 1000 km and an aft-looking swath of about 350 km. The fully integrated WindSat payload stands 10 feet tall and weighs approximately 675 lbs.

The retrieval was performed for Hurricane Isabel, which was a long-lived hurricane that reached Category 5 status on the Saffir–Simpson Hurricane Scale. It made landfall near Drum Inlet on the Outer Banks of North Carolina as a Category 2 hurricane. Isabel is considered to be one of the most significant tropical cyclones to affect portions of northeastern North Carolina and east-central Virginia since Hurricane Hazel in 1954 and the Chesapeake–Potomac Hurricane of 1933. Isabel formed from a tropical wave that moved westward from the coast of Africa on 1 September. Over the next several days, the wave moved slowly westward and gradually became better organized. By 0000 UTC 5 September, there was sufficient organized convection for satellite-based Dvorak intensity estimates to begin. Development continued, and it is estimated that a tropical depression formed at 0000 UTC 6 September, with the depression becoming Tropical Storm Isabel 6 h later.

Isabel turned west-northwestward on 7 September and intensified into a hurricane. Strengthening continued for the next 2 days while Isabel moved between west-northwest and northwest. Isabel turned westward on September 10 and maintained this motion until September 13 on the south side of the Azores–Bermuda High. Isabel strengthened to a Category 5 hurricane on September 11, with maximum sustained winds estimated at 145 kt at 1800 UTC that day. After this peak, the maximum winds remained in the 130–140 kt range until September 15. During this time, Isabel displayed a persistent 35–45 n mi diameter eye.

WindSat observed hurricane Isabel on September 13 as it entered its mature stage. As shown in Figure 7.3, Isabel surface vortex can be detected clearly by its third Stokes component as its magnitude changes from negative to positive. Retrieved wind vector from 10.7 GHz recovers all the wind vectors in relation to surface cyclonic circulation.

7.2.2 Simultaneous Retrieval of Sea Surface Temperature and Wind Speed

Microwave remote sensing of SST is primarily based on the measurements performed at lower frequencies where atmospheric scattering can be neglected. However, it is shown in Figure 7.4 that an increase in scattering and emission from large raindrops can result in an increase in brightness temperatures at 6.925 and 10.65 GHz by several degrees in Kelvin under severe weather conditions. These effects should be taken into account in the retrieval process.

The effects of nonspecular surface on the reflection term in Eq. (3.42) can also be included. Note that Ω is derived by Wentz [51] for AMSR-E instrument as

$$\Omega_v = [2.5 + 1.8 \times 10^{-2}(3.7 \times 10^1 - \nu)][\Delta^2 - 7.0 \times 10^1 \Delta\sigma^6]\Upsilon^{3.4}, \tag{7.10}$$

and

$$\Omega_h = [6.2 - 1.0 \times 10^{-3}(3.7 \times 10^1 - \nu)^2][\Delta^2 - 7.0 \times 10^1 \Delta\sigma^6]\Upsilon^2, \tag{7.11}$$

where

$$\Delta\sigma^2 = 5.22 \times 10^{-3}[1.0 - 7.48 \times 10^{-3}(3.7 \times 10^1 - \nu)^{1.3}], \quad \nu < 37 \text{ GHz}. \tag{7.12}$$

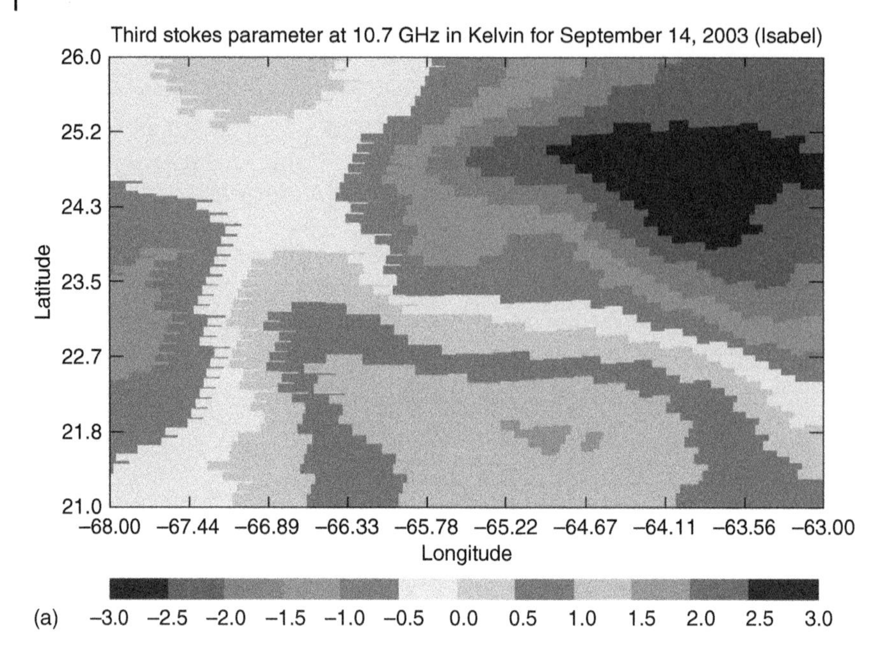

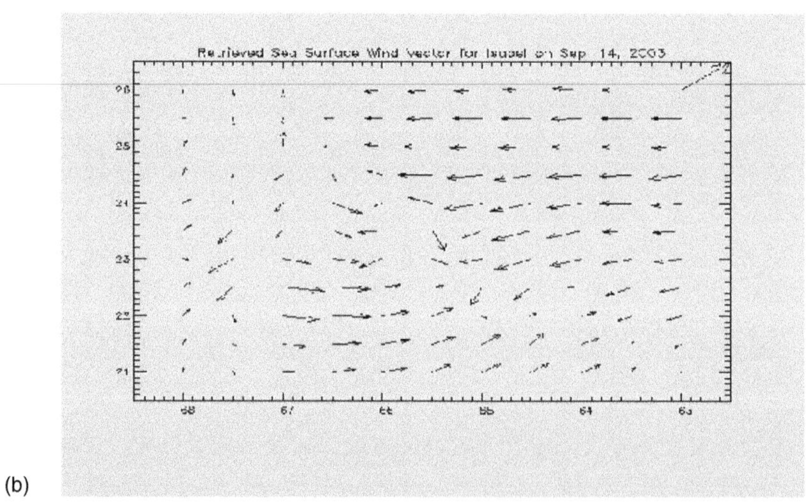

Figure 7.3 (a) WindSat third Stoke component for Hurricane Isabel and (b) derived wind field from all Stokes components at 10.7 GHz.

Using Eqs. (7.10–7.12), we can estimate the reflection magnitude contributed from nonspecular surfaces at 6.925 and 10.65 GHz. For a moderate wind of 10 m/s and relatively heavy raining atmosphere, Ω_v at 6.925 GHz is about 0.04 and Ω_h is 0.09, which result in an increase of about 0.40 and 1.28 K in V- and H-polarized brightness temperatures, respectively, while at 10.65 GHz, Ω_v, h are about 0.05 and 0.11, respectively, which result in an increase of about 0.89 and 3.18 K in V- and

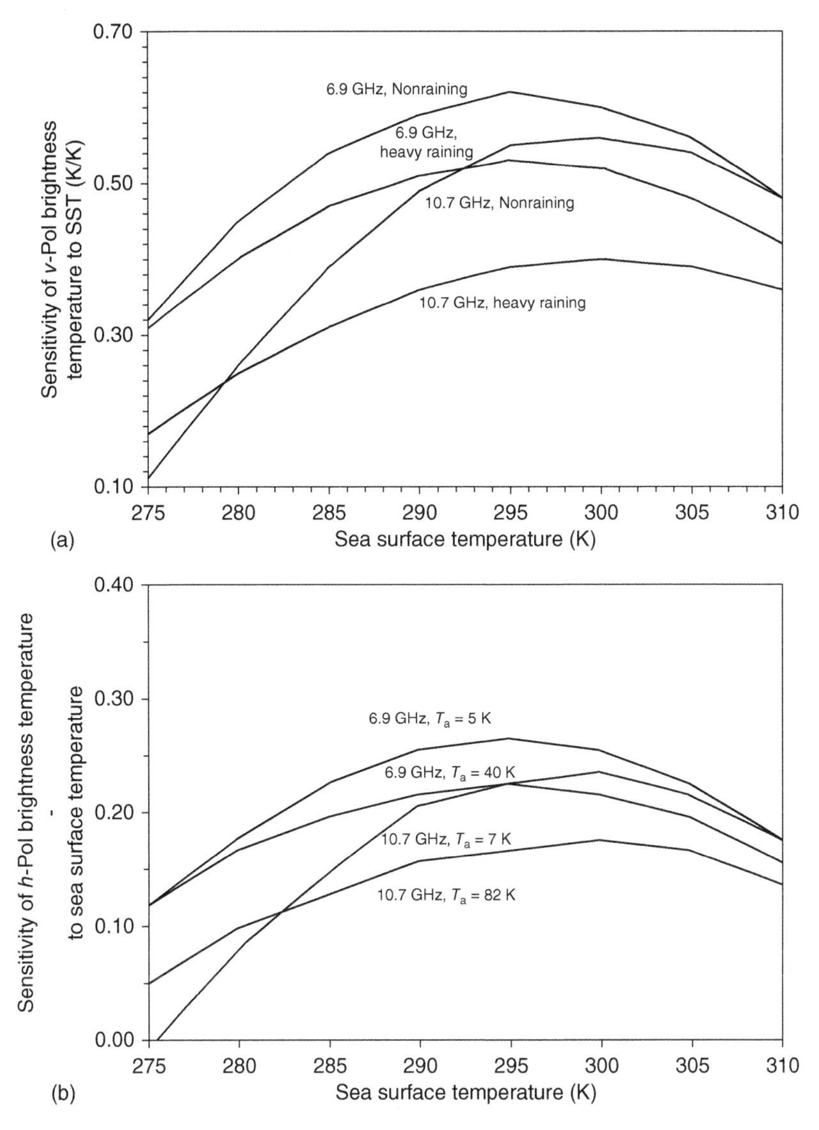

Figure 7.4 Sensitivity of brightness temperatures at 6.9 and 10.7 GHz of (a) v-Pol and (b) h-Pol for nonraining and heavy raining clouds versus sea surface temperature. Cloud liquid water path is 0.1 mm for nonraining clouds and 5 mm for heavy raining clouds. Surface wind speed is assumed to be 10 m/s. (Yan and Weng 2008 [198]. Reproduced with permision of Springer.)

H-polarized brightness temperatures, respectively. The contribution to brightness temperatures increases as the wind speed increases.

From Eq. (3.44), brightness temperatures at lower microwave frequencies are derived as a function of SST and emissivity. The emissivity is a function of surface wind and can be further used to retrieve wind speed. In general, brightness temperature increases as surface temperature increases. For example, the vertically polarized brightness temperature at 6.925 GHz can increase by 10–20 K as the surface temperature varies from 280 to 300 K. However, the sensitivity of brightness temperatures to SST decreases as atmospheric cloud water increases (see Figure 7.4). Thus, in SST retrieval, the correction for atmospheric emission at 10 GHz or higher is still significantly important for achieving a high accuracy.

To formulate a closure of equations for a set of retrieval unknowns, we utilize dual-polarization measurements from 6.925 and 10.65 GHz, which are the four AMSR-E channels. Using an approach similar to WindSat retrieval, we can derive

$$\begin{pmatrix} \Delta w \\ \Delta T_s \\ \Delta T_{a6} \\ \Delta T_{a10} \end{pmatrix} = (\mathbf{A}^t\mathbf{A} + \mathbf{E})^{-1}\mathbf{A}^t \begin{pmatrix} \Delta T_{v6} \\ \Delta T_{h6} \\ \Delta T_{v10} \\ \Delta T_{h10} \end{pmatrix}, \tag{7.13}$$

where $\mathbf{A}$ is a 4×4 matrix, and its elements are the partial derivatives of brightness temperatures relative to surface temperature and wind speed, atmospheric emission components at 6.925 and 10.65 GHz [198], which are

$$a_{11} = \varepsilon_{v6}\Upsilon_6 + \Upsilon_6[T_s - T_{a6}(1+\Omega_{v6})]\frac{\partial \varepsilon_{v6}}{\partial T_s}, \tag{7.14a}$$

$$a_{12} = \Upsilon_6[T_s - T_{a6}(1+\Omega_{v6})]\frac{\partial \varepsilon_{v6}}{\partial w} + (1-\varepsilon_{v6})\Upsilon_6 T_{a6}\frac{\partial \Omega_{v6}}{\partial w}, \tag{7.14b}$$

$$a_{13} = 1 + (1-\varepsilon_{v6})\Upsilon_6\left(1 + \Omega_{v6} + T_{a6}\frac{\partial \Omega_{v6}}{\partial T_{a6}}\right) + [\varepsilon_{v6}T_s + (1-\varepsilon_{v6})T_{a6}(1+\Omega_{v6})]\frac{\partial \Upsilon_6}{\partial T_{a6}}, \tag{7.14c}$$

$$a_{14} = 0, \tag{7.14d}$$

$$a_{21} = \varepsilon_{h6}\Upsilon_6 + \Upsilon_6[T_s - T_{a6}(1+\Omega_{h6})]\frac{\partial \varepsilon_{h6}}{\partial T_s}, \tag{7.14e}$$

$$a_{22} = \Upsilon_6[T_s - T_{a6}(1+\Omega_{h6})]\frac{\partial \varepsilon_{h6}}{\partial w} + (1-\varepsilon_{h6})\Upsilon_6 T_{a6}\frac{\partial \Omega_{h6}}{\partial w}, \tag{7.14f}$$

$$a_{23} = 1 + (1-\varepsilon_{h6})\Upsilon_6\left(1 + \Omega_{h6} + T_{a6}\frac{\partial \Omega_{h6}}{\partial T_{a6}}\right) + [\varepsilon_{h6}T_s + (1-\varepsilon_{h6})T_{a6}(1+\Omega_{h6})]\frac{\partial \Upsilon_6}{\partial T_{a6}}, \tag{7.14g}$$

$$a_{24} = 0, \tag{7.14h}$$

$$a_{31} = \varepsilon_{v10}\Upsilon_{10} + \Upsilon_{10}[T_s - T_{a10}(1+\Omega_{v10})]\frac{\partial \varepsilon_{v10}}{\partial T_s}, \tag{7.14i}$$

$$a_{32} = \Upsilon_{10}[T_s - T_{a10}(1+\Omega_{v10})]\frac{\partial \varepsilon_{v10}}{\partial w} + (1-\varepsilon_{v10})\Upsilon_{10} T_{a10}\frac{\partial \Omega_{v10}}{\partial w}, \tag{7.14j}$$

$$a_{33} = 0, \tag{7.14k}$$

$$a_{34} = 1 + (1 - \varepsilon_{v10})\Upsilon_{10}\left(1 + \Omega_{v10} + T_{a10}\frac{\partial \Omega_{v10}}{\partial T_{a10}}\right)$$
$$+ [\varepsilon_{v10}T_s + (1 - \varepsilon_{v10})T_{a10}(1 + \Omega_{v10})]\frac{\partial \Upsilon_{10}}{\partial T_{a10}}, \tag{7.14l}$$

$$a_{41} = \varepsilon_{h10}\Upsilon_{10} + \Upsilon_{10}[T_s - T_{a10}(1 + \Omega_{h10})]\frac{\partial \varepsilon_{h10}}{\partial T_s}, \tag{7.14m}$$

$$a_{42} = \Upsilon_{10}[T_s - T_{a10}(1 + \Omega_{h10})]\frac{\partial \varepsilon_{h10}}{\partial w} + (1 - \varepsilon_{h10})\Upsilon_{10}T_{a10}\frac{\partial \Omega_{h10}}{\partial w}, \tag{7.14n}$$

$$a_{43} = 0, \tag{7.14o}$$

$$a_{44} = 1 + (1 - \varepsilon_{h10})\Upsilon_{10}\left(1 + \Omega_{h10} + T_{a10}\frac{\partial \Omega_{h10}}{\partial T_{a10}}\right)$$
$$+ [\varepsilon_{h10}T_s + (1 - \varepsilon_{h10})T_{a10}(1 + \Omega_{h10})]\frac{\partial \Upsilon_{10}}{\partial T_{a10}}, \tag{7.14p}$$

where the derivatives of emissivity relative to wind speed and surface temperatures can be calculated using ocean emissivity models as discussed in the previous section, and the derivatives of Ω relative to wind speed and transmittance Υ are obtained as

$$\frac{\partial \Omega}{\partial T_a} = \frac{\partial \Omega}{\partial \Upsilon}\cdot\frac{\partial \Upsilon}{\partial T_a}, \tag{7.15}$$

where at lower frequencies,

$$\Upsilon = d_0 + d_1 T_a. \tag{7.16}$$

Since the retrieval is also a nonlinear process, initial values of atmospheric upwelling and downwelling radiation, T_a, at 6.9 and 10.7 GHz and surface wind are estimated from brightness temperatures, whereas the initial guess of SST is derived from dual polarization at 6.925 GHz with

$$T_s = \frac{1}{1 - C}T_v - \frac{C}{1 - C}T_h, \tag{7.17}$$
$$C = \frac{1 - \varepsilon_v}{1 - \varepsilon_h}.$$

The algorithm is applied for the AMSR-E data on board the EOS Aqua satellite, which was launched on May 4, 2002. AMSR-E provides microwave observations at frequencies ranging from 6.925 to 89 GHz (see Table 7.2). Its antenna has an aperture of 1.6 m and rotates continuously about an axis parallel to the local spacecraft vertical at 40 revolutions per minute (rpm). At an altitude of 705 km, it measures the upwelling scene brightness temperatures over an angular sector of $\pm 61°$ about

Table 7.2 AMSR-E instrument characteristics and parameters.

Frequency (GHz)	Channel	BW (MHz)	τ (ms)	NEDT	EIA (°)	IFOV (km)
6.925	v, h	350	2.6	0.3	55	43 × 74
10.65	v, h	100	2.6	0.6	55	30 × 51
18.7	v, h	200	2.6	0.6	55	16 × 27
23.8	v, h	400	2.6	0.6	55	18 × 31
36.5	v, h	1000	2.6	0.6	55	8 × 14
89.0	v, h	3000	1.3	1.1	54.5	4 × 6

the subsatellite track, resulting in a swath width of 1445 km. During a period of 1.5 s, the spacecraft subsatellite point travels 10 km. Even though the instantaneous field of view for each channel is different, active scene measurements are recorded at equal intervals of 10 km (5 km for the 89 GHz channels) along the scan. The half-cone angle at which the reflector is fixed is 47.4°, which results in an Earth incidence angle of 55.0°.

AMSR-E's calibration system has a cold mirror that provides a clear view of deep space (a known temperature of 2.7 K) and a hot reference load that acts as a blackbody emitter; its temperature is measured by eight precision thermistors. After launch, large thermal gradients due to solar heating developed within the hot load, making it difficult to determine the average effective temperature from the thermistor readings or the temperature displayed on the radiometer. The hot load temperature is not uniform or constant, and empirical calibration methods are developed. The radiometer calibration accuracy budget, exclusive of antenna pattern correction effects, is composed of three major contributors: warm-load reference error, cold-load reference error, and nonlinearity and errors with radiometer electronics. The total sensor bias error ranges from 0.66 K at 100 K to 0.68 K at 250 K.

NOAA receives AMSR-E level 1B data from NASA Distributed Active Archive Center (DAAC). The data was so far mainly used to test the new algorithm developments. Figure 7.5 displays the global SST retrievals from newly developed algorithms. Note that over high latitudes, SST retrievals are not reliable because of the persistent sea ice cover. Overall, global SST distribution is not affected by clouds and precipitation.

The AMSR-E SST algorithm is validated with *in situ* measurements from National Data Buoy Center (NDBC). The NDBC moored buoy stations are deployed mostly in the coastal and offshore water from the western Atlantic to the Pacific Ocean around Hawaii and from the Bering Sea to the South Pacific. NDBC's moored buoys measure barometric pressure; wind direction, speed, and gust; air and sea temperature; and other sea parameters. The time resolution of these measurements is 1 h. Since the continuous wind measurements associated with Coastal-Marine Automated Network (C-MAN) sites also provide six

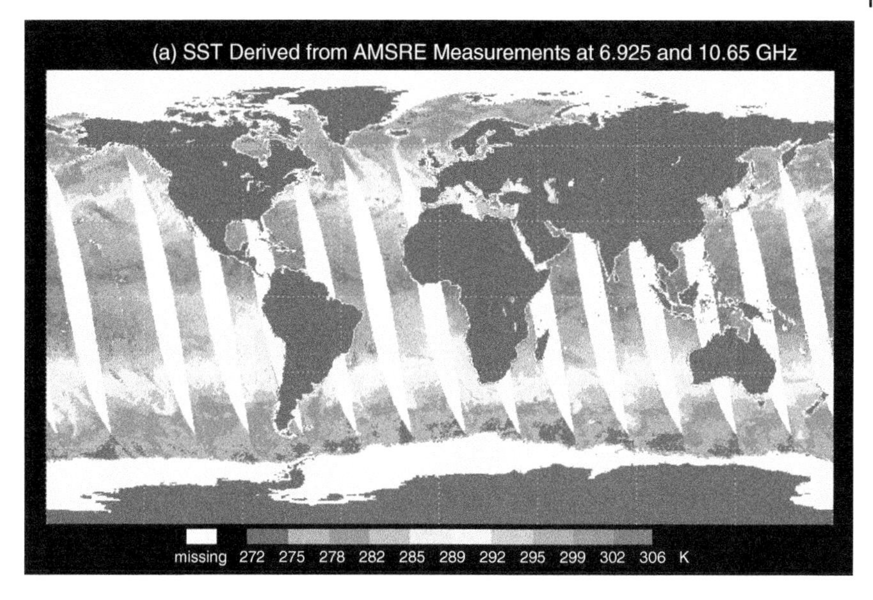

Figure 7.5 Sea surface temperature derived from EOS Aqua AMSR-E using brightness temperatures at 6.925 and 10.65 GHz.

10-min average values of wind speed and direction reported each hour, the time resolution of the wind speed used in our validation is 10 min. Buoy measurements of SST and SSW from May to October (a season with hurricane of high occurrence) of 2004 and 2005 are matched with AMSR-E observations of brightness temperatures. In the match process, the AMSR-E measurements near the coasts and contaminated by land are first removed. The differences in the buoy and AMSR-E time and location (latitude and longitude) are smaller than or equal to 10 min and 0.5°, respectively, where we assume that the ocean surface temperature is the same within 1 h. There are totally 21 754 of matchup data sets between AMSR-E and buoy measurements including clear and cloud weather conditions.

Figure 7.6 displays the comparisons of the SST retrieved from these AMSR-E matchup data under both clear and cloud weather conditions, respectively. The retrieved SST is plotted through its mean value. The sample interval of the buoy SST measurement is 0.5 K. The number of the retrieved SST corresponding to a certain buoy SST measurement varies from hundreds to thousands. It is noted that the algorithms show a better retrieval accuracy of SST at higher SST than at lower SST. For example, over areas where SSTs are greater than 295 K, the corresponding RMS error is around 1 K. This is primarily due to relatively higher sensitivity of brightness temperatures to SST over warm oceans. Overall, the algorithms produce an RMS error of 1.5 K in SST under all conditions including clear and clouds for the SST ranging between 275 and 300 K.

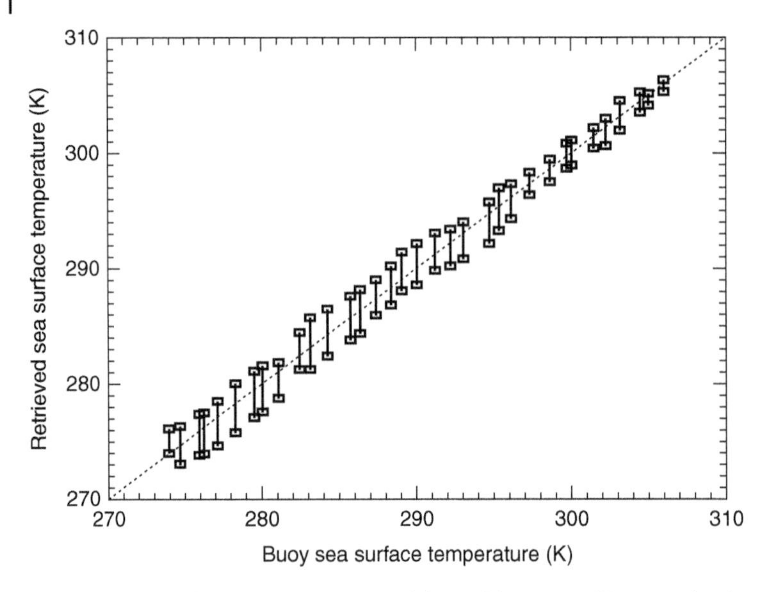

Figure 7.6 Sea surface temperature derived from EOS Aqua AMSR-E using brightness temperatures at 6.925 and 10.65 GHz validated against the NDBC Buoy data under all weather conditions. (Yan and Weng 2008 [198]. Reproduced with permision of Springer.)

7.3
Remote Sensing of Land Surface Parameters

7.3.1
Retrievals of Land Surface Temperature

Using the emission-based radiative transfer approximation (see Eq. 3.44), the brightness temperature measured from the satellite consists of three radiative components: (i) upwelling radiation emitted by the atmosphere; (ii) surface-emitted radiation attenuated by the atmosphere; and (iii) downwelling radiation from the atmosphere and cosmic background reflected by the surface and attenuated by the atmosphere. At frequencies when the atmospheric absorption resides in the lower troposphere, the atmosphere can be considered isothermal (T_m) so that brightness temperature can also be expressed as Eq. (3.44). For a specular surface,

$$T_b = T_s[1 - (1 - \varepsilon)\Upsilon^2] - \Delta T(1 - \Upsilon)[1 + (1 - \varepsilon)\Upsilon], \tag{7.18}$$

where $\Delta T = T_s - T_m$, which is the difference between the surface temperature and the isothermal atmospheric temperature. Responses of brightness temperatures to surface and atmospheric variables are shown in Figure 7.7. For a surface emissivity of 0.95, brightness temperature decreases slowly with increasing atmospheric transmittance. At a lower emissivity (0.90), brightness temperature

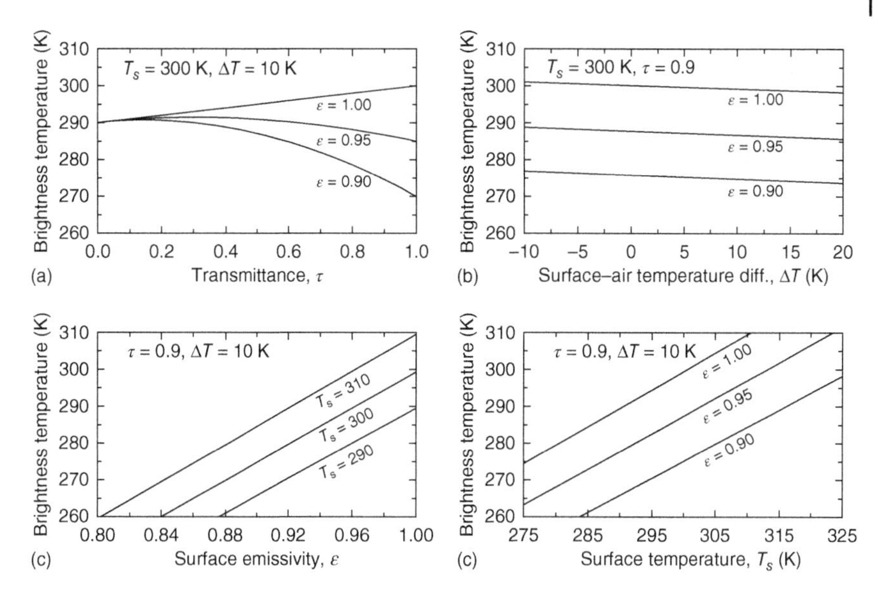

Figure 7.7 Brightness temperature versus (a) transmittance, (b) surface-air temperature difference, (c) surface emissivity, and (d) surface temperature. (Weng and Grody 1998 [194]. Reproduced with permission of Wiley.)

decreases rapidly with transmittance. Thus, microwave measurements under low emissivity conditions are primarily used for retrieval of atmospheric parameters such as cloud liquid and water vapor. However, for a blackbody, the brightness temperature increases with increasing transmittance. The minimal variation of brightness temperature occurs for an emissivity near 0.95, which is the average value of surface emissivity at 20 GHz for a dry land [118].

Figure 7.7b shows the sensitivity of the brightness temperature to ΔT. The variable ΔT depends on the surface–air temperature difference in the atmospheric boundary layer. Positive values of ΔT as large as 20 K correspond to strong superadiabatic conditions, while small or negative values are associated with temperature inversion that may be caused, for example, by the long-wave radiative cooling at night. For the three surface emissivity values (0.90, 0.95, and 1.00), the brightness temperature varies within a range of 5 K.

Figure 7.7c,d shows that the brightness temperature increases linearly with increasing surface emissivity and temperature, respectively. The brightness temperature variability over land is therefore primarily due to the changes in surface emissivity and temperature, with minimal effects due to transmittance variations. For example, over deserts where the soil moisture (i.e., emissivity) varies little with time, the brightness temperature variability is mainly driven by changes in the surface temperature. In general, brightness temperature measurements at a single frequency cannot differentiate between the changes due to surface temperature or emissivity. Multisensor techniques using microwave and infrared information measurements have been used to determine emissivity and surface temperature [199].

From radiative transfer simulations, it is found that the second term on the right-hand side of Eq. (7.18) is indeed insensitive to emissivity and can be parameterized primarily in terms of total precipitable water, V. The impact of clouds is also negligible for low microwave frequencies. This results in

$$T_b = T_s[1 - (1 - \varepsilon)\Upsilon^2] - \gamma V, \tag{7.19}$$

where γ is a constant that depends on the frequency. Apparently, brightness temperatures obtained at two nearby frequencies can be used to derive the surface temperature after the emissivity is eliminated. For an optimum retrieval of surface temperature, the surface emissivity should be similar, and the atmospheric emission should be significantly different for both frequencies. This requires the use of the SSM/I measurements at 19.35 and 22.235 GHz because both frequencies are located at different regions near the 22 GHz water vapor line. The emissivity values for both frequencies are very similar from the *in situ* measurements near 20 GHz [118]. Using the measurements at 19.35 and 22.235 GHz with the vertical polarization (TB_{19v}, TB_{22v}), we obtain the relationship

$$\frac{T_s - TB_{19v} - \gamma_{19} V}{T_s - TB_{22v} - \gamma_{22} V} = \left(\frac{\Upsilon_{19}}{\Upsilon_{22}}\right)^2, \tag{7.20}$$

where Υ_{19} and Υ_{22} are the atmospheric transmittances at 19.35 and 22.235 GHz, respectively; $\Upsilon_{19} = 0.047$; and $\Upsilon_{22} = 0.136$.

Owing to the stronger water vapor absorption at 22.235 GHz, its atmospheric transmittance is smaller than that at 19.35 GHz. The transmittance ratio in Eq. (7.20) is only a function of the total precipitable water. This can be illustrated as follows:

$$\Upsilon_\nu = \Upsilon_\nu(\text{oxygen})\Upsilon_\nu(\text{vapor})\Upsilon_\nu(\text{cloud}), \tag{7.21a}$$

where the transmittance of oxygen and clouds is nearly the same at both 19.35 and 22.235 GHz, and the transmittance of water vapor is exponentially related to the precipitable water

$$\Upsilon_{19} = \Upsilon_\nu(\text{oxygen})\Upsilon_\nu(\text{cloud}) \exp(-V/V_{19}), \tag{7.21b}$$

$$\Upsilon_{22} = \Upsilon_\nu(\text{oxygen})\Upsilon_\nu(\text{cloud}) \exp(-V/V_{22}), \tag{7.21c}$$

where $V_{19} = \cos(\theta)/\kappa_{19}$ and $V_{22} = \cos(\theta)/\kappa_{22}$. Here, θ is the local zenith angle and 53.1° for SSM/I. κ_{19} and κ_{22} are the mean water vapor mass absorption coefficients at 19.35 and 22.235 GHz, respectively, and they can be derived from the slopes of regression relationships between microwave optical thickness and precipitable water. Combining Eqs. (7.20) and (7.21) yields

$$\frac{T_s - TB_{19v} - \gamma_{19} V}{T_s - TB_{22v} - \gamma_{22} V} = e^{cV}, \tag{7.22}$$

where $c = 2(1/V_{22} - 1/V_{19}) = 1/42357 \times 10^{-2}$. Thus,

$$T_s = [e^{cV}(TB_{22v} + \gamma_{22}) - (TB_{19v} + \gamma_{19}V)]/(e^{cV} - 1). \tag{7.23}$$

Thus, for accurate remote sensing of surface temperature, it is important to know the atmosphere total precipitable water vapor, V. Typically, it can be approximated from numerical weather prediction (NWP) model. Notice that under some land conditions, the total precipitable water correlates well with the surface temperature. For example, the measurements from the First International Satellite Land Surface Climatology Project (ISLSCP) Field Experiment (FIFE) over Konza Prairie show these two variables can be fitted well [194]

$$V = 108.2be^{a(T_s-288)}, \tag{7.24}$$

where $a = 0.064\ \mathrm{K}^{-1}$ and b is dependent on the surface relative humidity and the shape of the water vapor profile. Under dry conditions (smaller b), precipitable water is less affected by T_s, when the temperature is less than 300 K. Under moist conditions (larger b), the precipitable water rapidly increases as the surface temperature rises. Thus, Eqs. (7.23) and (7.24) formulate a closure for retrieving the surface temperatures using microwave split-window technique.

It is also important to understand the retrieval accuracy when Eq. (7.23) is applied for *in situ* SSM/I measurements. Owing to the difficulty in obtaining global temperatures measured at the ground, we use the routine temperature observations made at 1.2-m shelter heights. Using the shelter air temperatures to test the algorithm results in errors due to the differences between the surface and shelter-height air temperatures. In particular, for the regions where a large lapse rate occurs near the surface (e.g., either superadiabatic or inversion), the shelter-height temperature can deviate significantly from the surface temperature. As the FIFE data indicated, the difference between the shelter and surface temperatures is found to be minimum in the morning and typically less than a couple of degrees [200], which is not strongly affected by the atmospheric conditions (dry and wet). This means that the shelter air temperature in the morning can be used to closely approximate the land surface temperature. To illustrate this fact, Figure 7.8 displays the seasonal variation of the difference between the surface and shelter air temperatures over the FIFE regions. The four time periods shown here correspond to the local times at which the current SSM/I sensor passes over the local area. At 0600 LT, the difference typically varies between −2 and 2 K. At 1000 LT, it becomes predominately positive, which means that the surface temperature becomes warmer than the air temperature. This is due to the initial heating of solar radiation on the surface. The difference becomes more negative at 1800 LT, which results from the long-wave radiation cooling at the surface. At 2200 LT, the difference is less negative and moves toward the smallest difference, which typically occurs in the morning. For this validation study, the SSM/I measurements were matched with the shelter-height air temperatures at 0600 LT. The entire 1993 hourly global surface observational

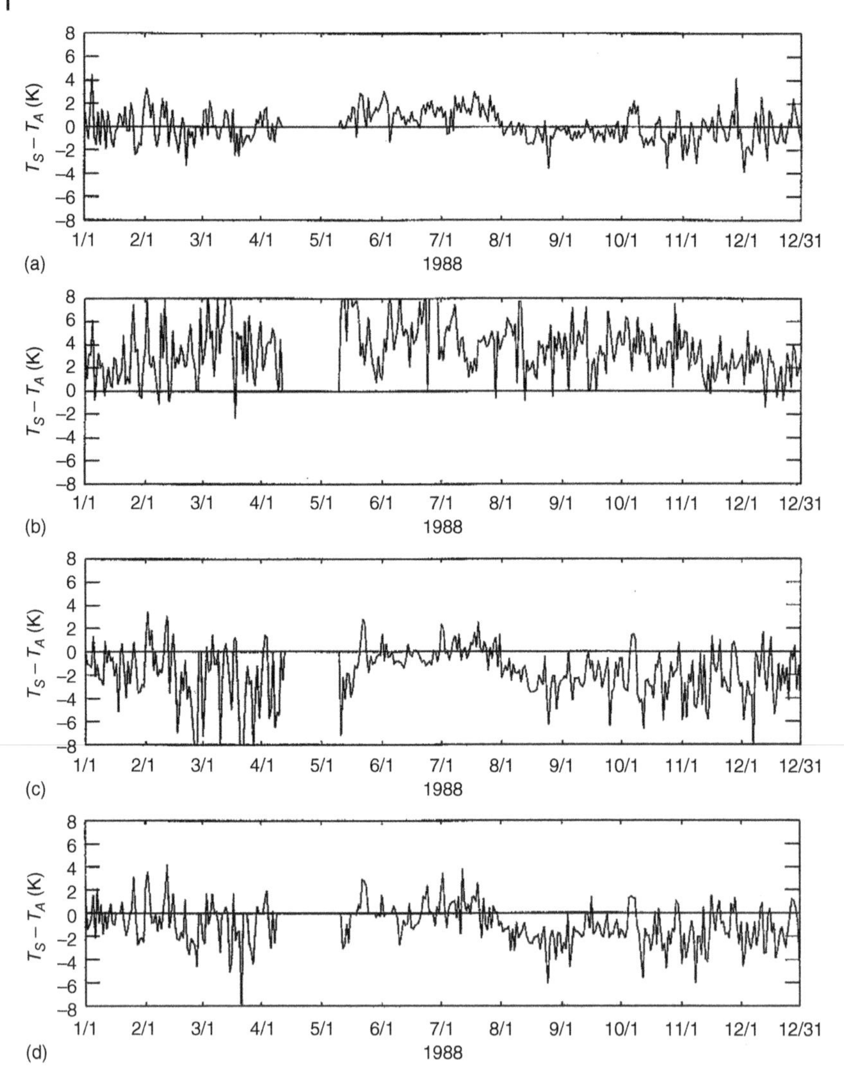

Figure 7.8 The difference between skin surface and shelter-height air temperatures at (a) 0600 LT, (b) 1000 LT, (c) 1800 LT, and (d) 2200 LT, for the FIFE region. The smallest difference occurs in the morning at about 0600–0800 LT. (Weng and Grody 1998 [194] Reproduced with permission of Wiley.)

data set was obtained from the National Climate Data Center. To maximize the use of surface and satellite observations, we allow certain spatial and temporal displacements in the matchup procedure. In particular, the geophysical locations between the surface and SSM/I measurements are limited to within 15 km, and the time difference should be less than 30 min.

The data are also filtered for heavy precipitation and deep snow-covered surfaces since these conditions can affect low-frequency measurements. These conditions are identified using a scattering index (SI), which is a brightness temperature difference between 22.235 and 85.5 GHz over land [201]. The SI exceeding 15 K indicates moderate-to-heavy precipitation or significant snow cover on the ground. As mentioned previously, the initial value (or the first guess) of surface temperature for the Newtonian iterative solution of Eq. (7.22) is obtained using the vertically polarized brightness temperature at 85.5 GHz.

With no correction being made for atmospheric emission and scattering, the first guess is given by

$$T_{s0} = TB_{85v}/\varepsilon_{85v} \tag{7.25}$$

where TB_{85v} is the brightness temperature and $\varepsilon_{85v} = 0.955$, which are similar to *in situ* measurements [118]. Figure 7.9a shows that the first guess correlates with the shelter air temperature, especially for the higher temperature conditions (>280 K), but has an RMS error of 6.22 K. The measurements with the triangle symbols are obtained from the scattering conditions excluding heavy precipitation and deep snow-covered surfaces, whereas those with the plus signs are from the nonscattering conditions. Apparently, the first guess estimates the surface temperatures very well for the nonscattering conditions, with large biases occurring under the scattering conditions.

The iterative solution of Eq. (7.22) converges very fast as soon as the first guess based on Eq. (7.25) has a bias less than 15 K from the actual surface temperature. As shown in Figure 7.9b, the retrieved surface temperatures correlate with the surface measurements very well. In general, the data points (plus signs) identified as nonscattering conditions converge more to the 1 : 1 line. In particular, the large errors of the first guess caused by the scattering (triangle) become much smaller for most cases. This significant improvement is achieved with the use of low frequencies, which have negligible scattering effects.

7.3.2
Retrieval of Land Surface Emissivity

Direct microwave emissivity measurements are not available under various land surface conditions. Thus, the "truth" emissivity must be also derived from the retrievals, which requires accurate knowledge of atmospheric emission, surface temperature, and surface types. From Eq. (3.44), we have

$$\varepsilon = \frac{T_b - T_u - (T_d + T_c)(1+\Omega)\Upsilon}{[T_s - (T_d + T_c)(1+\Omega)]\Upsilon}, \tag{7.26}$$

where the transmittance, upwelling, and downwelling radiation are calculated from the temperature and water vapor profiles. Here, we use NWP analysis of the profiles four times a day. Even over areas where the radiosonde measurements are sparsely distributed or not available, the Global Data Assimilation System (GDAS) still performs the analysis based on physical parameterization using the

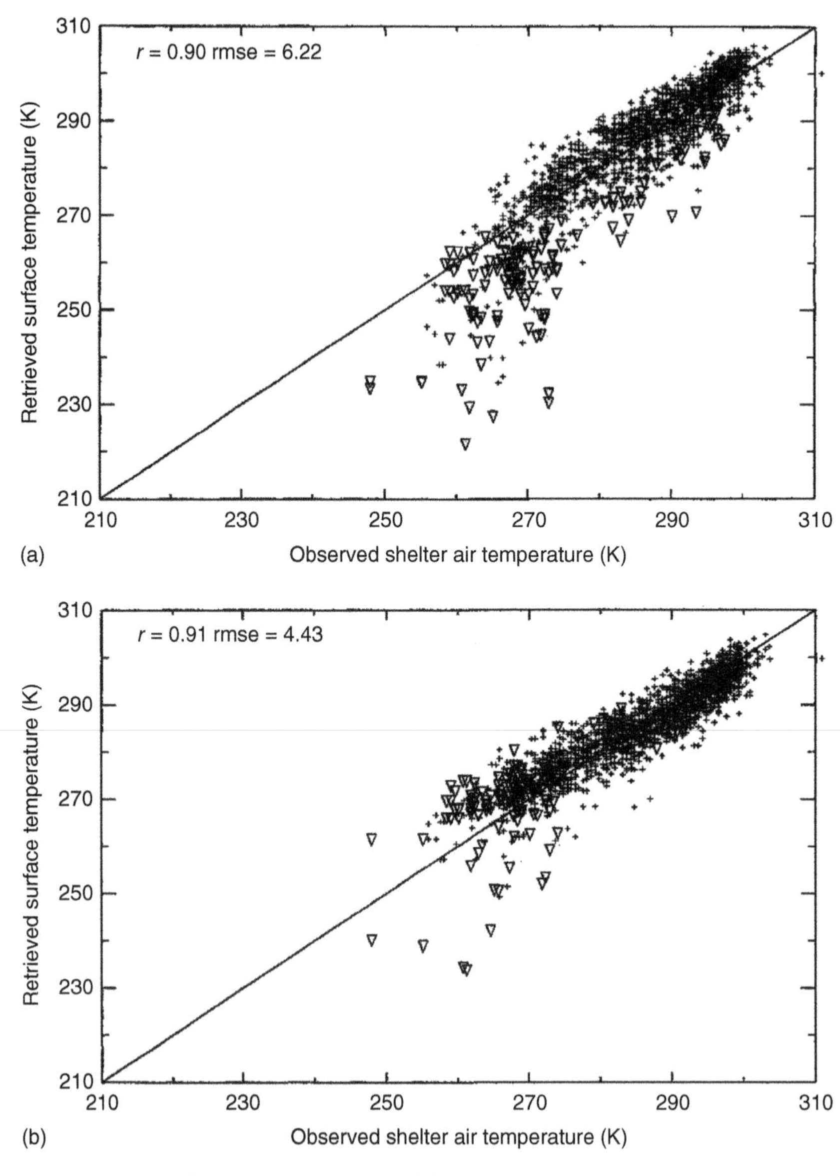

Figure 7.9 The surface temperature retrieved from the physically based retrieval using SSM/I brightness temperatures at 19.35 and 22.235 GHz in comparison with the shelter air temperatures in the morning (0600 LT): (a) the first guess based on vertically polarized brightness temperature at 85.5 GHz and (b) the iterative solution using the Newtonian method. The triangles in both figures represent the measurements having a scattering index greater than 5 K, which is most likely due to precipitation or surface snow. (Weng and Grody 1998 [194] Reproduced with permission of Wiley.)

numerical weather prediction model. However, since the GDAS does not produce cloud parameters such as liquid and ice water content, the computed parameters contain larger errors under cloudy conditions.

Brightness temperatures from DMSP SSM/I are utilized to estimate land surface emissivity using Eq. (7.26). The contributions to SSM/I measurements from the atmospheric emission and surface temperature are first removed using an emission-based radiative transfer equation. This is typically referred to as "truth" emissivity if the correction from atmosphere is performed very accurately.

The SSM/I instrument consists of an offset parabolic reflector of dimensions 24×26 in., fed by a corrugated, broadband, seven-port horn antenna. The reflector and feed are mounted on a drum that contains the radiometers, digital data subsystem, mechanical scanning subsystem, and power subsystem. The reflector–feed–drum assembly is rotated about the axis of the drum by a coaxially mounted bearing and power transfer assembly (BAPTA). A small mirror and a hot reference absorber are mounted on the BAPTA and do not rotate with the drum assembly. They are positioned off axis such that they pass between the feed horn and the parabolic reflector, occulting the feed once each scan. The mirror reflects the cold sky radiation into the feed, thus serving, along with the hot reference absorber, as calibration reference for the SSM/I. This scheme provides an overall absolute calibration that includes the feed horn. Corrections for spillover and antenna pattern effects from the parabolic reflector are incorporated in the data processing algorithms.

The SSM/I rotates continuously about an axis parallel to the local spacecraft vertical at 31.6 rpm and measures the upwelling scene brightness temperatures over an angular sector of 102.4° about the subsatellite track. The scan direction is from the left to the right when looking in the forward (F10, F11) or aft (F8) direction of the spacecraft with the active scene measurements lying ±51.2° about the forward (F10, F11) or aft (F8) direction. This results in a swath width of approximately 1400 km. The spin rate provides a period of 1.9 s during which the spacecraft subsatellite point travels 12.5 km. During each scan, 128 discrete uniformly spaced radiometric samples are taken at the two 85 GHz channels, and on alternate scans, 64 discrete samples are taken at the remaining five lower frequency channels. The antenna beam intersects the Earth's surface at an incidence angle of 53.1° as measured from the local Earth normal (see Table 7.3).

The retrievals of seven channel emissivity were performed from F13 SSM/I for each descending orbit and were then averaged in March of 1999. Retrievals

Table 7.3 SSM/I instrument characteristics and parameters.

Frequency (GHz)	Channel	BW (MHz)	τ (ms)	NEDT (K)	EIA (°)	IFOV (km)
19.35	v, h	250	7.9	0.6	53.1	43×69
22.235	v	250	7.9	0.6	53.1	40×60
37.0	v, h	1000	7.9	0.6	53.1	28×37
85.5	v, h	1500	3.89	1.1	53.1	13×15

are not performed over areas where atmospheres are identified with clouds and precipitation. An SI [201] is used to detect rain-bearing clouds.

$$SI = F - TB_{85v} \tag{7.27}$$

where F represents the brightness temperature from the scattering-free clouds and can be estimated from SSM/I lower frequencies as

$$F = 256.2 - 0.375 \times TB_{19} - [0.20 - 0.00217 \times TB_{22}]TB_{22} \tag{7.28}$$

The index used in Eq. (7.28) can also be very large for surface media such as snow, sea ice, and deserts. Some additional checks must be performed through the polarization index for the separation of surface scattering from that from the atmospheric clouds.

Figure 7.10 shows the mean monthly emissivity at three SSM/I frequencies (19.35, 22.235, and 85.5 GHz) for March 1999. Several pronounced features in global emissivity are seen from this SSM/I emissivity atlas. First, the emissivity over desert is highly polarized at all three SSM/I frequencies with the largest at 19.35 GHz. The emissivity decreases as frequency increases (Figure 7.10). Snow over Greenland and Antarctic also displays a similar characteristic.

Deserts over North Africa and Northwestern China display the largest polarization difference at lower frequencies (see Figure 7.11). A main reason may be smaller roughness. It is also shown that the polarization difference decreases as frequency increases.

7.3.3 Error Sensitivity of Land Surface Emissivity

Note that the retrieval is most accurate for the microwave window channels since these measurements are least affected by the atmospheric absorption, emission, and scattering (due to precipitation). However, as previously discussed [105], inaccurate profiles can still result in some retrieval errors. Additional errors can be associated with inaccurate surface temperature and inaccuracies of the water vapor absorption model. The total error in the retrieved emissivity at frequencies lower than 37 GHz was found to be generally less than 1% when using the Special Sensor Microwave Imager [105]. This error, however, increases as the surface wetness and atmospheric moisture content increase. The emissivity sensitivity can be generally derived from Eq. (7.26) as follows:

$$\frac{\delta\varepsilon}{\varepsilon} \approx -\frac{\delta T_b}{T_b}\left(1 + \frac{T_u + T_d \Upsilon}{T_b}\right), \tag{7.29}$$

where Ω effect is neglected in this analysis.

It is seen that the errors in brightness temperatures are directly related to the emissivity errors. The errors of brightness temperatures are most likely

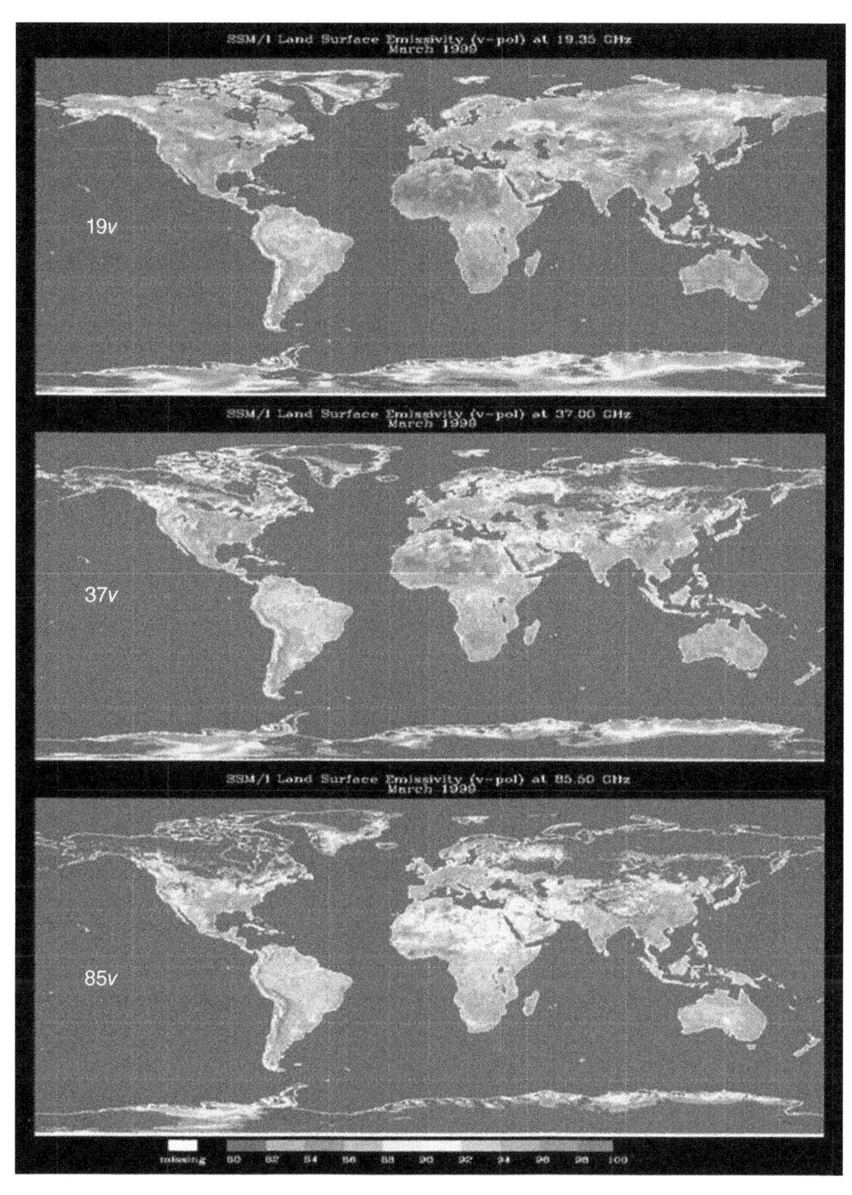

Figure 7.10 Global monthly (March 1999) mean emissivity at 19.35, 37 and 85.5 GHz retrieved from special sensor microwave imager (SSM/I).

related to the instrument calibration process. An increase in noises and biases in brightness temperatures may be from improper along-track averaging, inaccurate spillover correction, ignorance in nonlinear correction, and degradation of onboard calibration targets. This is an outstanding problem in conical scanning instruments because of a lack of end-to-end calibration. The conversion from

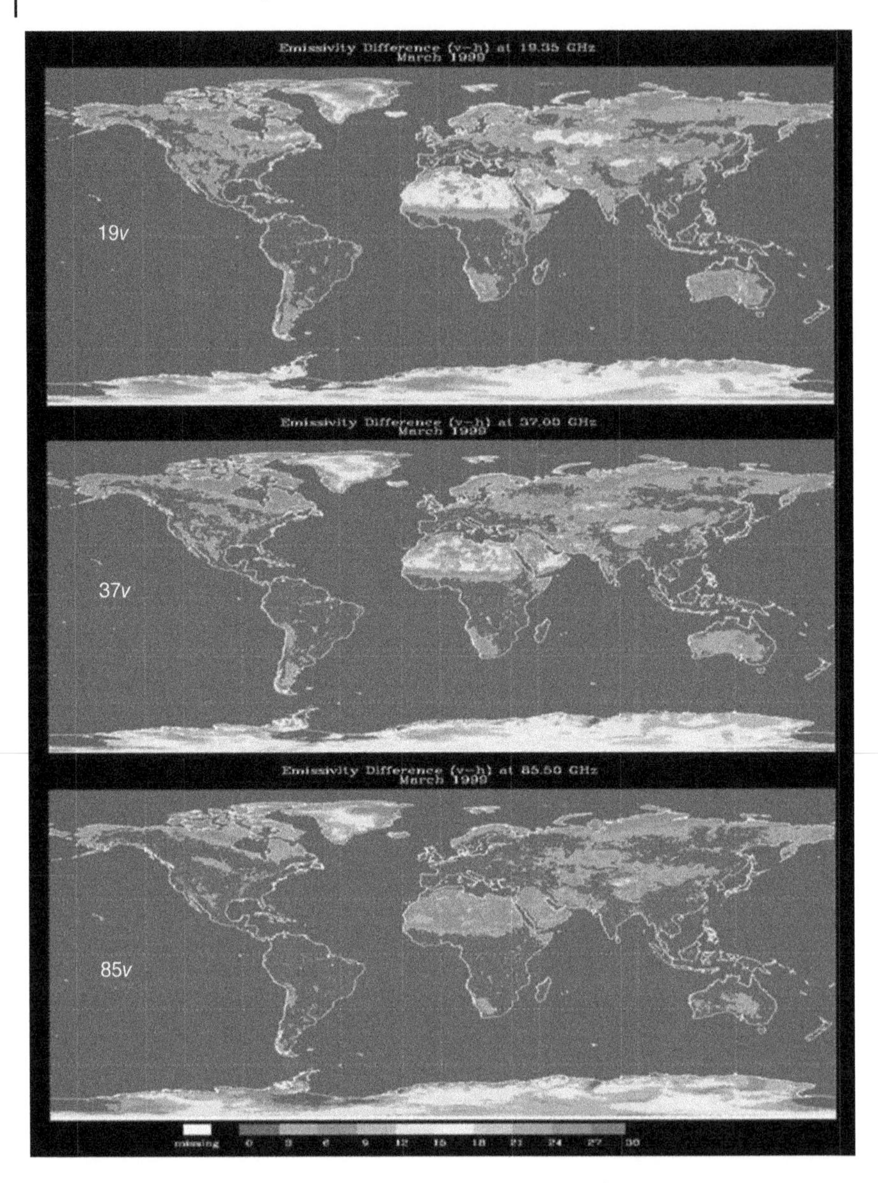

Figure 7.11 Global monthly (March 1999) mean polarization difference in emissivity at 19.35, 37, and 85.5 GHz retrieved from special sensor microwave imager (SSM/I).

antenna to sensor brightness temperatures is several degrees in magnitude at lower frequencies. Such conversion results in the spectra shift from the emission- to scattering-type brightness temperatures for most of land surfaces. Prigent *et al.* [105] estimated the range of these errors. A typical value of 0.6 K of radiometric noise in brightness temperature would induce small uncertainties in retrieved emissivity. Our analytic result shows that an error of 1% in T_b results in more than

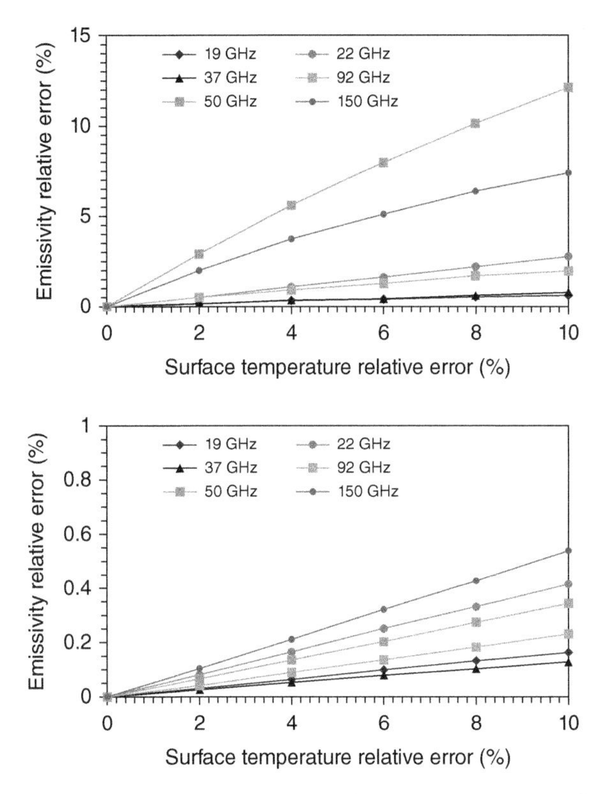

Figure 7.12 Emissivity error sources from water vapor and surface temperature.

1% error in emissivity, depending upon the upwelling and downwelling radiation. In tropical atmospheres where these two components are larger, the errors in T_b can be amplified (see Figure 7.12).

Under clear atmospheric conditions, most of the errors are related to uncertainty in surface temperatures and atmospheric temperature profiles that are used to calculate the atmospheric emission and transmittance.

$$\frac{\delta\varepsilon}{\varepsilon} \approx -\frac{\delta T_s}{T_s}\left(1+\frac{T_d}{T_s}\right), \tag{7.30}$$

and

$$\frac{\Delta\varepsilon}{\varepsilon} = -\frac{\Delta\Gamma}{\Gamma}\left[1+\frac{T_d+T'_u+T'_d\Gamma}{\left(T_s \quad T_d\right)\varepsilon}-\frac{T'_d\Gamma}{(T_s-T_d)}\right], \tag{7.31}$$

respectively. In particular, 1% error in the surface temperature would result in less than 1% error in land emissivity. The error from water vapor results in the largest error in emissivity at 22 V that is located at the water vapor absorption line.

In a cloudy atmosphere, the emission-based radiative transfer model becomes less accurate. As shown in Figure 7.13, the errors increase significantly at higher

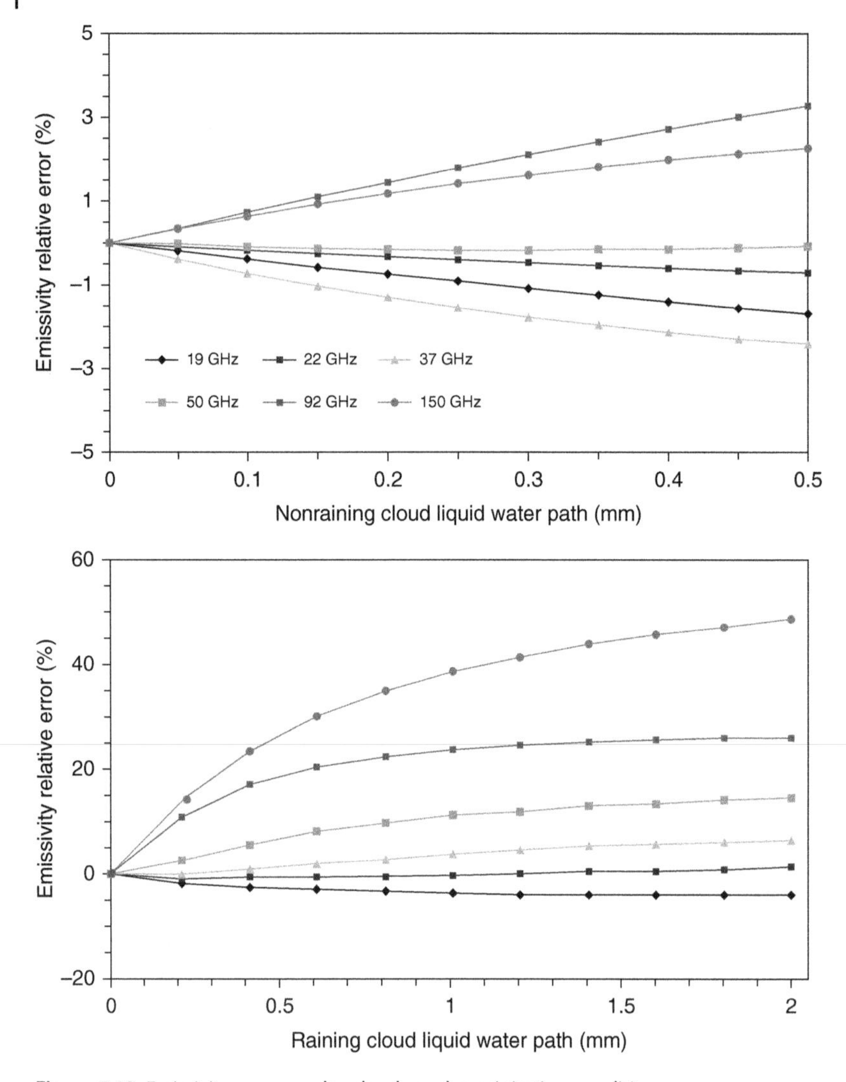

Figure 7.13 Emissivity errors under cloudy and precipitation conditions.

frequencies. The errors can be of several percent. In a raining atmosphere, the error can be as large as 50–60% at 150 GHz if the retrieval process does not consider the scattering from precipitating hydrometeors.

7.3.4 Fast Land Emissivity Algorithms

Since global temperature and water vapor profiles are not always available, the atmospheric upwelling and downwelling effects in Eq. (7.26) are not readily derived. Thus, the land surface emissivity algorithm can be derived from

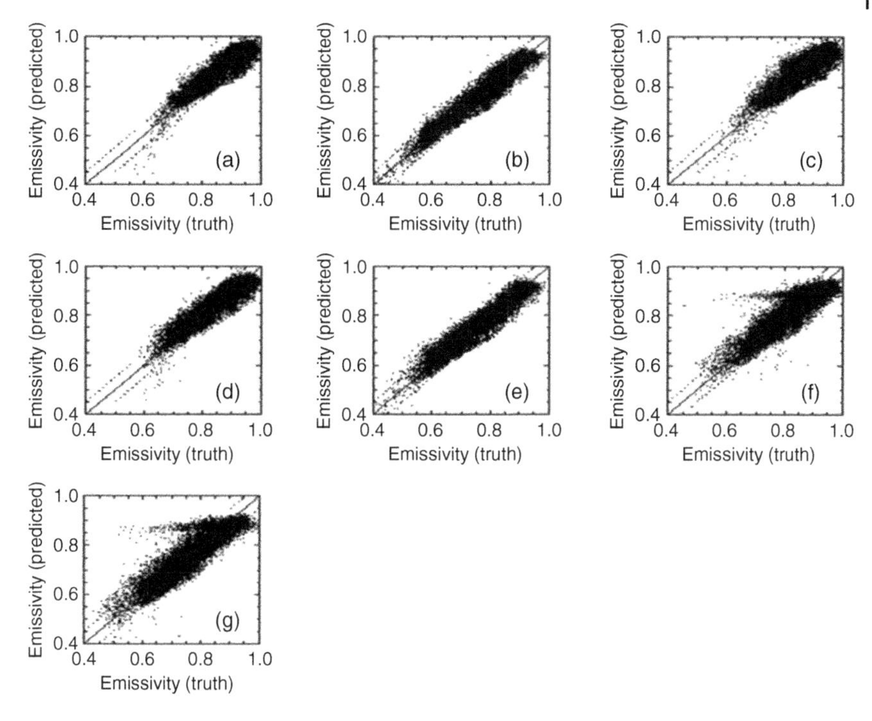

Figure 7.14 Regression-type emissivity retrieval from SSM/I brightness temperatures over land and its performance against the physical retrieval from the emission-based radiative transfer model.

regression against the "truth" (see Figure 7.14). The truth is a training data set from the analytical retrievals. From SSM/I seven channels, the land emissivity at each SSM/I channel is expressed as

$$\varepsilon_j = a_0 + \sum_{i=1}^{7}(a_{1,i}T_{b,i} + a_{2,i}T_{b,i}^2), \tag{7.32}$$

where $j = 1, 7$. This formula is applied to the prediction of the emissivity at all of seven SSM/I channels with a different set of coefficients as shown in Table 7.4. The emissivity at seven SSM/I channels can be retrieved under various land conditions.

Surface emissivity over land is modulated by many surface parameters such as vegetation, topography, flooding, and snow and, thus, contains information on the surface properties. The surface emissivity spectra express the differences in the surface properties. Here, the monthly mean land emissivity is computed using Eq. (7.32) at seven SSM/I channels for 10 years and a group of monthly mean emissivity spectra are analyzed over nine different surface conditions. The nine surface conditions are classified as (i) dense vegetation (jungle), (ii) agricultural/rangeland vegetation, (iii) arable soil (dry), (iv) soil (moist surface),

Table 7.4 Coefficients in SSM/I emissivity retrieval algorithms, channel order, 19*v*, 19*h*, 22*v*, 37*v*, 37*h*, 85*v*, 85*h*.

	ε_{19v}	ε_{19h}	ε_{22v}	ε_{37v}	ε_{37h}	ε_{85v}	ε_{85h}
a_0	−1.20E-01	−1.45E-01	−1.02E-01	−2.16E-01	−2.08E-01	−2.31E-01	−2.63E-01
a_{11}	5.57E-03	8.55E-04	−3.00E-03	−1.05E-03	−2.01E-03	−4.99E-03	−6.45E-03
a_{12}	2.98E-06	5.46E-06	1.93E-05	7.75E-06	1.07E-05	2.30E-05	3.02E-05
a_{13}	7.05E-01	1.15E-02	9.81E-03	7.21E-03	7.05E-03	9.97E-03	1.08E-02
a_{14}	−1.44E-05	−1.57E-05	−2.21E-05	−1.44E-05	−1.45E-05	−2.21E-05	−2.49E-05
a_{15}	−3.59E-03	−3.57E-03	1.54E-03	−3.98E-03	−3.20E-03	−3.19E-03	−2.28E-03
a_{16}	−9.27E-06	−9.88E-06	−1.87E-05	−8.78E-06	−1.10E-05	−1.92E-05	−2.55E-05
a_{17}	2.63E-03	2.25E-03	2.33E-03	9.97E-03	4.85E-03	5.86E-03	4.15E-03
a_{21}	1.17E-06	9.04E-07	1.84E-06	−4.00E-06	−3.37E-06	−4.48E-06	−1.13E-06
a_{22}	−3.78E-03	−3.19E-03	−3.83E-03	−3.96E-03	1.39E-03	−4.40E-03	−3.04E-03
a_{23}	7.37E-06	7.23E-06	9.78E-06	7.09E-06	6.08E-06	1.02E-05	9.00E-06
a_{24}	7.06E-03	5.16E-03	9.56E-03	7.85E-03	5.85E-03	1.24E-02	1.05E-02
a_{25}	−1.66E-05	−1.15E-05	−2.18E-05	−1.88E-05	−1.36E-05	−1.73E-05	−2.35E-05
a_{26}	−7.09E-03	−5.02E-03	−8.43E-03	−7.60E-03	−5.58E-03	−6.78E-03	−4.34E-03
a_{27}	1.40E-05	8.35E-06	1.59E-05	1.56E-05	1.02E-05	1.23E-05	1.73E-05

(v) semiarid surface, (vi) desert, (vii) composite vegetation and water, (viii) composite soil and water/wet soil surface, and (ix) snow, by using a classification algorithm [202] from SSM/I measurements. Table 7.5 lists mean emissivity spectra and polarization differences at SSM/I frequencies under these nine surface conditions that are calculated by averaging the monthly mean emissivity spectra for 10 years. It is shown that the polarization difference is sensitive to surface type. Vegetated land produces a smaller polarization difference. This is consistent with our knowledge that a smaller polarization difference implies that the region is heavily vegetated region and the most noticeable low maximum polarization

Table 7.5 Mean emissivity spectra for two polarizations and mean emissivity polarization difference under nine different surface conditions.

Type	19*v*	22*v*	37*v*	85*v*	19*h*	37*h*	85*h*	19*v*−*h*	37*v*−*h*	85*v*−*h*
1	0.927	0.900	0.930	0.911	0.920	0.927	0.907	0.007	0.003	0.003
2	0.930	0.909	0.936	0.918	0.913	0.924	0.909	0.017	0.012	0.009
3	0.940	0.921	0.942	0.914	0.908	0.919	0.900	0.032	0.023	0.013
4	0.930	0.910	0.935	0.919	0.898	0.910	0.899	0.032	0.025	0.021
5	0.940	0.924	0.930	0.906	0.875	0.881	0.871	0.065	0.048	0.035
6	0.942	0.933	0.941	0.923	0.823	0.841	0.857	0.119	0.099	0.067
7	0.918	0.891	0.920	0.908	0.900	0.907	0.898	0.018	0.013	0.010
8	0.888	0.876	0.880	0.887	0.787	0.799	0.827	0.100	0.081	0.060
9	0.907	0.893	0.833	0.760	0.805	0.751	0.705	0.103	0.082	0.056

1, dense veg; 2, agri veg; 3, dry soil; 4, moisture soil; 5, semiarid; 6, desert; 7, corn; 8, wet grnd; 9, snow.

difference areas are the forest regions. In contrast, a larger polarization difference implies that the region is over desert and/or snow, as summarized in Table 7.5.

7.4 Summary and Conclusions

Surface parameters over oceans and land can be retrieved from microwave imager data under all weather conditions. With the new polarimetric data from WindSat, the ocean wind vectors can be derived using a physical retrieval. A ratio of the third to the fourth Stokes component offers a robust first guess of the wind direction retrieval. In general, retrieval of oceanic SST requires use of data at lower microwave frequencies where the scattering from precipitation is minimal.

A microwave split-window algorithm is developed to retrieve the land surface temperatures. To form a closure without using any ancillary data, the total precipitable water vapor should be parameterized as a function of land surface temperature. Better accuracy can be derived if atmospheric total precipitable water is provided from some independent sources such as NWP model output.

For the land surface emissivity, we derive a retrieval form in an analytic form. The terms affecting the emissivity retrievals are all computed from NWP model outputs. The global emissivity atlas is derived and made available for all the NWP centers for data assimilation experiences.

8
Remote Sensing of Clouds from Microwave Sounding Instruments

8.1
Introduction

Clouds play a vital role in modulating the climate and Earth's radiation budget [203]. In the atmosphere, the latent heat release or consumption occurs either directly in clouds or in the precipitation produced by them. Clouds strongly affect the radiative fluxes through the atmosphere. Thus, the measurements on cloud hydrometeors in various phases critically affect the numerical weather prediction (NWP) models.

Global measurements of cloud liquid water path (LWP) are best determined by satellite-measured microwave brightness temperatures due to their direct response to the thermal emission of cloud particles. Early in the 1970s, Nimbus-6 scanning microwave spectrometer experience demonstrated the feasibility of the microwave measurement of cloud liquid water. A statistical relationship was first derived between the brightness temperatures at 21 and 31 GHz and cloud liquid water using Nimbus-6 scanning microwave spectrometer data [204], and large-scale distribution of cloud liquid water was obtained over the Pacific Ocean [205]. The capability was further displayed from Nimbus-7 scanning multichannel microwave radiometer (SMMR) data [206, 207]. However, more algorithms for cloud liquid water were developed for the special sensor microwave imager (SSM/I) flown on the defense meteorological satellite program (DMSP) (e.g., [148, 151, 208–210]). In the SSM/I algorithm for cloud liquid water algorithm [149], the liquid water in nonprecipitating and some precipitating clouds over oceans was estimated from the brightness temperature measurements at 19.35, 37, and 85.5 GHz. The algorithm was further revised as a full physical retrieval for the Advanced Microwave Sounding Unit (AMSU) measurements at 23.8 and 31.4 GHz [211].

With the millimeter-wavelength measurements from satellites, cloud ice water can also be retrieved [38, 39, 53]. Ice clouds, because of their high albedo in visible wavelengths, the reflection of short-wave radiation by ice clouds reduces the solar energy reaching the Earth's surface. On the other hand, ice clouds can trap the long-wave radiation emitted from the surface, resulting in less radiation to space compared to clear-sky conditions. The net radiative flux at the Earth's surface

Passive Microwave Remote Sensing of the Earth: For Meteorological Applications, First Edition. Fuzhong Weng.

resulting from the aforementioned two processes, however, depends on accurate description of the ice cloud parameters for radiative transfer calculations in climate models [212, 213]. Therefore, a quantitative measurement of microphysical parameters in ice clouds is important for both the validation of global climate models and understanding the nature variability of the Earth's climate [214–216].

In this chapter, we present the algorithms of remote sensing of cloud liquid and ice water using the microwave sounding instruments. In Section 8.2, we demonstrate the sensitivity of brightness temperatures to cloud liquid water at lower frequencies and then derive the physical retrieval of cloud liquid water. In Section 8.3, the algorithms of retrieving cloud ice water and particle size are derived using the brightness temperatures at high frequencies. In Section 8.4, we investigate the new methodology for remote sensing of hydrometeor profiles using the double oxygen bands in microwave regions.

8.2 Remote Sensing of Cloud Liquid Water

8.2.1 Principle of Microwave Remote Sensing of Clouds

As shown in Eq. (3.46), the brightness temperatures under these approximations are directly related by the layer mean temperature and atmospheric transmittance. Under a low emissivity condition, the brightness temperature increases as the atmospheric transmittance decreases. This physical principle drives the microwave remote sensing of clouds over oceans. The sensitivity of microwave measurements to cloud LWP over oceans is further analyzed using a vector radiative transfer model [46, 47]. The LWP is the sum of the vertically integrated liquid water content in nonraining and raining clouds. This vector radiative transfer model (RTM) is for a vertically stratified scattering and emitting atmosphere where the optical parameters for cloud drops, raindrops, and ice particles are calculated using the Mie theory. Cloud particles in nonraining clouds are distributed in a gamma distribution, and rain droplets in raining clouds are distributed in a Marshall and Palmer distribution [129]. In simulations, the atmosphere is divided into 20 layers, and the nonraining and raining clouds appear below the freezing level. The atmospheric temperature and water vapor profiles in NCEP Global Data Assimilation System (GDAS) are used as inputs to calculate the brightness temperatures at three frequencies at an angle of 55°.

Figure 8.1 displays the brightness temperatures against LWP under a condition of sea surface wind (SSW) of 10 m/s and sea surface temperature (SST) of 300 K. Note that the lower frequencies respond to cloud liquid water approximately linearly within a large dynamic range. When cloud liquid water is low (e.g., in nonraining clouds), the brightness temperature at 36.5 GHz is the most sensitive channel. The variation in brightness temperature under vertically polarization

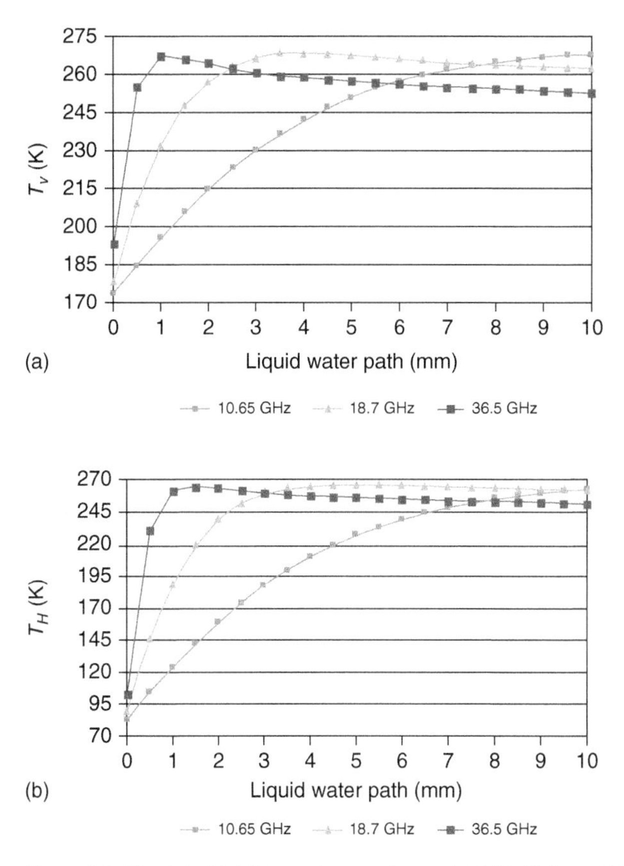

Figure 8.1 Simulated brightness temperatures at 10.65, 18.7, and 36.5 GHz as a function of cloud liquid water path for (a) vertically polarized and (b) horizontally polarized.

is about 75 K as cloud liquid water varies from 0 to 1.0 mm. However, the corresponding measurements at 36.5 GHz become saturated and further decrease when cloud liquid water is greater than about 1.0 mm. This is caused by scattering of raindrops and large cloud particles. Cloud liquid water corresponding to saturation point in brightness temperatures at 10.65, 18.7, and 36.5 GHz is about 8, 3, and 1 mm, respectively. Since the actual LWP can be up to 6 mm [149] in hurricane situation, it is necessary to use the brightness temperature at 10.65 GHz to retrieve the LWP for various raining clouds. Brightness temperatures at 6.925 GHz are the least sensitive to low cloud liquid water (not shown here). The brightness temperature at 6.925 varies by 4 K, when the cloud LWP varies from 0 to 2 mm. While the brightness temperatures at horizontally polarized channels are sensitive to cloud liquid water, they are also strongly affected by wind roughness [149]. Thus, in the cloud liquid water algorithm, we normally utilize the vertically polarized brightness temperatures.

8.2.2
Cloud Liquid Water Algorithm

Using Eq. (3.46), the brightness temperatures can be further linked to cloud LWP (L) and precipitable water path (V) [211] by further assuming an isothermal atmosphere ($\Delta T = T_s - T_m = 0$), that is,

$$T_b = T_s[1 - (1 - \varepsilon)\Upsilon^2], \tag{8.1}$$

where ε and T_s are the surface emissivity and temperature, respectively, and

$$\Upsilon = \exp[-(\tau_O + \tau_V + \tau_L)/\mu)], \tag{8.2}$$

where τ_O, τ_V, and τ_L are the optical thicknesses of oxygen, water vapor, and liquid, respectively.

$$\tau_L = \int_{\Delta Z} \kappa^{Ray}\text{LWC dz}, \tag{8.3}$$

where

$$\kappa^{Ray} = \frac{6\pi}{\lambda \rho_w}\text{Im}\left\{\frac{m^2 - 1}{m^2 + 2}\right\}, \tag{8.4}$$

and

$$\tau_V = \int_0^\infty \kappa^{H_2O}\rho_V \text{ dz}, \tag{8.5}$$

where κ^{H_2O} is the mass absorption coefficient of water vapor having a unit of m^2/kg, and ρ_v is the water vapor density in atmosphere. Let us assume κ^{Ray} and κ^{H_2O} to be independent of height. Then, we have

$$\tau_L = \kappa_L L, \tag{8.6}$$

where κ_L is the mass absorption coefficient of liquid-phase cloud, namely

$$\kappa_L = \frac{6\pi}{\lambda \rho_w}\text{Im}\left\{\frac{m^2 - 1}{m^2 + 2}\right\}, \tag{8.7}$$

Here, we use a different notation to indicate there is a further approximation being performed for cloud absorption coefficient, which can be derived from a mean cloud temperature in the complex dielectric constant. We also have

$$\tau_V = \kappa_V V, \tag{8.8}$$

where

$$V = \int_0^\infty \rho_V \, dz, \tag{8.9}$$

and

$$L = \int_{\Delta Z} \text{LWC} \, dz, \tag{8.10}$$

are the vertically integrated water vapor and liquid water, respectively. Thus, atmospheric transmittance becomes

$$\Upsilon = \exp[-(\tau_O + \kappa_V V + \kappa_L L)/\mu). \tag{8.11}$$

These deviations enable the fundamental microwave remote sensing of LWP and water vapor path (WVP) in an emission atmosphere. Normally, at least two channels are required, with one being more sensitive to liquid and other to water vapor. The emission regime can be identified from the relationship between brightness temperature and LWP at a saturation point (see Figure 8.1) where the brightness temperature no longer increases as the LWP increases. Of course, the saturation points are defined when the relationships are simulated over ocean surfaces. Over land where the emissivity is high, Eq. (8.1) most likely produces a monochromatic decrease as liquid water increases (say, emissivity is greater than 0.9). In such cases, the brightness temperature depression from clouds is very small, less than a few degrees, compared to cloud-free areas. Thus, it is difficult to detect liquid-phase clouds over land, given the same cloud microphysical distribution.

The primary channel used for liquid water remote sensing is dependent on the particular problem. For example, if we would like to detect the cloud liquid covering lower-to-moderate cloud liquid path, we use microwave brightness temperature at 30–40 GHz, which can measure a range up to 1.0 mm (or kg/m^2). The secondary channel used for the correction of water vapor effect would be near the 22 GHz absorption line. Two channels should have a frequency not far apart so that they can be valid in terms of the Rayleigh approximation. In general, a logarithmic function to the brightness temperature is obtained because the quantities we try to solve (e.g., V and L) are as exponents in the exponential function of the transmittance in Eq. (8.11). Thus, we now derive

$$\kappa_V V + \kappa_L L = -\frac{\mu}{2}\left\{\ln(T_s - T_b) - \ln\left[T_s(1-\varepsilon)\right] + \frac{2\tau_{O_2}}{\mu}\right\}. \tag{8.12}$$

Using two channel measurements, we can derive

$$L = a_0\mu[\ln(T_s - T_{b,1}) - a_1 \ln(T_s - T_{b,2}) - a_2], \tag{8.13}$$

and

$$V = b_0\mu[\ln(T_s - T_{b,1}) - b_1 \ln(T_s - T_{b,2}) - b_2], \tag{8.14}$$

respectively. $T_{b,1}$ is the channel sensitive to liquid, and $T_{b,2}$ is the channel sensitive to water vapor. The coefficients $ai, i = 0123$ and $bi, i = 012$ are related to water vapor and liquid water mass absorption coefficients as

$$a_0 = -0.5\kappa_{V2}/(\kappa_{V2}\kappa_{L1} - \kappa_{V1}\kappa_{L2}), \tag{8.15}$$

$$b_0 = 0.5\kappa_{L2}/(\kappa_{V2}\kappa_{L1} - \kappa_{V1}\kappa_{L2}), \tag{8.16}$$

$$a_1 = \kappa_{V1}/\kappa_{V2}, \tag{8.17}$$

$$b_1 = \kappa_{L1}/\kappa_{L2}, \tag{8.18}$$

$$a_2 = -2(\tau_{O,1} - a_1\tau_{O,2})/\mu + (1 - a_1)\ln[T_s(1 - \varepsilon_1)] - a_1 ln(1 - \varepsilon_2), \tag{8.19}$$

$$b_2 = -2(\tau_{O,1} - b_1\tau_{O,2})/\mu + (1 - b_1)ln[T_s(1 - \varepsilon_1)] - b_1 ln(1 - \varepsilon_2). \tag{8.20}$$

From Rayleigh's approximation, κ_L can be parameterized as a function of cloud layer temperature, $T_{L,}$ in Celsius as

$$\kappa_L = a_L + b_L T_L + C_L T_L^2. \tag{8.21}$$

Oxygen optical thickness is parameterized as a function of SST through

$$\tau_O = a_o + b_o T_s. \tag{8.22}$$

Table 8.1 lists some of the coefficients that can be used for various AMSU window channels. In the study by Weng *et al.* [211], AMSU measurements at 23.8 and 31.4 GHz were used for L and V retrievals. Figure 8.2 displays a global distribution

Table 8.1 The parameters calculated at four AMSU-A channels and used in liquid water and water vapor path algorithms.

	23.8 GHz	31.4 GHz	50.3 GHz	89 GHz
κ_V	4.80423E-3	1.93241E-3	3.76950E-3	1.15839E-2
$\kappa_1 - a_1$	1.18201E-1	1.98774E-1	4.53967E-3	1.03486E00
$\kappa_1 - b_1$	−3.48761E-3	−5.45692E-3	−9.68548E-3	−9.71510E-3
$\kappa_1 - c_1$	5.01301E-5	7.18339E-5	8.57815E-5	−6.59140E-5
$\tau_0 - a_0$	3.21410E-2	5.34214E-2	6.26545E-1	1.08333E-1
$\tau_0 - b_0$	−6.31860E-5	−1.04835E-4	−1.09961E-3	−2.21042E-4

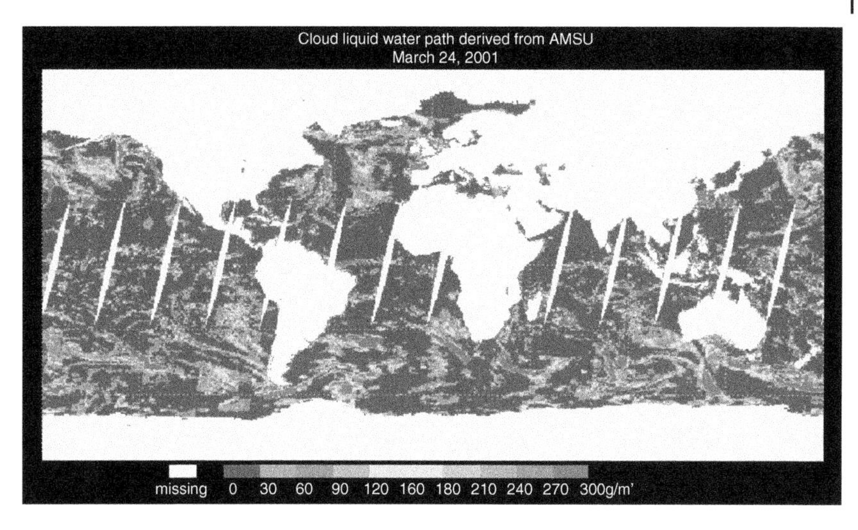

Figure 8.2 Global cloud liquid water path derived from AMSU onboard NOAA-16 satellite.

of cloud liquid water over oceans derived using the AMSU onboard NOAA-16 satellite. As discussed earlier, the AMSU antenna temperatures are first corrected for the asymmetric bias. The correction scheme is also obtained separately for individual satellites. Note that the AMSU descending measurements during a 24 h period do not completely cover the globe because of the orbital gaps. It is shown that the algorithm depicts cloud liquid water associated with various systems. The low clouds over oceans off the west coast of South America are detected although the amount of cloud liquid is low.

8.3
Remote Sensing of Cloud Ice Water

8.3.1
Microwave Scattering from Ice-Phase Cloud

The sensitivity of brightness temperatures to the ice water path (IWP) at higher frequencies or millimeter wavelengths is nearly independent of the cloud temperature. Vivekanandan *et al.* [217] studied the possibility of retrieving precipitation-sized ice water amount using simulated measurements at lower frequencies and found that the brightness temperature monotonically decreases as the cloud optical thickness increases. At higher frequencies, satellite microwave measurements provide estimates of cloud IWP (or vertically integrated ice water content) associated with both raining and nonraining ice clouds [39, 53]. Figure 8.3 displays the simulated brightness temperatures at 85 and 91 GHz for an ice cloud that is 2 km thick. The cloud base is set at an altitude of 330 mb, the cloud mean particle size is 0.5 mm, and all the particles constitute a bulk

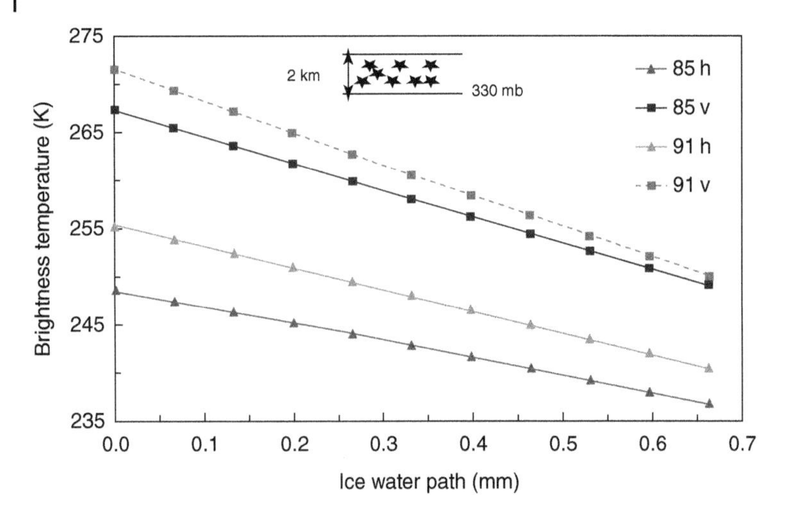

Figure 8.3 Brightness temperatures at 85 and 91 GHz simulated from scattering radiative transfer model for an ice cloud layer of 2 km thick, located at 330 hPa.

volume density of 400 kg/m^3. It is seen that a typical 10–20 K depression in the brightness temperature occurs as IWP increases from 0 to 0.7 mm (kg/m^2).

Ice cloud scattering is also demonstrated from a six-channel Millimeter-wave imaging radiometer (MIR), which was built for the National Aeronautics and Space Administration (NASA) by the Georgia Institute of Technology and flown on the NASA ER-2 aircraft to measure the atmospheric water vapor, clouds, and precipitation parameters [218, 219]. The MIR measures microwave radiation at three frequencies (183.6 $\pm$ 7, 183.6 $\pm$ 3, and 183.6 $\pm$ 1 GHz) near the 183.31 GHz water vapor line and at three frequencies (89, 150, and 220 GHz) in the window regions. It is a cross-track scanner that covers an angular swath of $\pm 50°$ centered at the nadir. Of particular importance to the National Oceanic and Atmospheric Administration (NOAA) is the fact that the MIR operates at the same frequencies as the NOAA AMSU-B module and offers a unique validation/calibration purpose for algorithm development. Measurements from the MIR were acquired during the Tropical Ocean Global Atmosphere Coupled Ocean Atmosphere Response Experiment (TOGA COARE) period (6 January–24 February 1993) in the western tropical Pacific. Accompanying the MIR data are radar measurements from the Airborne Rain Mapping Radar (ARMAR) that was installed on the NASA DC-8 aircraft, which provided detailed vertical profiles of hydrological parameters with a 60-m resolution. Radar backscatter measurements were performed at a frequency of 13.8 GHz and are thus most sensitive to hydrometeors of relatively large size. In addition to the MIR and ARMAR, measurements from several other instruments such as Advanced Microwave Precipitation Radiometer (AMPR) and Moderate-Resolution Imaging Spectrometer (MODIS) were collocated in both space and time for all scanning angles. However, this study only uses the measurements obtained at the nadir position. Figure 8.4

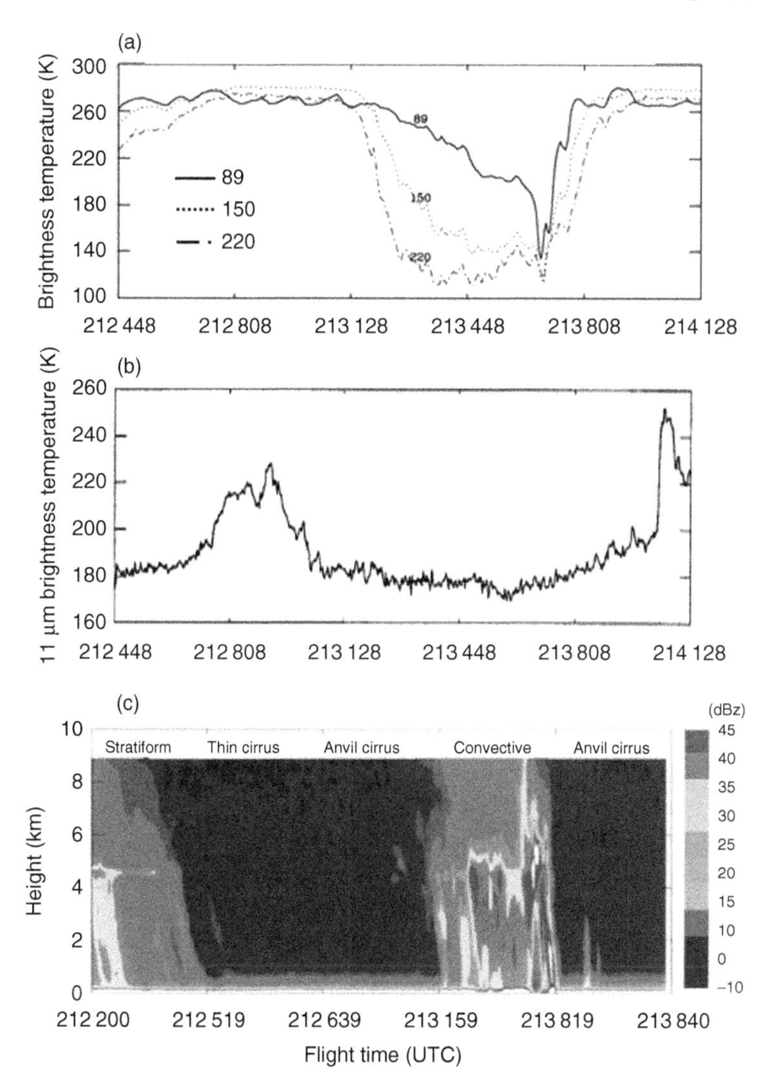

Figure 8.4 NASA ER2 and DC8 observations of convective systems over TOGA/COARE areas with (a) AMPR and MIR, and (b) MODIS-like channels, and (c) ARMAR.

shows a composite image from all the instruments for a tropical cloud system on February 22, 1993, over the equatorial Pacific to northeastern Australia. Clearly, several precipitating ice clouds are identified from radar vertical cross section of reflectivity. The precipitation between 21:22:00 and 21:25:19 is mainly stratiform because of an obvious melting bright band occurring near 4.5 km height. The rainfall type during 21:31:59 and 21:35:19 is likely convective due to high reflectivity values throughout the vertical column.

Figure 8.4 further displays and inter-compares various sensor responses to the cloud system. Note that the AMPR 10.7 and 19.35 GHz measurements increase rapidly from 21:33:48 to 21:38:08 of ER-2 flight time as a result of increasing precipitating liquid water amounts at low levels. Brightness temperatures at 19.35 and 37 GHz show depressions near 21:38:00 primarily due to ice particles in the deep convective region, although large raindrops under heavy rain conditions (rain rate greater than 15 mm/h) may also result in significant scattering [149]. It is important to recognize that the MIR is affected by both precipitating and nonprecipitating ice. This is clearly indicated at the ER-2 flight times between 21:31:28 and 21:34:48. Brightness temperatures at three window frequencies (89, 150, and 220 GHz) gradually decrease before entering the precipitating regions. An initial decrease before 21:32:00 is presumably due to thick anvil cirrus situated at higher altitudes. A detrainment of ice particles from the convective region is likely the primary process responsible for the generation of anvil clouds. Ice clouds are likely present during most of the study period. Note that the brightness temperatures at MODIS 11 μm are well below the 235 K threshold used for a typical cold cloud indicator [220]. The clouds between 21:28:08 and 21:34:48 are mainly identified as cirrus since they produce very little radar reflectivity. However, the cirrus between 21:31:28 and 21:34:48 must be very thick, since the IR temperatures are only 180 K, and the ice particles must be of millimeter size, since the MIR measurements are significantly depressed.

8.3.2
Cloud Ice Water Retrieval Algorithm

The several algorithms were developed and tested to retrieve IWP using aircraft millimeter-wavelength measurements [53, 221, 222]. Liu and Curry [221] presented a method to retrieve IWP using airborne MIR data at 89, 150, and 220 GHz channels. The algorithm was further modified to derive the IWP in the tropical cloud systems using the satellite microwave data [38]. Although the IWP algorithm works well for cirrus clouds in the Tropics, an uncertainty arises due to an unknown particle size. Weng and Grody [53] proposed an algorithm to derive both IWP and D_e using dual submillimeter-wavelength measurements. They found that for a given particle bulk volume density, the brightness temperature at microwave frequencies can be uniquely related to IWP and D_e through a two-stream radiative transfer model solution. The algorithm was also tested with the measurements obtained from the MIR. The retrieved IWP and D_e display a reasonable spatial distribution comparable to the radar and infrared measurements. Zhao and Weng [39] further improved the algorithm by Weng and Grody [53] and applied it for satellite measurements. Physical basis of retrieval algorithm is from a two-stream approximation. Weng and Grody [53] first derived a relationship between radiance and emanating from the ice clouds and ice cloud microphysical parameters by letting the interface reflection. Assuming optically thin cloud (i.e., $k\tau_1 \ll 1.0$), the upwelling and downwelling radiances from the ice cloud layer are

$$I(\tau_0, \mu) = \frac{I_1 + 2Ba^2\Omega(\mu) + I_0\Omega(\mu)(1 - a^2)}{1 + \Omega(\mu)(1 + a^2)}, \tag{8.23}$$

and

$$I(\tau_1, -\mu) = \frac{I_0 + 2Ba^2\Omega(\mu) + I_1\Omega(\mu)(1 - a^2)}{1 + \Omega(\mu)(1 + a^2)}, \tag{8.24}$$

where B is the Planck function; a is the similarity parameter; and I_1 and I_0 are the upwelling and downwelling radiances at the bottom and top of the ice cloud, respectively. The ice cloud scattering parameter, $\Omega(\mu)$, is defined as

$$\Omega(\mu) = \frac{\kappa\tau}{2a} = \frac{1}{2a}(1 - \omega g)\tau, \tag{8.25}$$

where τ is now the ice cloud optical thickness, g is the asymmetry factor, and μ is the cosine of the zenith angle. Note that the similarity parameter a is smaller than 1 at microwave frequencies [53]. Thus, Ωa^2 may be much less than 1. In addition, the contribution by I_0 is assumed to be negligible because it is typically close to the very low cosmic background radiation. As a result,

$$I(\tau_0, \mu) = \frac{I_1}{1 + \Omega(\mu)} \tag{8.26}$$

and

$$I(\tau_1, -\mu) = \frac{I_1\Omega(\mu)(1 - a^2)}{1 + \Omega(\mu)}, \tag{8.27}$$

The upwelling and downwelling radiances (or brightness temperature) at microwave frequencies are directly proportional to the incident radiation at the cloud lower boundary (i.e., I_1). From a space platform (satellite or aircraft), the upwelling radiance decreases as the scattering parameter increases. Conversely, the downwelling radiance, observed from a ground-based instrument looking up, increases as the scattering parameter increases. The variation of the scattering parameter may result from changes in the cloud IWP and particle size. The cloud optical thickness is defined as

$$\tau = \int_{z_b}^{z_t} \mathrm{d}z \int_0^{\infty} \frac{\pi}{4} D^2 \Omega_e(x, m) N(D)\mathrm{d}D, \tag{8.28}$$

where z_b and z_t are the cloud base and top heights, respectively; Ω_e is the extinction efficiency of ice particles; $N(D)$ is the particle size distribution function; x is the size parameter ($x = \pi D/\lambda$); m is the complex refractive index, which might vary with the ice particle bulk volume density. For spherical ice particles with the size distribution $N(D)$, the IWP can be expressed as

$$\mathrm{IWP} = \int_{z_b}^{z_t} \mathrm{d}z \int_0^{\infty} \frac{\pi}{6} D^3 N(D)\mathrm{d}D, \tag{8.29}$$

where ρ_i is the ice particle bulk volume density. For polydispersed ice particles, the scattering parameter can be expressed as a function of IWP and D_e [39]

$$\Omega(\mu) = \frac{\mathrm{IWP}}{\mu\rho_i D_e}\Omega_N(x, m), \tag{8.30}$$

where $\Omega_N(x, m)$ is the normalized scattering parameter defined as

$$\Omega_N = \frac{3}{4}(1 - \omega g)\Omega_N, \tag{8.31}$$

and the mean extinction efficiency of the ice particles is defined as

$$\Omega_N = \frac{\int_0^\infty D^2\Omega_e(x, m)N(D)\mathrm{d}D}{\int_0^\infty D^2N(D)\mathrm{d}D}. \tag{8.32}$$

If the brightness temperature at the cloud base is known, then Ω can also be determined through Eq. (8.26),

$$\Omega = \frac{I_1(\tau_b, \mu) - I(\tau_0, \mu)}{I_1(\tau_b, \mu)}, \tag{8.33}$$

and IWP can also be derived as

$$\mathrm{IWP} = \mu D_e\rho_i\frac{\Omega}{\Omega_N}. \tag{8.34}$$

As shown in Eq. (8.34), IWP is directly proportional to the scattering parameter. However, the relationship between Ω_N and D_e is nonlinear and may depend on the particular ice particle size distribution and bulk volume density. Therefore, measurements at two distinct frequencies are normally required to unambiguously determine both IWP and D_e for a given particle bulk volume density [31, 53]. Provided that the bulk volume density of ice particles can be determined independently from other sources (i.e., assumed to be either a constant or a function of ice particle size), the IWP essentially only depends on the scattering parameters, Ω and D_e. Since the most published bulk density–size relations are derived for ice particles in nonprecipitating cirrus clouds, there are some uncertainties in using these relationships in the retrieval because the AMSU measurements are primarily sensitive to the precipitating ice clouds. Thus, a constant density of 600 $\mathrm{kg/m^3}$ is used in this study.

Using two channel brightness temperature measurements (say 89 and 150 GHz), the scattering parameter ratio is directly related to the particle effective diameter [39, 53] by

$$r(D_e) = \frac{\Omega_{89}}{\Omega_{150}} = \frac{\Omega_{N89}(x, m)}{\Omega_{N150}(x, m)}, \tag{8.35}$$

Note that the scattering parameter ratio ideally varies between 0 and 1. For ice clouds having small ice particles, Ω_{89} nearly vanishes and the ratio approaches zero. For ice particles having a larger effective diameter, the scattering parameter ratio approaches unity when the scattering intensities at 89 and 150 GHz reach their geometrical optical limit. An empirical relationship between r and D_e is derived using simulated data from a radiative transfer model [46]. The Mie theory is applied to determine the scattering and absorption properties of the ice particles at 89 and 150 GHz. During the simulation, cloud ice water content is randomly generated within a range of 0–0.5 g/m^3, and the cloud ice particles are assumed to be spherical and distributed using the modified gamma distribution with an exponent of 2 [223]. The ice cloud base is at 9 km with a thickness of 1 km, which simulated a general atmosphere condition for ice clouds. The effective particle diameter randomly varies within the range of 0.1–3.5 mm. The incident radiation at the cloud base is set to a constant value corresponding to a brightness temperature of 280 K. Cumming's mixing formula is used to compute the refractive index of ice–air mixture [224]. The simulations were performed with various combinations of D_n and N_0 (a total of 62 500 gamma size distributions), and the results are shown in Figure 8.5. Note that the ratio initially increases as D_e increases and then approaches to constant value when the ice particle effective sizes become very large. For small $D_e < 0.4$ mm, the scattering parameter of 89 GHz is small due to its lack of sensitivity to the small-size ice particles. Therefore, the size information cannot be uniquely determined by the scattering parameter ratio of 89 and 150 GHz. A higher frequency pair (i.e., 150 and 220 GHz) is required for detecting these small-size ice particles [32]. For D_e between 0.4 and 2.5 mm, Ω_N at both 89 and 150 GHz linearly increases with D_e, and Ω_N at 150 is significantly higher than that at 89 GHz. However, for a larger D_e (greater 3.5 mm), Ω_N at both frequencies tends to approach the same constant value. The minimum detectable ice particle size is about 0.5 mm at 89 GHz. With the dual-frequency measurements at 89 and 150 GHz, reliable results are expected with the ratio ranging from 0.2 to 0.8.

Regression relations of $D_e \sim r$ and $\Omega_N \sim D_e$ are obtained as follows:

$$D_e = a_0 + a_1 r + a_2 r^2 + a_3 r^3, \tag{8.36}$$

and

$$\Omega_N = \exp[(b_0 + b_1 \ln(D_e) + b_2 (\ln D_e)^2], \tag{8.37}$$

where $a_i, i = 0, 1, 2, 3$ and $b_i, i = 0, 1, 2$ are the regression coefficients that may be dependent on the ice particle bulk volume density and assumed size distribution. A set of coefficients were derived for AMSU brightness temperatures at 89 and 150 GHz [39] as summarized in Table 8.2.

When the brightness temperatures at ice cloud base are estimated, the scattering parameter ratio can be computed using Eq. (8.22) with satellite measurements from two frequencies. Thus, IWP and D_e can be unambiguously determined from Eqs. (8.23)–(8.26) for a given bulk volume density. Over oceans, we can use an

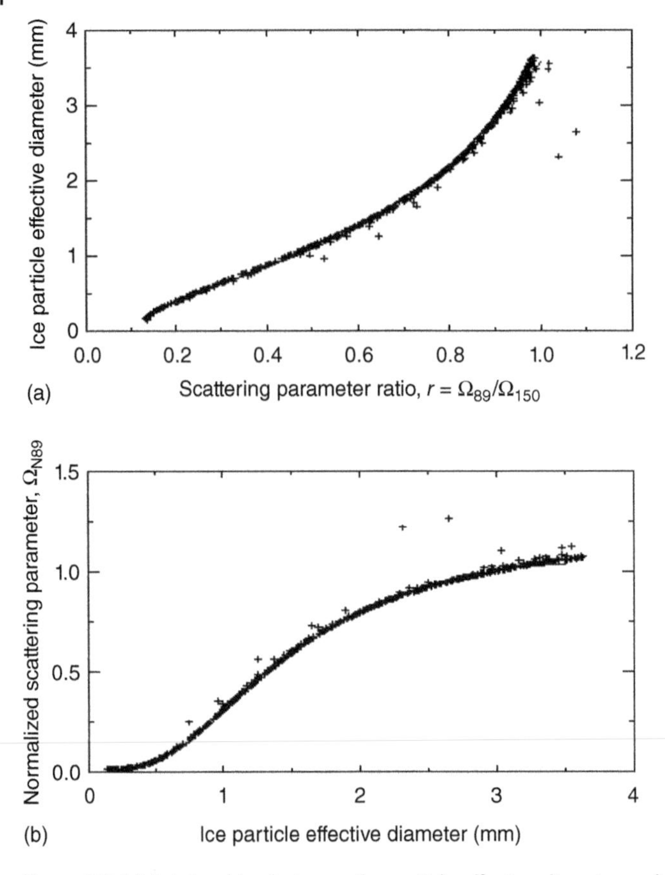

Figure 8.5 (a) Relationships between the particle effective diameter and the scattering parameter ratio, and (b) the relationship between the normalized scattering parameter and the particle effective diameter.

Table 8.2 The coefficients used in the IWP and D_e algorithms.

a_0	a_1	a_2	a_3
−0.24843	3.86726	−4.67150	4.70782
	b_0	b_1	b_2
$D_e < 2$ mm	−11.74663	1.90711	0.73029
$D_e \geq 2$ mm	−1.58571	1.52230	−0.52437

emission-based radiative transfer model to calculate the brightness temperatures at the ice cloud base. Alternatively, we can also use the measurements at lower frequencies to estimate the brightness temperatures at higher frequencies [201], assuming that lower frequencies can penetrate through ice clouds.

Over land, the cloud base brightness temperature is estimated using an empirical relationship between the AMSU lower and higher frequencies [39]. The AMSU brightness temperatures are collected under the scattering-free conditions and are then used to derive the relationships. The AMSU clear radiances are identified using infrared data from the advanced very-high-resolution radiometer (AVHRR)and surface temperature. The AVHRR is also onboard NOAA-15 and -16 satellites. The AMSU measurements corresponding to IR temperatures less than 275 K are excluded from the collocated data since the data possibly contain ice clouds [225]. The AMSU collocation data are also limited to within ±45° of scan angles to eliminate the effects of large footprints.

As shown in Figure 8.6, the brightness temperatures at 89 and 150 GHz at the ice cloud bases can be estimated with a root-mean-square (RMS) error of about 4.0 K,

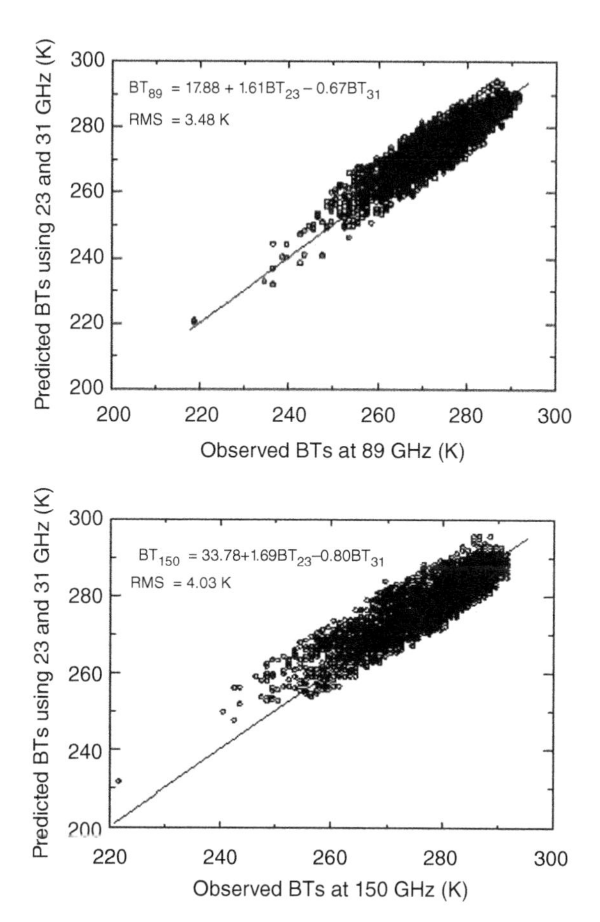

Figure 8.6 Regression relationship derived to estimate the upwelling brightness temperatures at 89 and 150 GHz at the ice cloud base over land using AMSU lower frequency measurements at 23 and 31 GHz.

which is equivalent to the results over oceans. This implies that the brightness temperature depressions at 89 and 150 GHz due to clouds must be greater than 4.0 K so that the clouds can be reliably identified. This RMS error corresponds to a minimum threshold of 0.01 and 0.02 for the scattering parameters at 89 and 150 GHz, respectively.

The scattering signatures resulting from desert, sea ice, and snow particles at higher microwave frequencies are quite similar to that of the ice particles because the dielectric constants among these scatters are almost the same. Therefore, for global application of the IWP and D_e retrieval algorithm, a procedure must be developed to discriminate the scattering signatures between atmospheres and various surface materials [39]. However, the satellite measurements alone provide very limited information on surface types. Other data sets such as AVHRR infrared and GDAS surface temperature and surface types are used as part of the proposed screening procedure.

Surface scattering from snow and sea ice can be largely removed using the measurements at lower AMSU-A frequencies [39]. AMSU-A-derived products of snow cover and sea ice concentration are first used to indicate their presence. The GDAS surface temperature (T_s) of less than 269 K is used as an additional threshold to identify the scattering signatures of frozen surfaces. The retrieval of atmospheric ice is not performed under these surface conditions. Furthermore, there is no retrieval over high terrains such as Tibetan plateau, where the surface temperatures are usually less than 273 K. Deserts also scatter at AMSU 89 and 150 GHz [106]. However, the scattering from clouds can be easily separated from the surface using the satellite infrared measurements and GDAS surface temperatures. If the atmosphere is free from ice clouds, the IR temperature is close to the surface temperature, and therefore, the scattering at 89 and 150 GHz must be from the surface. More specifically, for desert scatters, the temperature difference between the frequencies is less than 10 K and Ω is greater than 0.01. In the case that the satellite infrared data are not available, AMSU measurements at 183 ± 7 GHz can be used as a substitute because the channel peaks in the lower troposphere and is less affected by the surface. Note that the upwelling brightness temperatures at 89 and 150 GHz are corrected to the two-stream brightness temperatures before they are used in the retrieval.

Figure 8.7 displays a tropical cyclone (TC) IR, IWP, and D_e over an oceanic cyclone system. All images illustrate some interesting features such as spiral rain bands associated with larger IWP and particle size. In particular, IWP and cloud top temperature are well correlated, especially for precipitation clouds with cloud top temperature colder than −40 °C. Large IWP values correspond to colder cloud top temperature, which is a typical feature of convective precipitating systems. This result is also similar to that found in the previous studies for nonraining cirrus clouds [38, 226].

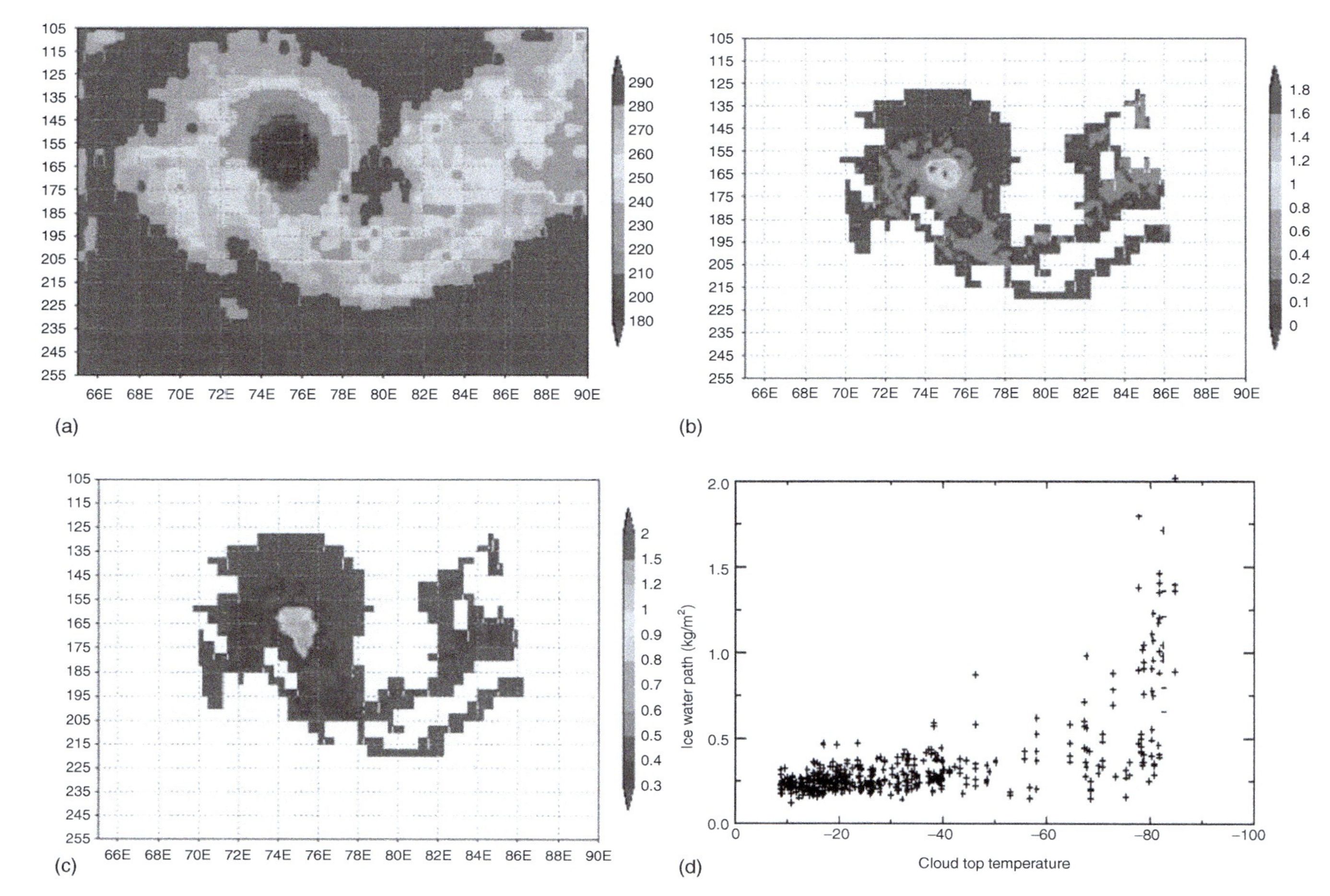

Figure 8.7 (a) IR temperature measurements for a tropical cyclone system occurred on February 28, 2000; (b) retrieved cloud ice water path; (c) ice particle effective diameter; and (d) cloud ice water versus cloud top temperature.

8.4 Cloud Vertical Structures from Microwave Double Oxygen Bands

The microwave radiation with channels near the oxygen absorption lines at 50–70 GHz and 118.75 GHz was explored for profiling the hydrometeors in precipitation systems using their differential responses to absorption and scattering [227–229]. There are two main advantages of using these sounding channels over window channels: (i) lesser sensitivity to surface contributions and, therefore, more global applicability; (ii) the possibility of cloud slicing caused by the increasing height of weighting function peaks with increasing clear-sky absorption [230]. In this section, we further study the impacts of clouds and precipitation on upwelling radiation at microwave sounding channels and quantify the relationship in terms of the hydrometeor size and condensed water content. Observational atmospheric profiles including temperature, vapor, and clouds are used as inputs to the radiative transfer model.

The oxygen absorption band at 50–60 GHz has been used in the first-generation microwave sounding unit (MSU) for probing the atmospheric temperature from space since the launch of NOAA TIROS-N satellite on October 13, 1978. Initially, MSU only had four channels within the band and can provide four deep-layer mean temperatures within the troposphere and low stratosphere. Since the launch of NOAA-15 satellite on May 13, 1998, MSU has been replaced by AMSU-A. AMSU-A has 12 channels (channels 3–14) located at 50–60 GHz oxygen absorption band, and 3 window channels 1, 2, and 15 are located at 23.8, 31.4, and 89 GHz, respectively. The three AMSU-A window channels can be used to detect clouds and precipitation, providing vertically integrated cloud LWP within the TCs through physical retrievals [211]. The eight AMSU-A sounding channels (channels 5–14) can be utilized to detect three-dimensional (3D) warm core structures of TCs [231, 232]. On October 28, 2011, Suomi National Partnership Polar-Orbiting (NPP) satellite was launched with the Advanced Technology Microwave Sounder (ATMS) on board. ATMS not only inherited all the channels from AMSU-A for profiling atmospheric temperature but also added a new channel with its central frequency located at 51.76 GHz to provide temperature information in the lower troposphere. With a higher spatial resolution and broader swath, ATMS can depict much more details of cloud and warm core structures [147, 232]. A direct assimilation of ATMS radiance measurements in Hurricane Weather Research and Forecast (HWRF) model system has a large positive impact on the prediction of 2012 superstorm Sandy track and intensity [233].

A substantial investment has recently been made in China to build and enhance meteorological satellite capabilities. The FY-3 satellite series is a new generation of polar-orbiting satellite series and consists of seven satellites, with about a 2-year interval between two subsequent launches [234]. The first two experimental satellites FY-3A and FY-3B were successfully launched on May 27, 2008 and December 5, 2010, respectively. The third one in the FY-3 series, FY-3C, was launched on September 23, 2013.Both FY-3A and -3C are configured in the mid-morning orbits with their local equator crossing times (LECTs) around 10 am, and FY-3B is in an afternoon orbit with its LECT around 2 pm. There are 11 instruments on board all the missions, providing the measurements in ultraviolet, visible, infrared, and microwave wavelengths [235]. Of particular interest for NWP applications and atmospheric remote sensing is two sounding instruments: Microwave Temperature Sounder (MWTS) and Microwave Humidity Sounder (MWHS). The MWTS on board the first two missions are similar, but not identical, in channel specification to MSU, and the MWHS on board FY-3A and -3B are similar to MHS on board the NOAA's Polar-Orbiting Environmental Satellites (POES) series, which started in 1998 [236]. MWHS and MHS have three channels at 183 GHz and the other two at 89 and 157 GHz. However, MWTS and MWHS on board FY-3C are enhanced with more channels, compared to the previous missions. The FY-3C MWTS has 13 channels and is located at 50–60 GHz oxygen absorption band, whereas the FY-3C MWHS includes additional eight sounding channels located near the 118 GHz oxygen absorption band.

It is the first time for an operational agency to explore the applications of 118 GHz for profiling atmospheric temperature and humidity by a space-borne cross-track microwave sounding instrument. The dual use of 60 and 118 GHz measurements from airborne platforms for assessing cloud and precipitation could be found in Blackwell *et al.* [227] and Bauer and Mugnai [228]. Based on the data from the NPOESS Aircraft Sounder Testbed-Microwave (NAST-M), it was pointed out that brightness temperatures at 118 GHz were much lower than those at 54 GHz due to strong frequency dependence of ice particle scattering in convective areas [227]. Two oxygen sounding channel data at 50–57 and 118.75 GHz during NAST-M were used in the Baysian retrieval algorithm to derive the hydrometeor profiles [228]. This study presents a new capability of using these FY-3C dual oxygen bands to detect the vertical structures of cloud and precipitation associated with hurricane and typhoon systems.

8.4.1 FY-3C Microwave Sounding Instruments and Their Channel Pairing

FY-3C MWHS is a 15-channel cross-track scanning instrument. The MWHS channels 1, 10, 11, 13, and 15 have their central frequencies located at 89, 150, 183.31 ± 1, 183.31 ± 3, and 183.31 ± 7 GHz, respectively, which are similar to those of the five MHS channels (e.g., 89, 157, 183.31 ± 1, 183.31 ± 3, 183.31 ± 7 GHz). The MWHS channels 2–9 are located in the oxygen band near 118.75 GHz. Specifically, the weighting functions of MWHS channels 5–7 are located near 230, 340 hPa and the Earth's surface, respectively. The antenna scans within the angles of ±53.35°, leading to a swath width of about 2600 km. There are a total of 98 field of views (FOVs) along each scan line. The nominal spatial resolution of FY-3C MWHS is 15 km at the nadir for channels 10–15, and that for channels 1–9 is 33 km.

FY-3C MWTS has 13 channels in the frequency range from 50.3 to 57.6 GHz for profiling the atmospheric temperatures from the Earth's surface to about 1 hPa. These 13 channels have the same central frequencies as those of the 13 AMSU-A sounding channels 3–14 [237] and respond to the thermal radiation of the atmosphere at various altitudes. The FY-3C MWTS has an instantaneous FOV of 2.2° and a nominal spatial resolution of 33 km at its nadir. It scans to ±52.725° from the nadir with a total of 90 FOV over a swath of 2600 km. Since MWTS has a smaller beam width, compared to AMSU, the temperature structures within typhoons and hurricanes can be better depicted [5]. Since MWTS does not have the window channels at 23.8, 31.4, and 89 GHz, it is very difficult to detect the clouds and precipitation from MWTS alone. However, the cloud information is contained in the sounding channel measurements and can also be detected using a cloud emission and scattering index (CESI) [2]. In this study, a combination of FY-3C MWTS and MWHS dual oxygen bands are investigated for more details of the cloud and precipitation associated with typhoon systems.

Figure 8.8 shows the weighting functions calculated by the Community Radiative Transfer Model (CRTM) from a standard US atmospheric profile for all channels of both MWHS and MWTS on board FY-3C [29, 238]. Five microwave humidity sounding channels (i.e., channels 11–15) with their frequencies centered around 183.31 GHz have their weighting functions evenly distributed in the middle and low troposphere. The weighting functions of these channels are narrowly peaked and are similar to those of MHS. The eight new microwave humidity sounding channels (i.e., channels 2–9, color) with frequencies centered near 118.75 GHz were added to FY-3A/B MWHS to allow profiling both temperature and water vapor from a single instrument since the 118.75 GHz is located at the O2 absorption band and its absorption intensity is temperature dependent.

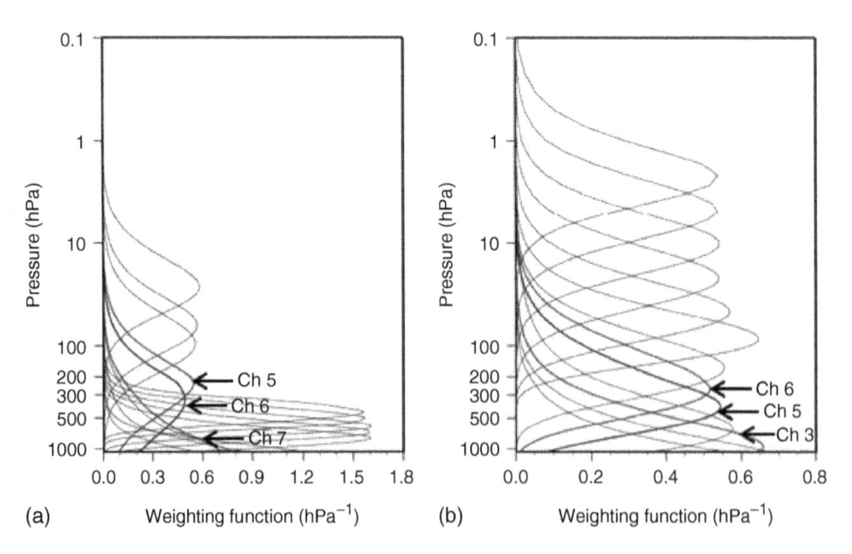

Figure 8.8 Weighting functions (WFs) for the (a) FY-3C MWHS and (b) MWTS channels. The WFs of paired channels of MWHS and MWTS are indicated by colored curves.

These new MWHS O2 channels have broader weighting functions compared to MWHS channels 11–15 and extend from the surface to about 20 hPa. The weighting functions at MWTS channels are distributed from the Earth's surface to about 1 hPa and are identical to those of AMSU-A sounding channels. Thus, the two instruments (MWTS and MWHS) can be paired in channels for remote sensing of vertical structures of clouds. Specifically, the three paired MWHS and MWTS channels indicated in color bars – MWHS channel 5 and MWTS channel 6; MWHS channel 6 and MWTS channel 5; and MWHS channel 7 and MWTS channel 3 – are chosen for this study. The three paired MWHS and MWTS channels have similar oxygen absorption intensity in magnitude. Here, the variation of the total optical depth with respect to frequency is computed by the line-by-line radiative transfer model [11, 239] using the US standard atmosphere. Each of the three MWTS channels has a single band located in the valley of the oxygen absorption line. Each of the three MWHS channels has two bands located symmetrically on the two sides of the absorption peak near 118.7 GHz. Since the absorption intensities at both sides are similar, the addition of receiver signals at both sides allows reducing the random noise. In addition, the shift in the receiver carrier frequency does not significantly affect the average absorption intensity, and thus, the radiometer has a stable performance.

Antenna brightness temperatures from both FY-3C MWHS and MWHS for Super Typhoon Neoguri that occurred over the Pacific Ocean on July 6, 2014, have been used in this study. Calibration details on the conversion from radiance to antenna brightness temperature can be found in Gu *et al.* [240] and You *et al.* [241]. The European Center for Medium-Range Weather Forecasting (ECMWF) analysis profiles of temperature and water vapor within 55S–55N on July 1, 2014, are used for generating a linear relationship for each of the three paired MWHS and MWTS channels with the same weighting function peak altitude. Detailed structures of a CESI calculated from the three paired MWHS and MWTS channels with their weighting function peaks located at three different altitudes are calculated and shown. As a reference, the cloud liquid and ice water contents from the ECMWF analysis are provided.

8.4.2
Typhoon Neoguri Observed by MWHS and MWTS

The spatial distributions of brightness temperatures observed from double oxygen band microwave sounding instruments on board FY-3C within Super Typhoon Neoguri at 1236 UTC July 6, 2014, are shown in Figure 8.9. The following three paired channels are selected: MWHS channel 5 and MWTS channel 6; MWHS channel 6 and MWTS channel 5; and MWHS channel 7 and MWTS channel 3. The center of Super Typhoon Neoguri was located at (130.1°E, 19.1°N) at this time. Since both MWHS and MWTS are the cross-track radiometers, the brightness temperatures have an obvious scan-dependent feature. For the six selected tropospheric channels shown in Figure 8.9, the weighting functions peak at a lower altitude at a smaller scan angle compared to that at a larger

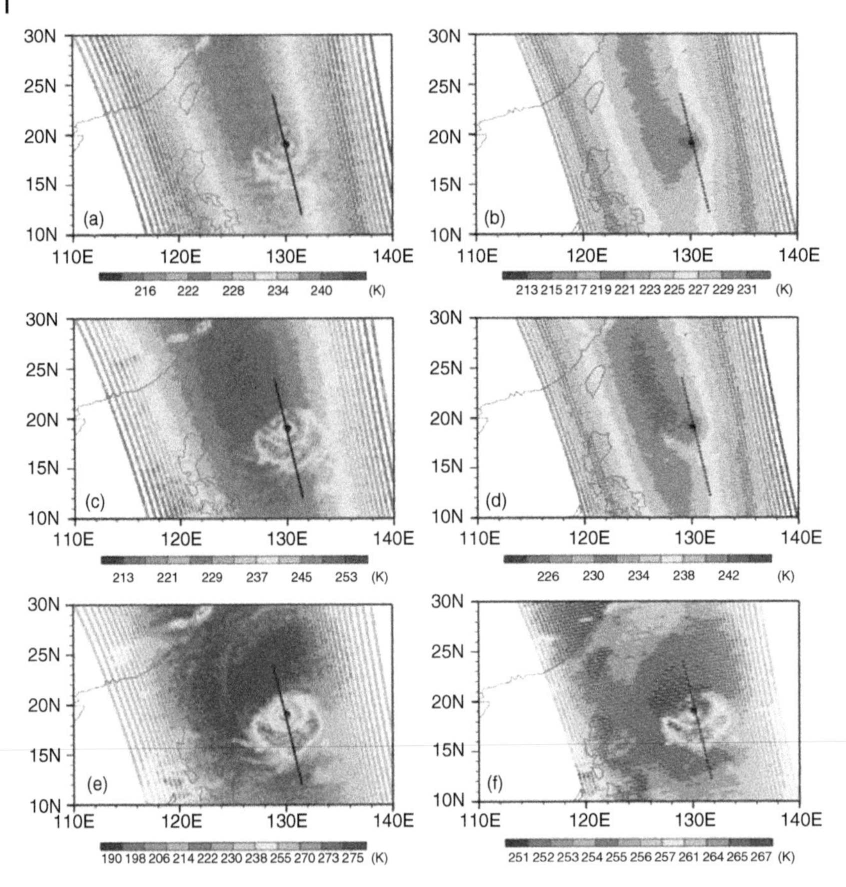

Figure 8.9 Spatial distributions of brightness temperatures for the three paired channels: (a) MWHS channel 5 (118 ± 0.8 GHz) and (b) MWTS channel 6 (54.94 GHz); (c) MWHS channel 6 (118.75 ± 1.1 GHz); (d) MWTS channel 5 (54.40 GHz); (e) MWHS channel 7 (118.75 ± 2.5 GHz); and (f) MWTS channel 3 (52.80 GHz) at 1236 UTC July 6, 2014. The center of Super Typhoon Neoguri is located at (130.1°E, 19.1°N) and indicated by a hurricane symbol in black.

scan angle. The weighting function is the lowest at the nadir (e.g., zero scan angle), and thus, the brightness temperatures in the atmospheres under clear-sky conditions are the highest as expected. However, typhoon perturbations to the brightness temperature fields are visibly present in the FY-3C measurements of all six channels shown in Figure 8.9. A typhoon rain-band-like cold brightness temperature distribution is located in the southeast half of the ring surrounding the typhoon center at MWHS channels 5–7 (Figure 8.9a, c, d) and MWTS channel 3 (Figure 8.9f). For MWTS channels 5 and 6, whose weighting functions are located near 400 and 250 hPa, respectively, a warm perturbation can be noticed at the center of Typhoon Neoguri.

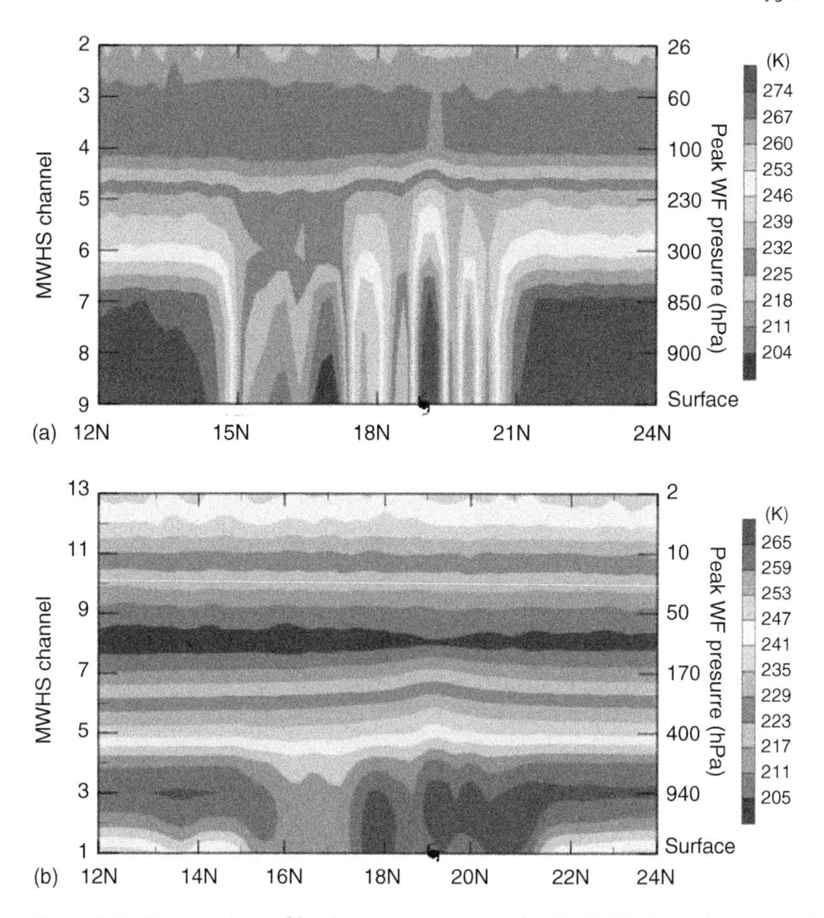

Figure 8.10 Cross sections of brightness temperatures for (a) MWHS channels 2–9 and (b) MWTS channels 1–13 in the along-track direction through Super Typhoon Neoguri's center (see the black line in Figure 8.9).

Figure 8.10 shows two cross sections of antenna temperatures for MWHS channels 2–9 (Figure 8.10a) and MWTS channels 1–13 (Figure 8.10b) in the along-track direction that passes through the center of Super Typhoon Neoguri. The limb effect of cross-track radiometer on brightness temperature mentioned before is avoided in such cross sections along a fixed scan angle. The brightness temperature is as low as 204 K for MWHS window channel 9 to the south of the Super Typhoon Neoguri center (Figure 8.10a). MWHS channel 5 with its peak WF located at 850 hPa is the warmest. A warm anomaly is found near the hurricane center throughout the troposphere. The eye of Super Typhoon Neoguri is characterized by a warm brightness temperature of a similar magnitude to that in the Neoguri's environment for all eight MWHS channels. The measured MWHS brightness temperature is as high as 274 K at the typhoon center below

850 hPa. In other words, the brightness temperatures in the rain-band regions are more than 100 K lower than those in the hurricane eye and its environment. There are two narrow bands of low brightness temperatures next to the warm center that are associated with the upward motion, cloud, and precipitation in the eyewall region. Outside the eyewall region, there are two narrow bands of warm brightness temperatures, which reflect the locations of clear streams within Super Typhoon Neoguri. Such a large brightness temperature contrast between rain bands and clear streaks (including the eye) suggests the robustness for MWHS to capture the thermal and cloud features of typhoons. The asymmetric structures of Super Typhoon Neoguri are fully captured by MWHS observations.

The cross section of MWTS brightness temperature distribution (Figure 8.10b) does not provide as much detailed structures as that of MWHS within Neoguri due to a weaker cloud emission and scattering effect at lower frequency oxygen band. The brightness temperatures within the eye and clear streaks are about 6 K higher than those in the rain bands and the environment at the same altitudes, which are of a similar magnitude to the known warm temperature anomaly in the eye of typical typhoons or hurricanes. This number is much smaller than the difference in the brightness temperatures between the troposphere and stratosphere, which is more than 60 K (see Figure 8.10b).

8.4.3 The Cloud Emission and Scattering Index (CESI)

Early studies [53, 211] show that microwave radiometers, such as AMSU-A and MHS, offer limited information regarding cloud vertical structures. Microwave measurements at lower frequency window channels could be used to derive the total cloud LWP of nonraining clouds. By using the measurements from multiple microwave window channels, the atmospheric WVP and cloud LWP can be resolved simultaneously. Specifically, the cloud LWP can be derived from two AMSU-A window channels at 23.8 and 31.4 GHz [211]. Microwave radiation interacts with ice particles primarily through scattering. Since ice clouds are above the absorbing part of the atmosphere, the emission and cloud temperatures simply modulate the upwelling microwave radiation from below. Therefore, the cloud IWP can be derived from two MHS window channels at 89.0 and 157.0 GHz [53, 211].

Instead of relying on the two low-frequency surface channels for obtaining cloud information, FY-3C involved an additional higher frequency oxygen band centered around 118.75 GHz with eight new MWHS channels. By combining these channels with the low-frequency oxygen band of MWTS, it is expected that cloud LWP from the top of the atmosphere to different pressure levels can be derived. Figure 8.11 provides four scatter plots of CRTM simulated and FY-3C observed brightness temperatures at MWHS channel 6 and MWTS channel 5. The brightness temperature simulations are obtained by using the diverse profile data sets from the ECMWF 91-level short-range forecasts [242] as input to CRTM. Only the clear-sky data at the nadir within a latitude range of 55S–55N

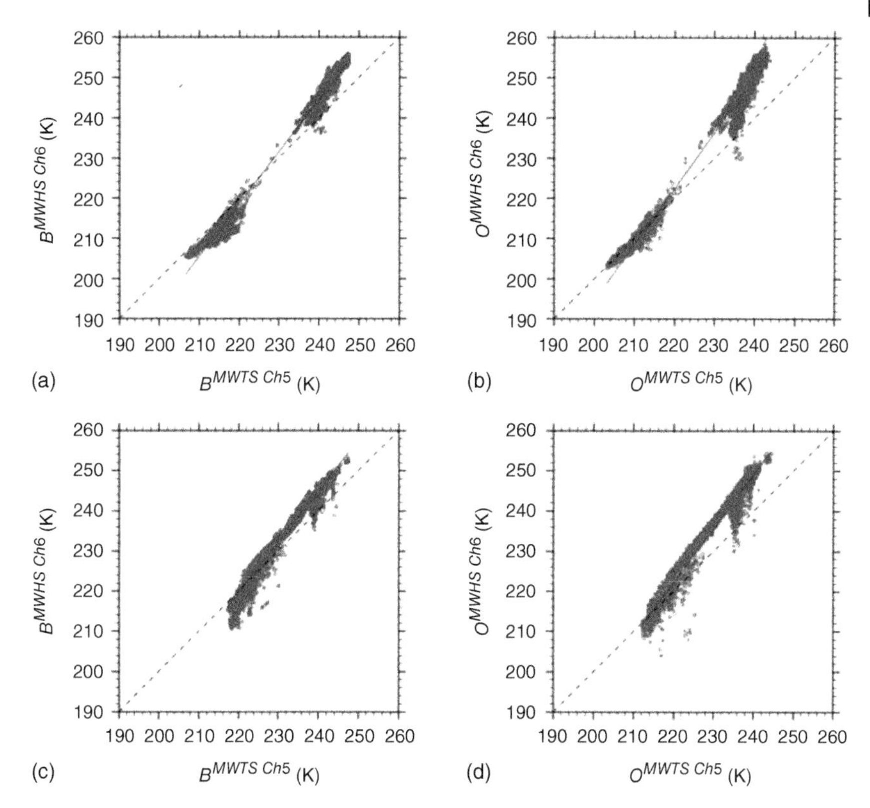

Figure 8.11 Scatter plots of CRTM calculated brightness temperatures using ECMWF analysis as input (left panels) and FY-3C observations (right panels) for the paired MWHS channel 6 (118.75 ± 1.1 GHz) and MWTS channel 5 (54.40 GHz) using all clear-sky data at the nadir over (a)–(b) land and (c)–(d) oceans within a latitude range of 55S–55N, on July 1, 2014.

on July 1, 2014, are shown in Figure 8.11. A linear relationship between the two paired channels is found for both model simulations and FY-3C observations over both land (Figure 8.11a,b) and ocean (Figure 8.11c,d). The slopes of the variation in the MWHS brightness temperatures with respect to MWTS brightness temperatures are all greater than 1. More variability and larger slopes are found in the observations. A linear regression relationship is also obtained for each of the other two selected paired MWHS and MWTS channels.

As a conceptual demonstration for using microwave radiometers to probe the vertical structures of cloud, a CESI can be defined [243]:

$$\mathrm{CESI} = T^{reg}_{b,\,H} - T^{obs}_{b,\,H}, \tag{8.38}$$

where $T^{obs}_{b,H}$ is the level-1b MWHS brightness temperature, and $T^{reg}_{b,H}$ is calculated from the level-1b MWTS brightness temperature ($T^{obs}_{b,T}$) using the following linear regression model:

$$T_{b,H}^{reg} = \alpha T_{b,T}^{obs} + \beta, \tag{8.39}$$

where α and β are the regression coefficients calculated through a linear fit between the MWHS brightness temperature observations ($T_{b,H}^{obs}$) and simulations ($T_{b,H}^{ECMWF}$) using ECMWF analysis profiles; we have atmospheric profiles including temperature within a latitude zone of 55S–55N on July 1, 2014. Since the ECMWF model does not have hydrometer size and size distribution, a fixed size of all-phase particles with a gamma size distribution is used in radiative transfer model simulation.

Figure 8.12 shows the spatial distributions of CESI within Super Typhoon Neoguri at 1236 UTC July 6, 2014, derived from the three selected paired channels. The ECMWF integrated water paths from the top of the atmosphere to three different pressure levels close to the peaks of the weighting function of the three paired channels at 1200 UTC July 6, 2014 are also shown in Figure 8.12. Super Typhoon Neoguri's eye, eyewall, and rain bands are clearly seen in the CESI distributions. A vertical continuity of the cloud structures is reflected in the CESI distributions derived from the three paired channels located at three different altitudes. On the contrary, except for a broad region of cloud located to the southeast of the Typhoon Neoguri center, the detailed eye and eyewall structures are not well defined in the ECMWF forecasts.

8.5 Summary and Conclusions

Microwave sounding instruments have some channels that can be uniquely utilized to derive cloud liquid water and IWPs. The cloud liquid algorithms are physically based using two channels at 23.8 and 31.4 GHz. In the past, cloud liquid water algorithms for SSM/I and special sensor microwave imager/sounder (SSMIS) were regression-based and thus required some *in situ* measurements to generate the regression coefficients. The *in situ* measurements are rarely available for the algorithm developments of new mission, and thus, the new instrument data need to be calibrated to the heritage sensors prior to the generation of the cloud products from the new mission. With the developments of the physical retrieval algorithms for AMSU-A and ATMS instruments, the cloud liquid water products can be developed right after the satellite launch.

For cloud IWP retrieval, we developed a physical retrieval algorithm that determines both cloud IWP and ice particle size simultaneously using 2-mm wavelength channels from MHS and ATMS WG band measurements. It has been shown from both aircraft and satellite observations that the cloud ice water depicted at the millimeter wavelengths is mostly associated with the convection and surrounding anvil cirrus clouds. In the deep convective regions, the retrieval

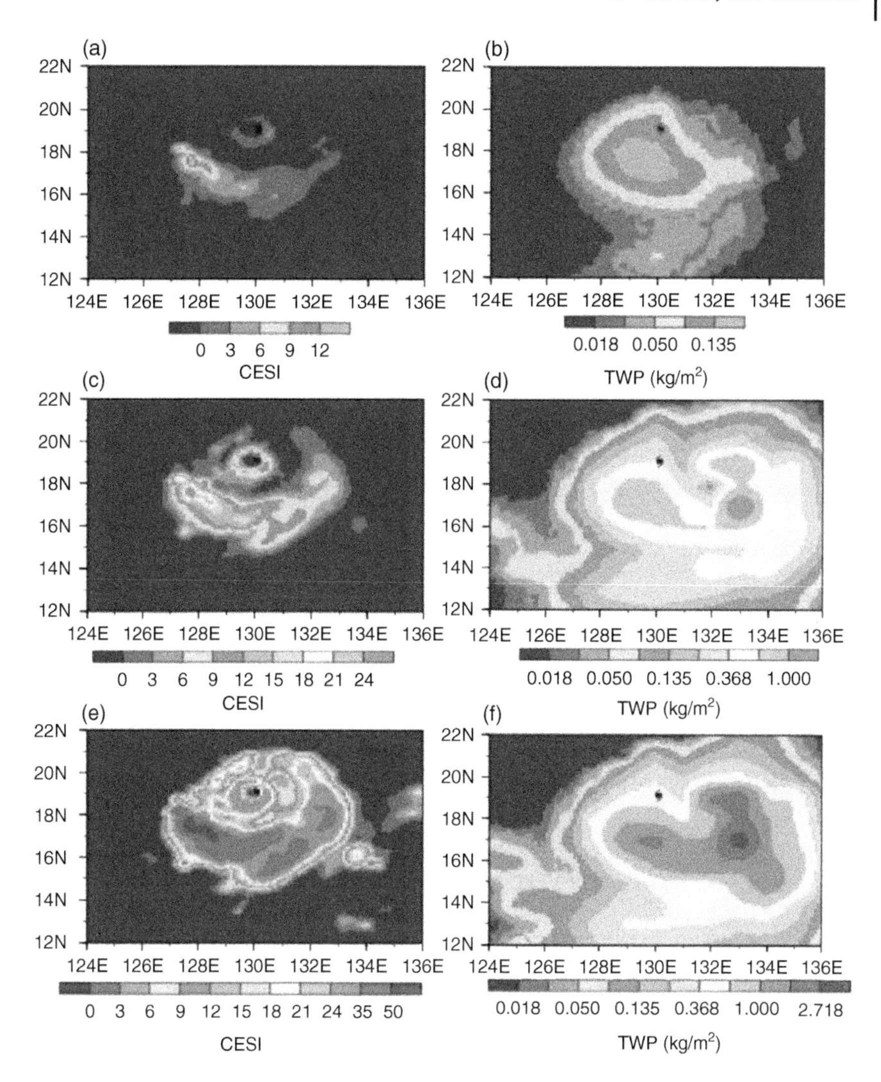

Figure 8.12 Cloud emission and scattering index (CESI) derived from FY-3C (a) MWHS channel 5 (118 ± 0.8 GHz) and MWTS channel 6 (54.94 GHz); (c) MWHS channel 6 (118.75 ± 1.1 GHz) and MWTS channel 5 (54.40 GHz), and (e) MWHS channel 7 (118.75 ± 2.5 GHz); and (f) MWTS channel 3 (52.80 GHz) for Typhoon Neoguri at 1236 UTC July 6, 2014, and the vertically integrated liquid and ice (total) water path (TWP) from the top of the atmosphere to (b) 200 hPa, (d) 500 hPa, and (f) 850 hPa at 1200 UTC July 6, 2014. The center of super typhoon Neoguri is indicated by a hurricane symbol in black. TWP is calculated from ECMWF global model analysis field.

algorithm becomes insensitive to the total amount of ice since the scattering from large ice particles is in the geometrical optical regime.

The launch of Chinese FengYun-3C satellite is remarkable since there are two unique microwave sounders on board. MWHS and MWTS have provided us new tools for monitoring the atmospheric temperature, water vapor, and cloud due to their unique capability through the dual oxygen absorption band sounding channels. It is demonstrated that the vertical and horizontal structures of cloud and precipitation within Typhoon Neoguri (2014) can be detected using a simple CESI that is computed from the MWHS and MWTS paired channel measurements.

It is pointed out that the frequency location of the MWHS channels can be made more evenly distributed around 118.75 GHz compared to the current MWHS on board FY-3C so that more paired channels can be made available with MWTS. Currently, there are almost no high-frequency oxygen band channels in the middle and low troposphere, yet there are three MWHS channels (i.e., channels 7–9) that have the peak weighting functions located near the surface. It is anticipated that an improved channel design will be explored for the future missions of other space agencies.

9
Microwave Remote Sensing of Atmospheric Profiles

9.1
Introduction

Microwave sounding measurements carried out under cloudy and precipitation conditions carry a wealth of information on the temperature and water vapor profiles as well as cloud hydrometeors. The effect of hydrometeors on the brightness temperatures measured by the microwave sensors may be negligible, significant, or something in between, depending on the spectral region considered and on the type and intensity of the precipitation, making these microwave and millimeter-wave sensors an ideal tool to probe the atmosphere in weather active areas. Kidder *et al.* [244] have provided a comprehensive overview of the utilization of advanced microwave sounding unit (AMSU) data in estimating tropical cyclone (TC) intensity, retrieving upper tropospheric temperature anomaly and gradient wind, and determining the TC precipitation potential. Knaff *et al.* [245] and Zhu *et al.* [231] have retrieved and analyzed the atmospheric temperatures in hurricane systems. Spencer and Braswell [246] estimated TC's maximum sustained wind (MSW) using the temperature gradient derived from AMSU-A measurement. Demuth *et al.* [247, 248] have developed regression algorithms to estimate TC MSW, minimum sea-level pressure (MSLP), and TC size (radii of winds). In addition to monitoring TCs, microwave observations can be easily utilized in numerical weather prediction (NWP) models because microwave radiances respond linearly to atmospheric temperatures. Zhu *et al.* [231] have developed a scheme to construct the TC's initial vortex for a mesoscale model simulation of Hurricane Bonnie (1998). The atmospheric temperatures were retrieved from AMSU-A data, and then a nonlinear balance equation was used to derive the 3D wind field of Hurricane Bonnie. Zhu and Gelaro [249] have indicated that AMSU-A measurements have the largest positive impact on global medium-range forecasts among all the satellite observations in the Gridpoint Statistical Interpolation (GSI) data assimilation system.

In this chapter, we first present a regression-based algorithm to derive the temperature profiles under hurricane conditions from microwave oxygen sounding measurements. Then, we propose a physical algorithm to simultaneously retrieve the vertical profiles of temperature, water vapor, and hydrometeor parameters.

Passive Microwave Remote Sensing of the Earth: For Meteorological Applications, First Edition. Fuzhong Weng.

The surface boundary layer is also treated dynamically by including the surface emissivity spectrum and the skin temperature as part of the control parameter vector. Including the hydrometeors in the retrieved state vector increases the number of degrees of freedom in the solution-finding process. It is important to note that these degrees of freedom are also due to the limited number of channels available.

9.2 Microwave Sounding Principle

As illustrated in Figure 2.2, a number of channels within the oxygen absorption band can be used to profile the atmospheric temperature. The sounding principle at the microwave frequency can be proved through a radiative transfer equation in which single- and multiple-scattering terms are neglected and there is no azimuth-dependent terms. For a channel that is not affected by the surface emission, the brightness temperature is the same as the upwelling radiation in Eqs. (3.44) and (3.45), namely

$$T_b = \int_{\Upsilon_s}^{1} T(\Upsilon)\ \mathrm{d}\Upsilon, \tag{9.1}$$

where Υ is the atmospheric transmittance with the reference to the top of the atmosphere, and T is the atmospheric temperature. The subscript s denotes the parameter at the surface. The transmittance is related to the atmospheric optical thickness such that

$$\Upsilon = \exp\left[-\frac{(\tau_s - \tau)}{\mu}\right]. \tag{9.2}$$

The atmospheric weighting function is defined as

$$W = \frac{\partial \Upsilon}{\partial \ln p}, \tag{9.3}$$

where p is the atmospheric pressure. Here, a logarithmic function is used in the pressure coordinate. Thus, the brightness temperature for a channel i can be written as

$$T_{b,i} = \int_{p_s}^{0} T(p) W_i d \ln p. \tag{9.4}$$

The integration in Eq. (9.4) can also be discretized as

$$T_{b,i} = \sum_{j=1}^{L} c_i T_j W_{i,j}, \tag{9.5}$$

where L is the number of layers for atmospheric vertical stratification, and c_i is the coefficient relating the temperature to the Planck function, which is dependent on the wavelength.

Equation (9.5) illustrates that the microwave brightness temperature is determined from a linear combination of vertical temperature profile with the weight determined by $W_{i,j}$. Since atmospheric sounding channel is selected with its weighting function having peaked at a certain height, the brightness temperature measured from satellites can directly reflect the physical temperature roughly at that height. However, since the function W is of a particular shape as seen in Figure 9.1, the radiation at a given channel is also contributed from a layer of atmosphere.

Microwave temperature sounding instruments including AMSU-A/B, Microwave Humidity Sounder (MHS), and Advanced Technology Microwave Sounder (ATMS) have been launched on board US and European satellites. Both AMSU-A and ATMS measure the thermal radiation at microwave frequencies ranging from 23.8 to 89.0 GHz and are mainly designed to provide information on the atmospheric temperature profiles. In particular, AMSU-A channels (3–14) and ATMS channels (3–16) respond to the thermal radiation at various altitudes because of their weighting function distributions. Several window channels at frequencies of 31.4, 89, and 150 GHz have been used as primary channels to determine the liquid water and ice water content of the clouds because they

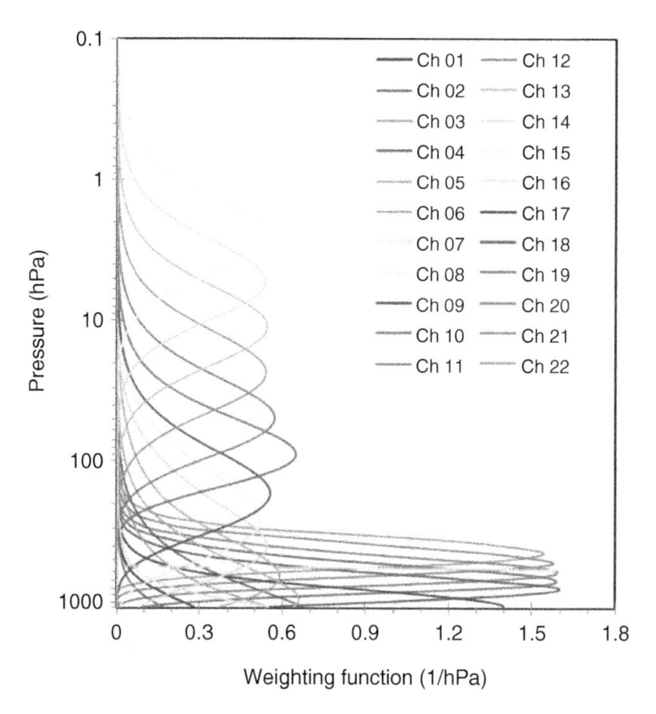

Figure 9.1 Vertical distribution of the ATMS weighting function at nadir computed from the mid-latitude standard atmospheric profile.

directly respond to the thermal emission of liquid droplets and scattering of ice particles, as discussed in Chapter 8.

When a sounding channel becomes semitransparent, the contributions to the brightness temperature from surface emission and reflected downwelling radiation can be significant. Typically, an uncertainty in the surface emissivity model can result in major errors in probing the lower tropospheric profile near the surface boundary. Using Eq. (3.44), we can study the impacts of the surface emissivity on the sounding channel through

$$\Delta T_b = (T_s - T_d)\Upsilon\Delta\varepsilon. \tag{9.6}$$

Thus, an error in emissivity is directly translated into an error in brightness temperature. Table 9.1 summarizes the brightness temperature perturbations at a few selected frequencies to a surface emissivity variation of 0.04. Obviously, at a window channel where the transmittance is relatively large and T_d is small, the emissivity uncertainty has a much larger impact on the brightness temperatures. For example, at 150 GHz, T_b is ~7.0 K for a surface pressure of 1000 hPa (mb). For a surface pressure of 600 hPa, T_b increases to ~8.0 K. At the sounding channels near the 50–60 GHz oxygen absorption band, T_b decreases as the frequency approaches the center of the absorption band. At 52.8 GHz, T_b increases from 0.2 to 2.3 K as the surface pressure decreases from 1000 to 600 mb.

At the sounding channels near the 183.3-GHz water vapor absorption band, T_b strongly varies with the water vapor amount, surface pressure, and frequency. At 183.3 ± 7 GHz, which is furthest from the band center, T_b increases from 1.8 to 6.0 K as the water vapor amount increases from 0.5 to 2.0 mm. When the surface pressure decreases from 1000 to 600 mb, T_b reaches up to 7.9 K. At 183.3 ± 1 GHz,

Table 9.1 Brightness temperature responses to surface emissivity of 0.04.

Frequency (GHz)	T_s = 230 K and TPW = 0.5 mm					
	P_s = 600 (mb)			P_s = 1000 (mb)		
	T_d (K)	τ	ΔT_B (K)	T_d (K)	τ	ΔT_B (K)
6.925	1.50	0.99	9.08	4.00	0.98	8.87
10.65	1.60	0.99	9.07	4.40	0.98	8.84
18.7	2.30	0.99	9.02	6.20	0.97	8.70
23.8	3.30	0.98	8.93	8.50	0.96	8.51
36.5	7.10	0.97	8.63	19.10	0.91	7.69
50.3	49.30	0.77	5.59	112.50	0.49	2.29
52.8	111.20	0.49	2.34	188.60	0.15	0.25
89	8.20	0.96	8.54	22.30	0.99	7.46
150	4.40	0.98	8.84	12.50	0.94	8.21
183.3 ± 7	16.60	0.93	7.89	43.50	0.81	6.02
183.3 ± 3	55.30	0.75	5.24	104.10	0.54	2.71
183.3 ± 1	134.60	0.39	1.50	160.10	0.29	0.81

the impact of surface emissivity on the brightness temperature is the smallest (0.01 K) for a water vapor amount of 2 mm. This implies that the uncertainty in surface emissivity over a high-elevation terrain and in a moisture-deficient atmosphere will significantly increase the uncertainty in simulating the brightness temperatures at microwave sounding channels.

9.3 Regression Algorithms

From a set of surface blind channels, Eq. (9.5) illustrates a much simplified approach for atmospheric temperature sounding. Essentially, the retrieval of the temperature profile from microwave measurements is a linear problem because the weighting functions at various temperature sounding channels are relatively stable and therefore are the fixed coefficients. Alternatively, temperatures at any pressure level can be expressed as a linear combination of brightness temperatures measured at various sounding channels, namely

$$T(p) = C_0(p,\mu) + \sum_{j=4}^{11} C_j(p,\mu) T_{b,j}(\mu), \tag{9.7a}$$

where C is derived using collocated radiosonde and satellite data, and j is the AMSU-A channel index. For AMSU-A, C is derived separately at each pressure level and at the cosine of the local zenith angle [231]. In collocating AMSU and rawinsonde observations, two observations within 1 h in time and 100 km in space are collected. At each angle, the data pairs should be large enough (100–200) to make the regression results statistically reliable. A similar algorithm was also developed for ATMS applications, but the coefficients are independent of the local zenith angle as a predictor [232], namely

$$T(p) = C_0(p) + \sum_{j=5}^{12} C_j(p) T_{b,j}(\mu) + C_{sz}(p)\mu^{-1}, \tag{9.7b}$$

where $T_{b,j}$ is the ATMS antenna brightness temperature at channel j, and all the regression coefficients are as listed in Table 9.2.

The performance of the AMSU-A temperature retrieval can be illustrated from a vertical profile of the root-mean-square (RMS) error. The RMS is computed on the basis of all the rawinsonde observations used in the regression. As shown in Figure 9.2, the RMS increases from 700 hPa downward due partly to the lack of sharp weighting function and partly to the atmospheric structures being more variable closer to the boundary layer. Large errors also occur above 250 hPa where the reversal of the temperature lapse rate results in small changes in the brightness temperature.

In general, the vertical layers of temperature profiles from a regression-type of algorithm should be set close to the number of sounding channels available.

Table 9.2 Coefficients of the ATMS regression-based temperature retrieval algorithm.

Level (hPa)	C_0	C_5	C_6	C_7	C_8	C_9	C_{10}	C_{11}	C_{12}	C_{sz}
50	−19.380	−0.010	−0.173	0.071	0.592	−0.325	0.061	1.115	−0.240	−0.259
70	21.016	0.197	−0.634	0.209	0.493	−0.371	0.852	0.495	−0.274	−7.434
100	52.627	−0.069	0.508	−0.578	−0.870	0.641	2.034	−0.690	−0.140	−12.392
125	92.005	−0.074	0.664	−0.975	−1.070	1.610	1.249	−0.706	−0.050	−10.029
150	112.761	0.168	0.229	−1.015	−0.848	2.173	0.320	−0.611	0.137	−9.430
175	106.053	0.460	−0.384	−0.878	−0.248	2.197	−0.520	−0.307	0.257	−7.987
200	105.279	0.681	−1.076	−0.516	0.720	1.474	−1.155	0.196	0.242	−5.098
225	118.178	0.657	−1.415	0.007	1.420	0.494	−1.473	0.668	0.109	1.775
250	105.111	0.520	−1.350	0.534	1.644	−0.373	−1.350	0.848	−0.008	10.234
275	60.489	0.368	−1.012	0.983	1.393	−0.882	−0.962	0.756	−0.041	17.970
300	19.265	0.248	−0.649	1.296	0.987	−1.116	−0.577	0.562	−0.010	23.244
350	−34.306	0.109	−0.108	1.573	0.353	−1.199	−0.076	0.265	0.023	29.276
400	−62.373	−0.026	0.433	1.555	−0.197	−1.042	0.221	0.062	0.047	32.267
450	−65.056	−0.201	0.966	1.398	−0.692	−0.763	0.380	−0.119	0.097	34.310
500	−60.065	−0.409	1.470	1.207	−1.081	−0.486	0.462	−0.245	0.126	36.971
550	−46.377	−0.588	1.872	1.009	−1.326	−0.290	0.492	−0.295	0.116	38.936
600	−29.615	−0.651	2.063	0.831	−1.405	−0.176	0.484	−0.307	0.095	38.910
650	−14.066	−0.621	2.124	0.649	−1.410	−0.063	0.436	−0.310	0.084	37.242
700	−10.486	−0.541	2.146	0.440	−1.344	0.051	0.357	−0.256	0.043	35.241
750	−19.197	−0.440	2.131	0.247	−1.205	0.129	0.262	−0.153	−0.013	33.533
800	−39.417	−0.293	2.001	0.113	−0.942	0.123	0.157	−0.031	−0.058	31.782
850	−74.453	−0.015	1.736	−0.036	−0.611	0.141	0.064	0.078	−0.105	28.477
1000	23.847	1.189	−0.279	−0.147	0.545	−0.001	−0.430	0.248	−0.092	1.447

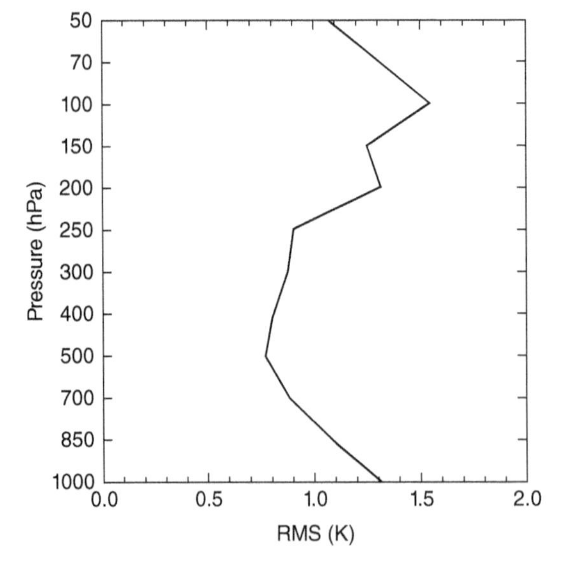

Figure 9.2 Vertical distribution of RMS errors of the AMSU-derived temperatures.

The retrieval accuracy for a temperature profile with an inversion is generally poor because of the coarse vertical resolution of the microwave sounding channels. Similarly, it is also difficult to determine the temperature profile near the tropopause because of a sign reversal of the temperature lapse rates from the troposphere to the stratosphere. In addition, some sounding channels are affected by surface radiative and thermal properties, which are functions of the geophysical parameters such as canopy water content, terrain height, desert constituents, and snow and glacial age. In addition, some channels may be strongly affected by the emission and scattering from raindrops and large ice particles during storms.

Using the ATMS retrieval algorithm, we first studied several TCs that occurred in the Southern Hemisphere TC season of 2012 and compared the results with the retrieval results from the NOAA-15 AMSU-A observation using the old algorithm. TC Giovanna is the second intense TC in the 2011–2012 Southern Indian Ocean cyclone season. TC Giovanna developed off the eastern coast of Madagascar during February 9–21, 2012. Figure 9.3a shows the vertical cross-section of the ATMS temperature anomaly (temperature minus environmental temperature at each level) retrieved at 2130 UTC on February 11, 2012. The temperature anomaly is computed with respect to the environmental temperature, which is an averaged temperature within 15° lat/lon of the storm but without the temperature perturbations near the core region depending on the size of the storm. Giovanna was a category-3 hurricane according to the Saffir–Simpson scale, with an MSW of 100 mph and MSLP of 948 hPa. A maximum warm core of ~6 K can be found at the 250 hPa level. The cold anomalies below 500 hPa correspond to the heavy precipitation regions in the storm eyewall, which indicated that the radius of maximum wind was ~100 km at that time. As a

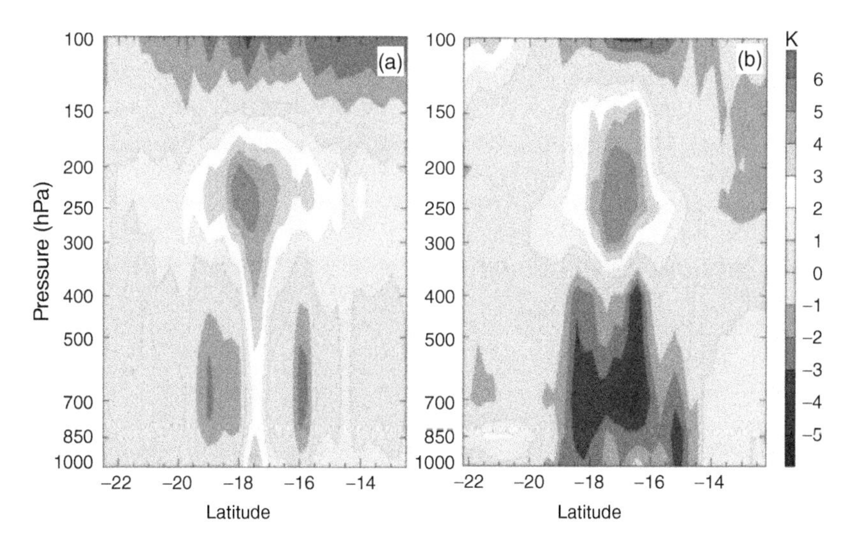

Figure 9.3 Vertical cross sections of temperature anomalies for tropical cyclone Giovanna retrieved from (a) Suomi NPP ATMS along 58.2E longitude at 2130 UTC, (b) NOAA-15 AMSU-A along 59.2E longitude at 1300 UTC, 11 February 2012.

comparison, NOAA-15 AMSU-A-retrieved temperatures (Figure 9.3b) were derived at 1300 UTC on February 12, 2012, when the satellite passed over Giovanna, which was about 8.5 h earlier than the ATMS observations. Giovanna stayed as category-3 hurricane at this time, with an MSLP of 944 hPa. ATMS measurements show a wider scan swath and finer horizontal resolution compared to AMSU-A. Over 850 hPa, ATMS retrievals clearly depict the cold temperature anomalies in TC's spiral rain bands and the warm core of the storm. The TC's warm core feature retrieved from ATMS extended from the 200 hPa level to the ocean surface. However, the AMSU-A retrievals could not fully resolve the storm's eye at lower levels because of its coarse resolution. AMSU-A-retrieved warm core over 250 hPa and lower level cold anomalies are stronger than those of ATMS. It can be seen that the warm core heights and cold eyewall locations are similar for the two retrievals. The differences in the intensity of temperature anomaly are due partially to the different algorithms used for the retrievals and also to the difference between the overpassing times of NOAA-15 and Suomi NPP satellites.

Hurricane Sandy was the most devastating storm that occurred in 2012 in the Northern Hemisphere. It started as a tropical wave in the Eastern Caribbean Sea on October 19, 2012, and strengthened into the 18th Atlantic tropical depression at 1500 UTC 22 October 2012. Figure 9.4a indicates that there was a weak, warm core with a maximum of about 4.0 K at the incipient stage at 0630 UTC 24 October. The maximum warm anomaly is located at the 400 hPa level. The lower level cold temperature anomalies were located at ~100–200 km to the north and south of the storm center, corresponding to a broad range of precipitation surrounding the storm. At 1500 UTC 24 October, Hurricane Sandy intensified as the 10th hurricane of the 2012 Atlantic hurricane season, with an MSLP of 973 hPa and MSW of 80 mph. During the period from 0600 UTC 25 October to 0000 UTC 26 October, Sandy reached category-2 hurricane intensity with a maximum wind exceeding 100 mph. Later, on October 25, Sandy encountered an upper-level trough from its west. The large vertical wind shear weakened Sandy and created asymmetric structures of the storm. Precipitation was mostly concentrated on the northwest side of the storm, which was consistent with the temperature anomaly feature retrieved from ATMS (Figure 9.4b). The first peak of maximum warm anomaly of the storm, 8.9 K over the 400 hPa level, was found at 0710 UTC 26 October. Weak upper-level warm anomalies could also be found on the south side of the storm, where less rainfall occurred. The cold anomalies developed on the north side of the storm correspond to the rainfall bands over there. The asymmetric pattern of Sandy lasted for about 2 days until October 28, during which another mid-latitude cold frontal system encountered with Sandy. The ATMS-retrieved temperature anomaly captured the asymmetric structure at 1810 UTC 27 October (Figure 9.4c). There was a partial eyewall (cold anomaly) on the northwest side of the storm. The warm core tilted northwestward with height on upper levels. After the interaction with the cold frontal system, Sandy experienced a second intensification period and reached a peak of 940 hPa for MSLP on October 29. Figure 9.4d shows an 8.5 K maximum warm core at ~450 hPa level at 1730 UTC 29 October,

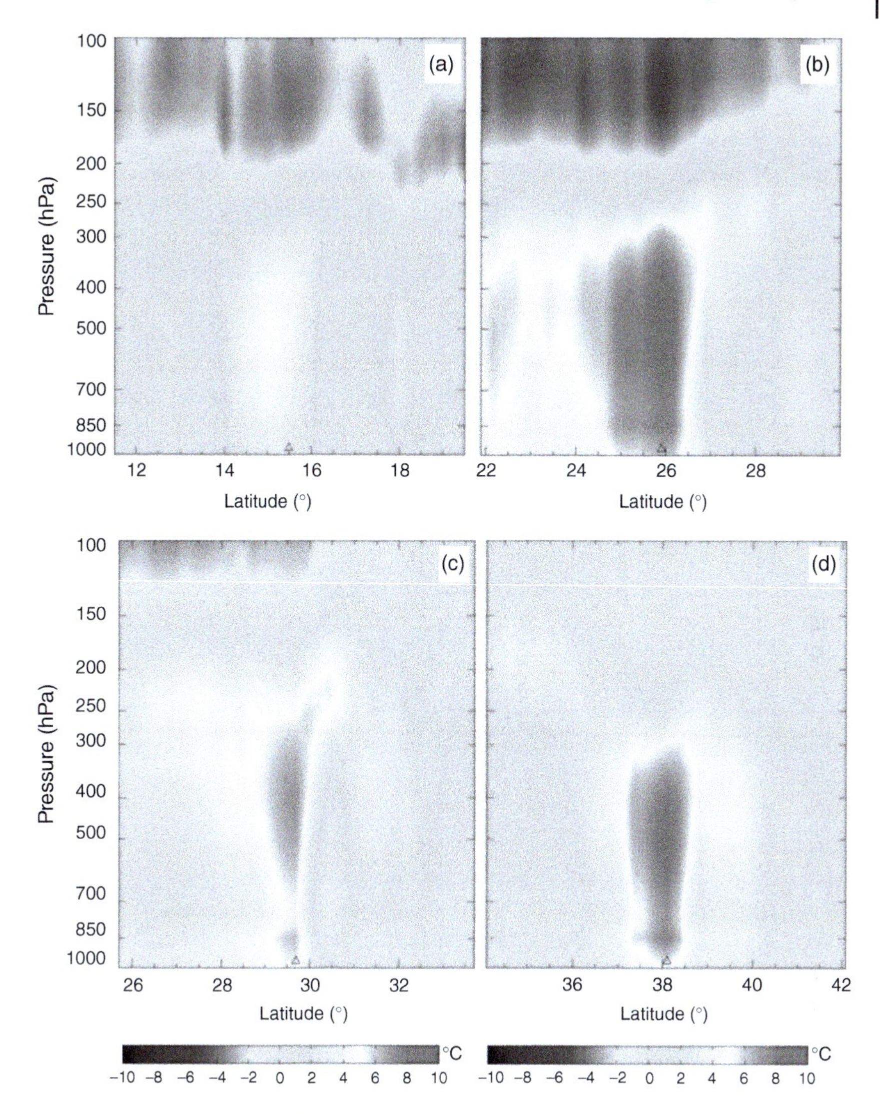

Figure 9.4 Vertical cross section of Hurricane Sandy temperature anomaly structures retrieved from ATMS observations at (a) 0630 UTC 24 October along longitude 77.0W, (b) 1710 UTC 26 October along longitude 76.6W, (c) 1810 UTC 27 October along longitude 75.5W, (d) 1730 UTC 29 October along longitude 72.9W.

when the storm reached a maximum wind speed of 90 mph. Sandy made landfall at 0200 UTC 30 October at about 8 km southwest of Atlantic City. The warm core feature quickly diminished after the landfall.

Throughout the lifetime of Sandy, the maximum warm core was located at ~400 hPa level, which is much lower than that of a typical TC (~250 hPa, to be discussed). It was mostly because Sandy took place in the fall season and middle latitudes. The averaged 850 hPa temperature retrieved from ATMS within 15° radius of Hurricane Sandy center was calculated and compared with that of another nine TCs in 2012, that is, hurricanes Ernesto, Isaac, and Kirk in the Atlantic Ocean; typhoons Guchol, Bolaven, Tembin, and Sanba in the Northwest Pacific Ocean; TC Giovanna in the South Indian Ocean; and TC Jasmine in the South Pacific Ocean. It is found that the averaged 850 hPa temperatures for Sandy and the other nine storms are ~11.7 and 15.8 K, respectively. It may suggest that the relative lower supply of thermal energy in the lower levels is one of the main factors that prevented the development of deep convections and vertical circulations in Hurricane Sandy. Therefore, the height of maximum warm core is lower than that of a typical TC. The lower level cold anomalies indicated that Sandy's radius of maximum wind was generally ~200 km. Sometimes, there was no strong cold temperature anomaly or heavy rainfall seen within 400 km from the center at a certain quadrant. The large size of Hurricane Sandy was likely a result of interactions with an upper-level trough and an extratropical frontal system. These processes enhanced Sandy's asymmetric structures and strong convections far away from the center but restrained the development and concentration of the storm near the core region. The variation of maximum temperature anomaly reflects the evolution of storm intensity. The retrieved warm core data on October 24 evening and on October 25 are not used, because the centers of Hurricane Sandy were located at the very edge of the observation swath at these three times as shown in Figure 9.4, and they were over or close to the Caribbean islands. The hurricane's warm core and eyewall structures cannot be resolved at a very large field of view (FOV) because the signal is smoothed out under coarse resolution. The correlation coefficient between the maximum warm core and MSW is 0.86, which passes the 99% confidence level. However, the correlation between the maximum warm core and MSLP is just 0.38, which fails in the 95% significance test. One possible reason is that Hurricane Sandy was a very large sized storm, so the intensity of the warm core at the center cannot represent the mass (pressure) features of the whole system very well.

9.4 One-Dimensional Variational (1DVAR) Theory

The mathematical basis of one-dimensional variation retrieval (1DVAR) is a proven and widely used variational approach [250]. We briefly review it here for the purpose of showing that it is valid for general applications in atmospheric sounding. There are three important assumptions made for this type of retrievals: the forward problem is locally nonlinear; both the geophysical state vector and the errors associated with the forward model and the instrument noise are Gaussian; and, finally, the measurements and the forward operator are nonbiased to each other. It is important to keep in mind that the variational, Bayesian, optimal estimation theory, maximum probability all give the same solutions (if the same assumptions are made), although reached through different paths.

The following aspect links the probabilistic approach to the variational solution, which seeks to minimize the cost function. Intuitively, the retrieval problem amounts to finding the geophysical vector x that maximizes the probability of simulation of the measurements vector y using x as the input and using H as the forward operator. The Bayes theorem states that the joint probability $P(x/y)$ could be written as

$$P(x, y) = P(y/x)P(x) = P(x/y)P(y). \tag{9.8}$$

Therefore, the retrieval problem amounts to maximizing

$$P(x/y) = \frac{P(y/x)P(x)}{P(y)}, \tag{9.9}$$

where x is assumed to follow a Gaussian distribution:

$$P(x) = \exp\left[-\frac{1}{2}\left(x - x_b\right)^T B^{-1}\left(x - x_b\right)\right], \tag{9.10}$$

where x_b and B are the mean vector (or background) and covariance matrix of x, respectively. Ideally, the probability $P(y/x)$ is a Dirac delta function with a value of zero except for x. Modeling errors and instrumental noise all influence this probability. For simplicity, it is assumed that the PDF of $P(y/x)$ is also a Gaussian function with $y(x)$ as the mean value (i.e., the errors of modeling and instrumental noise are nonbiased), which can be written as

$$P(y/x) = \exp\left[-\frac{1}{2}\left(y - H(x)\right)^T R^{-1}\left(y - H(x)\right)\right], \tag{9.11}$$

where R is the measurement and/or modeling error covariance matrix. In a vector form of $\mathbf{x}$, maximization of $P(x/y)$ is the minimization of $-\ln(P(x/y))$, which can be computed from the given equations as

$$J(\mathbf{x}) = \frac{1}{2}(\mathbf{x} - \mathbf{x}_b)^T\mathbf{B}^{-1}(\mathbf{x} - \mathbf{x}_b) + \frac{1}{2}[H(\mathbf{x}) - \mathbf{y}]^T\mathbf{R}^{-1}[H(\mathbf{x}) - \mathbf{y}], \tag{9.12}$$

where $J(\mathbf{x})$ is called the cost function, which we want to minimize. The first term on the right-hand side, J_b, represents the penalty in departing from the background value (*a priori* information), and the second term, J_r, represents the penalty in departing from the measurements. The solution that minimizes this two-term cost function is sometimes referred to as a constrained solution. The minimization of this cost function is also the basis for the variational analysis retrieval. In theory, one can also find another optimal cost function for a non-Gaussian distribution and nonlinear problems. It is just not a straightforward problem. The solution that minimizes this cost function is easily found by solving

$$\frac{\partial J(\mathbf{x})}{\partial \mathbf{x}} = 0, \tag{9.13}$$

and assuming local linearity around x, which is generally a valid assumption if there is no discontinuity in the forward operator

$$H(\mathbf{x}_\mathrm{b}) = H(\mathbf{x}) + K(\mathbf{x}_\mathrm{b} - \mathbf{x}), \tag{9.14}$$

where K, in this case, is the Jacobian or derivative of H with respect to x. This results in the following departure-based solution:

$$\Delta\mathbf{x} = \mathbf{x} - \mathbf{x}_\mathrm{b} = \{(\mathbf{B}^{-1}\mathbf{K}^T\mathbf{R}^{-1}\mathbf{K})^{-1}\mathbf{K}^T\mathbf{R}^{-1}\}[\mathbf{y} - H(\mathbf{x}_\mathrm{b})]. \tag{9.15}$$

If the given equations are ingested into an iterative loop, each time assuming that the forward operator is linear, we end up with the following solution to the cost function minimization process:

$$\mathbf{x}_{n+1} = \{(\mathbf{B}^{-1}\mathbf{K}^T\mathbf{R}^{-1}\mathbf{K})^{-1}\mathbf{K}^T\mathbf{R}^{-1}\}[\mathbf{y} - H(\mathbf{x}_\mathrm{b})] + \mathbf{K}_n\Delta\mathbf{x}_n, \tag{9.16}$$

where n is the iteration index. The previous solution can be rewritten in another form, after matrix manipulations, as

$$\mathbf{x}_{n+1} = \{\mathbf{B}\mathbf{K}_n^T(\mathbf{K}_n B\mathbf{K}_n^T + \mathbf{R})^{-1}\}\{[\mathbf{y} - H(\mathbf{x}_n)] + \mathbf{K}_n\Delta\mathbf{x}_n\}, \tag{9.17}$$

The latter is more efficient, as it requires the inversion of only one matrix. At each iteration n, we compute the new optimal departure from the background, given the derivatives as well as the covariance matrices. This is an iterative-based numerical solution that accommodates moderately nonlinear problems or/and parameters with moderately non-Gaussian distributions. This approach to the solution is generally labeled under the general term of physical retrieval and is also employed in NWP assimilation schemes along with horizontal and temporal constraints. The whole geophysical vector is retrieved as one entity including the temperature, moisture, and hydrometeor profiles as well as skin surface temperature and emissivity vector, ensuring a consistent solution that fits the radiances.

In the 1DVAR inversion process, a forward operator is called to simulate the brightness temperatures or radiances including multiple-scattering effects due to ice, rain, snow, graupel, and cloud liquid water at all microwave frequencies and to generate the corresponding Jacobians for all atmospheric and surface parameters. The forward operator is the Community Radiative Transfer Model (CRTM) developed at the Joint Center for Satellite Data Assimilation (JCSDA) [251]. CRTM produces radiances as well as Jacobians for all geophysical parameters. It is valid under clear, cloudy, and precipitating conditions. Derivatives are computed using the K matrix developed by tangent linear (TL) and adjoint (AD) approaches. This is ideal for retrieval and assimilation purposes. The different components of CRTM, briefly, are the OPTRAN fast atmospheric absorption model [252], a microwave emissivity model [53, 66], and the advanced doubling–adding radiative transfer solution for the multiple-scattering modeling [44].

The covariance matrix plays an important role in variational algorithms. Lopez (2001) estimated an error covariance matrix of cloud and rain from the French global model ARPEGE [253] and simply defined an empirical covariance matrix of clouds with large errors. Moreau *et al.* [254] used the regular covariance matrix of temperature and humidity, which they convolved with moist convection and large-scale condensation schemes to produce an ensemble of rain water and cloud profiles. In their study, the covariance was computed for each grid point. Boukabara *et al.* [255] related the part of the covariance matrix B to temperature and humidity using a set of globally distributed radiosondes (known as the NOAA-88 set) and the part related to the cloud parameters, which is built independently. The statistics of clouds are generated from multitude runs based on fifth-generation mesoscale model (MM5) simulations, corresponding to Hurricane Bonnie (1998), with 4 km resolution and 23 vertical levels, extrapolated to the internal pressure grid at 100 layers. The ability of these runs to represent the hydrometeor global variability is not fully established, but this is believed to be accurate enough for the case of hurricanes and tropical storms. Impact studies (not shown) were also performed, which show that the system is able to reach convergence (therefore, a radiometric solution) under many conditions that are independent of the set that was used to generate these covariances.

9.5 Multiple 1DVARs for All-Weather Profiles

The 1DVAR system can begin with the background information that has already been retrieved from the previous 1DVAR. Each time, the new 1DVAR retrieval is built on robust background information and by concentrating on additional profiles with newly added channels. In the case of microwave 1DVAR design, we can first retrieve the atmospheric temperature profiles from oxygen sounding channels near 50–60 GHz. The problem is almost linear, and the forward model can be purely emission-based, which is applied for most of the atmospheric conditions. The products can be less affected by precipitation ice scattering.

Next, the 1DVAR adapts the temperature profiles from the previous retrievals and uses them as background information. At this time, the unknowns are the water vapor profiles. Cloud ice water path was derived prior to this process (Figure 9.5) [256].

A good first guess can allow fast converge to a local minimum of the cost function $J(x)$ in Eq. (9.12). In the temperature 1DVAR process, Liu and Weng [256] also developed a regression algorithm [257] to retrieve temperature as a first guess (i.e., x_0) in Eqs. (9.16) and (9.17). The first guesses are obtained at a maximum of 42 levels from the surface to 0.1 hPa. Here, the regression coefficients for temperature and water vapor profiles are predetermined using our collocated radiosondes and satellite measurements. Note that the first guess for cloud water profile is set to zero above the freezing level and to a small value of 0.005 for the rest of the profile, since the *in situ* cloud liquid water profile is

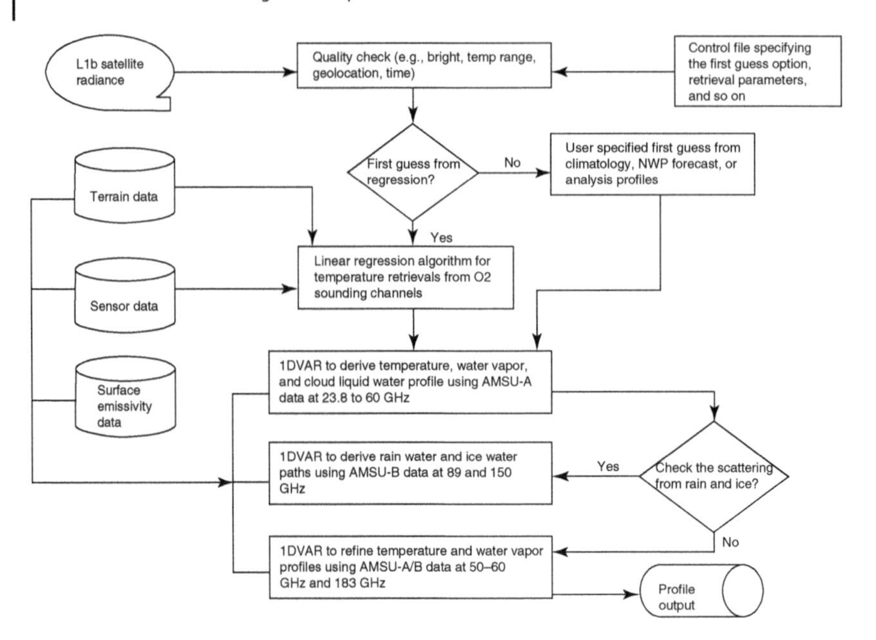

Figure 9.5 Flowchart of the microwave one-dimensional variation algorithm. The core module describes the retrieval procedure.

not sufficient for the development of the regression algorithm. For the rain water profile, we need to identify the precipitating atmospheres. In doing so, the rain is determined through an algorithm that uses AMSU-B measurements at 89 and 150 GHz [258]. While the rain pixel is identified, a vertically uniform distribution of rain water below the freezing level is assumed.

The first guess of ice water content is assumed to be uniformly distributed above the freezing level if the scattering signature is larger than a threshold [211]. The ice particle's effective diameter is calculated from the scattering parameter ratio at 89 and 150 GHz, which in turn is defined as the ratio of the difference between the predicted and measured brightness temperatures to the measured brightness temperature. The predicted brightness temperature is computed from AMSU-A channels 1 and 2 brightness temperatures based on a regression technique. The ice water path is then calculated from the effective diameter and the scattering parameter. As required by 1DVAR, the surface pressure and ocean wind speed are taken from the 6-h forecast data from the National Centers for Environmental Prediction (NCEP) global forecast system (GFS). A high-resolution dataset having a 1/6 degree resolution is used to define surface elevation and 24 surface types. The surface types are as follows: water, old snow, fresh snow, compacted soil, tilled soil, sand, rock, irrigated low vegetation, meadow grass, scrub, broadleaf forest, pine forest, tundra, grass soil, broadleaf pine forest, grass scrub, oil grass, urban concrete, pine brush, broadleaf brush, wet soil, scrub soil, broadleaf 70 – pine 30, and new ice.

It should be pointed out that an individual AMSU-A channel is not sensitive to the water vapor profile. However, AMSU-A measurements at channels 1 and 2

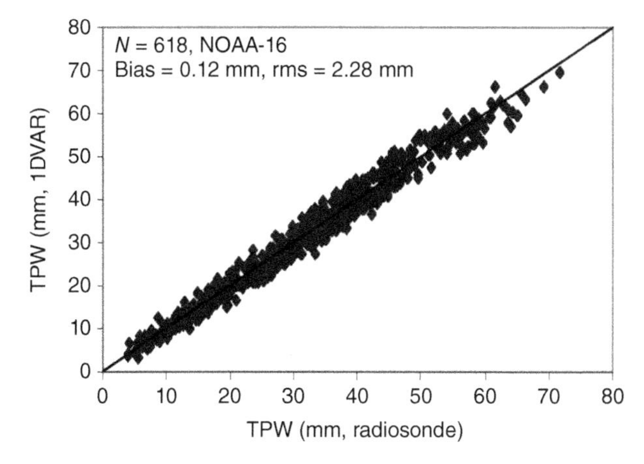

Figure 9.6 Comparison of the total precipitable water between radiosondes and retrievals using the data from AMSU on the NOAA-16 satellite.

are sensitive to the total precipitable water vapor (e.g., 23.8 GHz), and those from channels 3 to 15 are more sensitive to the vertical temperature profile. Thus, the initial water vapor profile can still be estimated from the AMSU-A because of the correlation between the temperature and water vapor profiles and the AMSU-A channel 1 sensitivity to the total precipitable water. Thus, in the temperature 1DVAR, water vapor profiles are also treated in the state vector; however, the convergence is controlled only by the difference between two consecutive retrievals in temperatures.

The 1DVAR retrieval algorithm is also validated with collocated satellite measurements and radiosonde data. The radiosonde stations are distributed globally and are shown in the website http://raob.fsl.noaa.gov/. All AMSU and radiosonde data are matched with a spatial distance of less than 50 km and a temporal difference of less than 2 h. Figure 9.6 compares the total precipitable water from radiosondes with that retrieved from NOAA −15, −16, and −17 AMSU data. Note that the entire matchup data during 2002 are used in this analysis, and the comparisons are only performed with the radiosonde data under clear conditions. Overall, the biases are relatively small, and the RMS error is stable and nearly 2.5 mm (or 2.5 kg/m^2), which is better than the results from our previous study [211].

It is also important to assess the performance of the 1DVAR under cloudy and precipitation conditions. In this study, the AMSU observations in Hurricane Isabel are used for our tests. Hurricane Isabel was one of the most powerful storms that affected the eastern portion of the United States in 2003. From the AMSU data at 150 GHz, we can clearly define the hurricane center near 21.75°N and 56.55°W and the spiral rainfall bands surrounding the eye (Figure 9.7). Figure 9.7a–d shows the retrieved cloud ice water path, surface rain rate, cloud liquid water path, and total precipitable water. Notice that the area covered by the cloud ice water is much broader than that covered by surface precipitation. The nonprecipitating ice

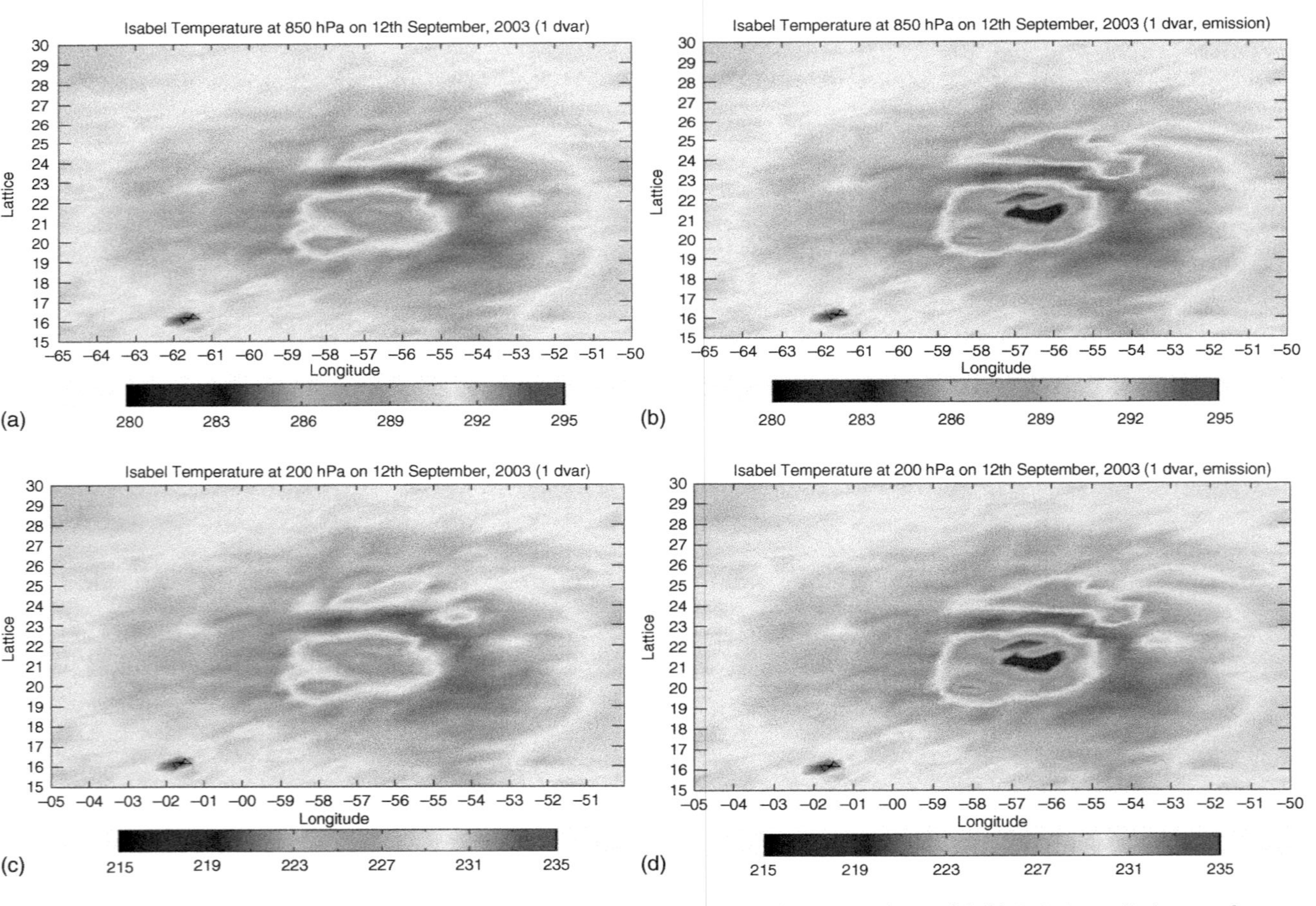

Figure 9.7 (a) Retrieved atmospheric temperature at 850 hPa through a scattering radiative transfer model. (b) Emission radiative transfer model; (c) Retrieved atmospheric temperature at 200 hPa through the scattering radiative transfer model. (d) Emission radiative transfer model for Hurricane Isabel on September 12, 2003.

clouds are primarily depicted by the AMSU at 150 GHz. In addition, both cloud liquid water and total precipitable water fail to detect the hurricane eye regions because of the poor spatial resolution of the AMSU data, although the amounts are generally higher near the center.

The effects of cloud and precipitation scattering on the temperature retrievals can be demonstrated by comparing the results from a forward model including scattering with those from the emission-based model. In Figure 9.7a, the magnitude of the cold temperature anomaly at 850 hPa derived from the scattering model is more agreeable with the other results [231, 259] compared to that from the emission-based model shown in Figure 9.7b. Because of the scattering of liquid-phase hydrometeors, the microwave brightness temperatures at sounding channels are strongly depressed. If this scattering effect were not properly taken into account in the retrieval process, the physical temperatures at 850 hPa would have been forced to the lower values, which becomes highly nonphysical (see Figure 9.7b). Using the scattering model, the retrieved structure of cooling at 850 hPa is smooth, and the anomaly of $\sim$3 K is realistic according to Hawkins and Rubsam.

The quality of the temperature distribution at 200 hPa has also been improved significantly using the scattering model. Note that the temperature at the center is $\sim$10 K higher than that of its environment (Figure 9.7c) and compares favorably with other early observations [231, 259]. It is also interesting to see that the retrieved temperature from the emission model (Figure 9.7d) is higher than that from the scattering model (Figure 9.7c), which is opposite to the pattern at 850 hPa. This can be explained as follows: for the clouds at high altitudes, the transmittance in the emission model is smaller, attenuating the high radiance from lower warm and humid/cloudy atmosphere. Since the measured brightness temperatures are higher than those predicted by the emission-based model, the retrieval algorithm must enforce a higher temperature there when the emission model is applied.

9.6 Microwave Integrated Retrieval System (MIRS)

The microwave integrated retrieval system (MIRS) performs the retrieval in the empirical orthogonal function (EOF) space so that the system can handle a large array of the state vector and can converge when all microwave channels are utilized simultaneously. An integrated retrieval of atmospheric temperature, water vapor, and hydrometeor profiles has run into a number of technical difficulties. (i) Microwave measurements are obtained from a limited number of channels. We have many more unknowns than measurements. (ii) All channels can be affected by the scattering of large rain and ice particles. If we like to perform the retrievals under precipitating atmospheres, how can we run a scattering forward model at an affordable CPU time? (iii) When the products are developed independently of any background information, from where can we obtain the background information and covariance matrices for all state variables?

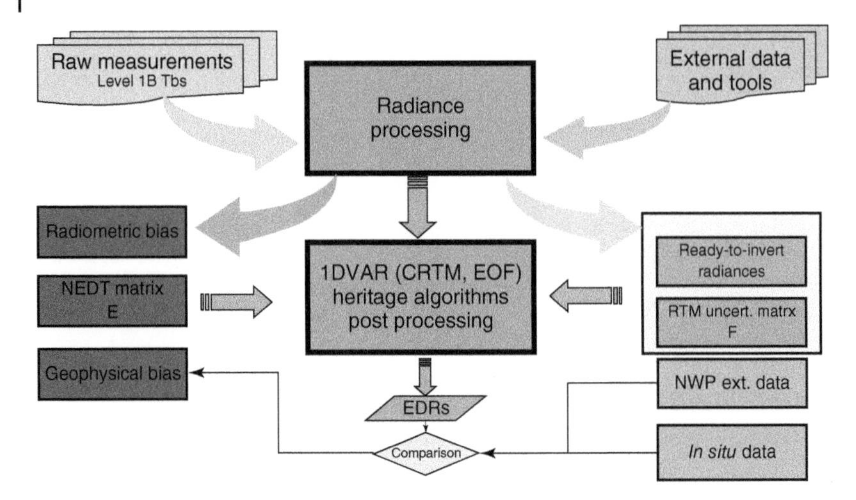

Figure 9.8 Microwave integrated retrieval system (MIRS) flowchart.

MIRS addresses all these technical issues by integrating the CRTM, which produces both radiances under all-weather conditions and the corresponding Jacobians for all parameters including cloud and hydrometeor parameters (Figure 9.8). The MIRS methodology described here is based on consistently treating all the parameters that impact the measurements. It is also independent of NWP-related information. The ill-posed nature of the inversion is handled through the use of the eigenvalue decomposition technique, which renders the inversion very stable and results in a high convergence rate. It was shown, in an ideal simulation case, that the null space is a limiting factor. This translates into cases where the retrieval process reaches a solution that satisfies the measurements but is different from the actual, in terms of hydrometeor and cloud profiles. Because of this and the limited information content of the radiances, the aim of this retrieval was essentially to target the temperature and moisture profiles as well as the surface parameters in very active weather regions. The effects from the scattering and emission of hydrometeors on measurements could be similarly produced by other parameters which are not explicitly accounted in the control variables. Improvement in the cloud and hydrometeor profiling is, however, expected if the temperature and moisture profiles are provided externally, for example, through accurate NWP forecasts. Designing the retrieval of cloud and hydrometeors in the profile form presents a number of advantages including the avoidance to explicitly account for the cloud top pressure and the cloud thickness, which could, in certain cases, cause instability or oscillation. The designed system could also, in theory, provide information about the multilayer nature of the clouds and mixture of phases within the cloud/precipitating layers, provided enough information on the radiances exists. The retrieval system is used under clear, cloudy, and precipitating conditions. It was shown by simulation and confirmed with real data that the performances, when applied to clear skies, are not degraded and that the retrieval algorithm is able to reach a zero-amount solution for all cloud and hydrometeor parameters if the radiances indicate so.

The 1DVAR retrieval is performed in the EOF space through projections back and forth, at each iteration, between the original geophysical space and the reduced space. This method has been routinely used in operational centers as a standard transform approach for control variables [260]. It has also been used in the context of retrieval of trace gases, sounding, and surface properties [256, 261–263]. Applying it in the context of our 1DVAR retrieval is therefore not very original, except maybe for its extension to cloud and precipitation profiles, which is, to our knowledge, new. Only a limited number of eigenvectors/eigenvalues are maintained in this reduced space. The selection of the number of EOFs to be used for each parameter is somehow subjective but depends on the number of channels available that are sensitive to that parameter. Other approaches exist, such as in [264] where an objective way for choosing the parameters to be included in the control parameters is suggested, using the ratio between the background covariance matrix and the *a posteriori* covariance (ratio of diagonal elements). This ratio, however, depends on the Jacobian, which is only known at the end of the iterative process unless the problem is purely linear (not the case when cloud and precipitation as well as high-frequency channels are involved). The advantages of performing the retrieval in the EOF space are as follows: (i) handling the strong natural correlations that sometimes exist between parameters that usually create a potential for instability (or oscillation) in the retrieval process (small pivot), which is reduced significantly by performing the retrieval in an orthogonal space, and (ii) time saving by manipulating and inverting smaller matrices. The projection in the EOF space is performed by diagonalizing the *a priori* covariance matrix:

$$B \times L = L \times \Theta, \tag{9.18}$$

where L is the eigenvector matrix, also called the *transformation matrix*, and Θ is the eigenvalue diagonal matrix, which contains independent pieces of information. The retrieval can therefore be performed using the original matrices as stated before (retrieval in original space), or alternatively, by using the matrices/vectors (retrieval in reduced space). The transformations back and forth between the two spaces are performed using the transformation matrix L. It is important to note that, at this level, no errors are introduced in these transformations. It is merely a matrix manipulation. However, the advantage of using the EOF space is that the diagonalized covariance matrix and its corresponding transformation matrix can be truncated to retain only the most informative eigenvalues/eigenvectors. By doing so, we are bound to retrieve only the most significant features of the profile, leaving out the fine structures. How much truncation to use depends on how much information the channels contain. In the AMSU configuration, six EOFs are used for temperature, four for humidity and surface emissivity, one for skin temperature, one for nonprecipitating cloud, and two for both rain and frozen precipitation (20 in total).

Several criteria have been reported for deciding on the convergence of variational methods, among which are (i) testing that the increment in the parameter values at a given iteration is less than a certain threshold (usually a fraction of the associated error of that particular parameter), (ii) testing that the cost function $J(x)$ decrease is less than a preset threshold, or (iii) checking that the obtained geophysical vector $\mathbf{x}$ at a given iteration produces radiances that fit the measurements within the noise level impacting the radiances. We have chosen the last criterion, as it maximizes the radiance signal extraction. A convergence criterion based on $J(x)$ while being mathematically correct would produce an output that carries more ties to the background and therefore would be more inclined to present artifacts due to it. The convergence criterion adopted is when

$$\phi^2 = |(y^m - y(x))^T E^{-1}(y^m - y(x))| \leq N, \quad (9.19)$$

where N is the number of channels used for the retrieval process. Mathematically, this means that the convergence is declared to have reached if the residuals between the measurements and the simulations at any given iteration are less than or equal to one standard deviation of the noise that is assumed in the radiances. Note that fitting the radiances within the noise level is a necessary but not sufficient condition. We should note here that the convergence criteria do not alter the balance of weights to the radiances (or to the background) in the cost function minimized by 1DVAR.

The evolution of the humidity profile is monitored for supersaturation in the iterative process. A maximum of 130% relative humidity is allowed. Currently, it is set in an ad hoc manner at each step. This has the potential to nonlinearly steer the convergence from its mathematical path and should, in general, be avoided, but our experience has shown that this does not significantly increase the divergence rate.

Microwave imaging and sounding data from the NOAA-18 satellite were used to validate the retrieval system described under both clear and extreme weather conditions, in the eye and within the eyewall of hurricane Dennis in the summer of 2005. This was performed by comparing the retrievals of temperature and humidity profiles to the measurements performed by GPS dropsondes. Before retrieval is performed, the brightness temperatures of the two sensors are collocated and corrected for any bias when compared to the forward model simulations. The collocation is performed in two different ways: (i) averaging of 3×3 MHS footprints is performed to fit the AMSU spatial coverage (low resolution), or (ii) it is assumed that the AMSU footprint is valid within all subpixel MHS footprints (high resolution). In the latter case, the subpixel heterogeneity is computed from the MHS footprints and translated into AMSU channels but only for those that are sensitive to the same geophysical parameters, namely channels 23.8, 31.4, 50.3, and 89 GHz. Bias removal is effected by simulating the brightness temperatures over ocean using the NCEP global data assimilation system (GDAS) analyses as inputs. These biases were found to be scan-dependent.

The instrumental/modeling error covariance matrix E is also built partly during this process by using the variances of the same comparisons. These variances are subjectively scaled down to account for the uncertainties in the GDAS inputs and collocation errors. The diagonal elements (in standard deviation, in kelvin) of the forward modeling error matrix E for the AMSU and MHS channels from 1 to 20 are the following:

$$E = (1.9, 1.7, 1.2, 0.6, 0.3, 0.2, 0.3, 0.4, 0.4, 0.3, 0.8, \\ 0.0, 0.0, 0.0, 2.1, 2.2, 1.4, 1.6, 1.3, 1.1)_{diagonal} \tag{9.20}$$

where channels 12–14 peak above the maximum altitude reported by GDAS, so the comparison to GDAS simulation is not very meaningful; therefore, the variances for these channels were deemed unreliable, and the channels were disabled. These modeling errors are used more predominantly than the instrumental errors (NEΔT values), which are computed exclusively from the raw AMSU/MHS Level-1B data available from NOAA using the approach in [165]. For window channels, modeling errors are dominant over instrumental errors. These values are slightly lower than those found in the previous studies [264, 265]. They, however, allow stable convergence in most cases. Note that these modeling errors are computed over ocean under clear-sky conditions. The same values are used under cloudy/rainy conditions.

Figure 9.9a shows the field of 157 GHz MHS brightness temperature for Hurricane Dennis, which occurred in July 2005. The dropsonde launch location is near the eye and within the eyewall of the hurricane and close in time to the satellite measurements. MIRS retrieved the vertically integrated graupel-size ice amount computed from the retrieved profile (Figure 9.9b). This is shown as a qualitative validation. Although the retrieval is performed in profile form, the resulting integrated value displays physically plausible features and values. The retrieval corresponds to the same hurricane Dennis on July 8, 2005 (the same descending orbit shown before). First, where no activity is present (from the 157 GHz Tb field), the retrieval reports no ice or rain even though the first guess used is actually a nonzero profile (the same used everywhere). Second, the large values of graupel amount are concentrated in the middle of the active area and decrease gradually at the edges. One can even see that in what seems to be the eye of the hurricane, the value of the integrated ice amount is actually very small compared to the surrounding pixels.

Figure 9.9c,d shows the comparison of MIRS retrievals to a few selected sondes that were dropped within the eye and eyewall of the hurricane. GDAS is also represented as reference. These figures correspond to temperature and moisture, respectively. Both time difference and distance between the space-based measurement and the dropsonde are shown on the plots. Note that the vertical extent reaches 700 hPa only for this particular aircraft that dropped the sondes. GDAS and MIRS still report retrievals up to 20 and 0.1 hPa. It is found that these comparisons show a rather good agreement between MIRS and the dropsondes, at

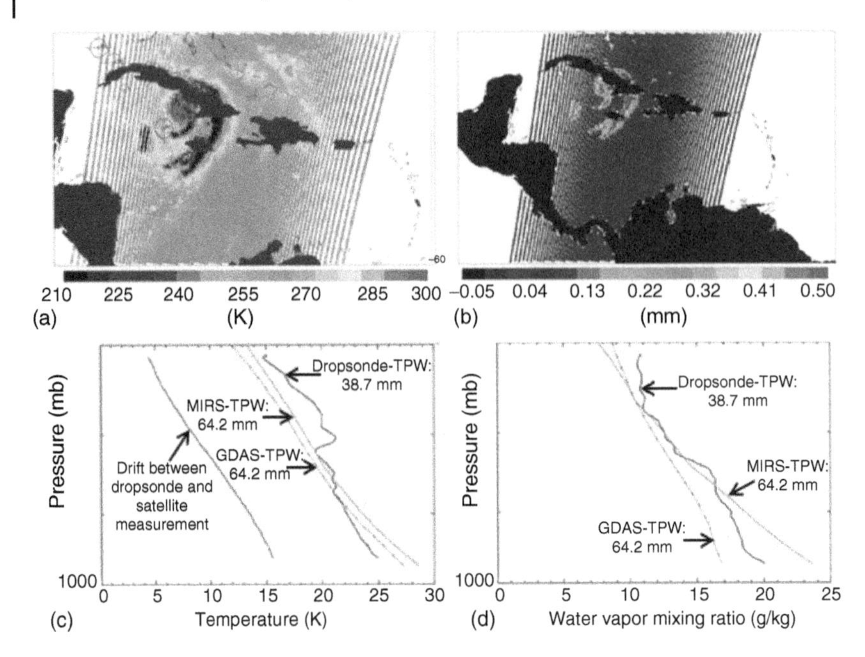

Figure 9.9 NOAA-18 MHS 157 GHz and MIRS retrieved graupel-size ice content, temperature, and water vapor profiles within the eyewall of Hurricane Dennis, 2005, passing through Cuba Island. Retrievals were performed at MHS resolution (roughly 20 km).

least for temperature. The differences are indeed well within the intra-variability of the sonde measurements [266]. In addition to the intra-variability and the representativeness issues reported before, the vertical descent of the sondes apparently tends to drift horizontally more drastically within very active regions. Although the distance at the launch location was reported to be 2.6 km for the first sonde, for instance, we can see that when reaching the surface, the distance was around 10 km. Again, in fast-moving features such as hurricanes, this factor could make a significant difference. For the closest collocation (less than 12 min and less than 3 km in distance), the difference in water vapor is actually also within the previously reported intra-variability. When the time and distance differences are larger, the moisture differences are larger. But the errors of representativeness and the vertical drift of the sondes could at least, in part, explain the remaining differences. It is worth mentioning that NCEP GDAS does ingest the dropsonde measurements within its assimilation cycle but not the rain-impacted AMSU/MHS radiances. It is interesting to notice in this case that GDAS analyses exhibit similar differences with the dropsondes as with MIRS retrieval, although the latter is based solely on microwave radiances measured from AMSU and MHS.

9.7
Summary and Conclusions

Satellite microwave sounding measurements are used for deriving the TC's inner core dynamics and constructing the initial vortex for the storm simulation. With AMSU-A and ATMS observations, a regression algorithm was developed by using the collocated microwave data and radiosonde data. Since these algorithms are trained with the clear-sky data, they are ideal for uses in the regions where satellite data are not affected by the scattering from precipitation. Thus, for AMSU-A and ATMS, those channels with the weighting functions having peaked below 500 hPa are not used in the regression algorithms.

The 1DVAR retrieval algorithm has been developed to retrieve a comprehensive suite of geophysical parameters from spaceborne microwave measurements. The suite of parameters consists of those that most directly impact the measurements: atmospheric profiles of temperature, moisture, liquid and ice cloud, liquid precipitation, and surface emissivity spectrum and its skin temperature. The design of the MIRS is generic, and it can accommodate any microwave sensor that can be handled by the forward operator CRTM. It is currently operational for AMSU-A/MHS, SSMI/S, and ATMS. It also significantly reduces the amount of time needed to develop an algorithm for a new sensor. It is noteworthy that, in MIRS, the same code is used for all sensors, as well as the same atmospheric covariance, background, and forward model. Therefore, applying MIRS to a new sensor comes with a high degree of confidence stemming from the previous tuning, improvement, and assessment performed for the previous sensors.

10
Assimilation of Microwave Data in Regional NWP Models

10.1
Introduction

Effective and accurate environmental prediction based on numerical modeling requires several elements. One of these is a set of observations that is accurate and comprehensive enough to help establish the initial conditions for the prediction model. Since most environmental prediction problems are of a global nature, space-based observing systems are the only practical means of achieving the required data coverage, and satellite sensors are, therefore, by far the largest and most important data source. Currently, satellite data comprise 99% of all observations received by operational weather and climate prediction centers. Satellite data that have been used in global numerical weather prediction (NWP) models include primarily the sounding measurements observed from microwave and infrared instruments on board the National Oceanic and Atmospheric Administration (NOAA) and EUMETSAT satellites. In addition, satellite-estimated parameters such as ocean winds, atmospheric cloud and water vapor track winds, and atmospheric temperature and water vapor information inferred from global position satellite (GPS) radio occultation (RO) are also effectively assimilated. The relative importance of all the sensors in the NWP system can be quantified through an adjoint sensitivity, which calculates the 24-h forecast error (the so-called energy norm) reduction by adding the data of each category of instrument into the data assimilation system. Although this kind of analysis is not a full observing system experiment, it has drawn the same conclusion on the current observing systems, that satellite microwave sounding data are critical and as effective as that from the conventional radiosondes in improving the global medium-range forecasts [249]. The full observing system experiments with the National Centers for Environmental Prediction (NCEP) global forecast system (GFS) were conducted to demonstrate the impacts from the use of each specific sounding system on the global medium-range forecast.

In this chapter, the radiance data from cross-track and conical-scanning microwave sounders are used for assimilation in NWP models. The impacts from assimilating the two sounder data on global forecast skills are compared [267, 268]. After launching the new generation of the polar-orbiting satellite, Suomi National

Polar-Orbiting Partnership Satellite (NPP), all the NWP centers have reported large positive impacts from the assimilation of Suomi NPP Advanced Technology Microwave Sounder (ATMS). It has been shown that the use of ATMS data in US GFS alone results in an increase of 70% in forecast score as measured by the global 500-hPa geopotential height anomaly correlation coefficient (ACC) [268]. ATMS radiances are also directly assimilated in the Hurricane Weather Research and Forecast (HWRF) system, and the impacts from assimilating ATMS data on hurricane tack and intensity forecasts are significantly positive [233].

10.2 NCEP GSI Analysis System

The NCEP uses a gridpoint statistical interpolation (GSI) analysis system in its global and regional NWP models. GSI is a three-dimensional variational (3D-Var) data assimilation system, and detailed description of the theory and development of the GSI system can be found in [269]. The GSI analysis system is more advantageous than the earlier NCEP Spectral Statistical Interpolation (SSI) analysis system [270] in that it has more flexibly in situations where observations are greatly inhomogeneous in terms of their data density and quality. Through the application of recursive filters, the spectral definition of background errors in the SSI analysis system is replaced with a nonhomogeneous gridpoint representation of background errors [269, 271, 272]. The GSI system was made available for the NWP community, along with a GSI User's Guide, providing a step-by-step procedure for users to install, compile, and run the GSI system on different local computer systems.

The GSI data assimilation system finds optimal analysis fields from forecast fields, conventional observations, some retrieval products as observations, and satellite radiances under dynamic constraints following a set of physical laws. The Community Radiative Transfer Model (CRTM) discussed in the earlier chapters is used in GSI for the direct radiance assimilation [251].

The variational approach in deriving the NWP initial conditions uses both satellite measurements and an initial guess. Specifically, assuming that the errors in the observations and in the *a priori* information are unbiased, uncorrelated, and have Gaussian distributions, the best estimate of atmospheric state vector ($\mathbf{x}$) will minimize the cost function

$$J(\mathbf{x},\beta) = (\mathbf{x}-\mathbf{x}_b)^T\mathbf{B}^{-1}(\mathbf{x}-\mathbf{x}_b) + (\beta-\beta_b)^T\mathbf{B}_\beta{}^{-1}(\beta-\beta_b) + \left[H(\mathbf{x})-\mathbf{y}-\mathbf{b}(\mathbf{x},\beta)\right]^T\mathbf{R}^{-1}\left[H(\mathbf{x})-\mathbf{y}-\mathbf{b}(\mathbf{x},\beta)\right] + J_c, \quad (10.1)$$

where $\mathbf{B}$ is the error covariance matrix associated with the background state vector $\mathbf{x}_b$; $\mathbf{R}$ is the error covariance matrix associated with observations and forward models, respectively; and H is the so-called observation or forward operator. For satellite radiance assimilation, the radiance vector is simulated for a set of channels (or frequencies) at the state vector $\mathbf{x}$, and $\mathbf{y}$ denotes the observations. The parameter $\boldsymbol{\beta}$ denotes a set of coefficients for the dynamic bias correction $\mathbf{b}$. In general, the element of $\mathbf{b}$ is related to a set of predictors chosen from the state vector $\mathbf{x}$. For AMSU-A bias correction, the predictors are the satellite viewing angle, cloud liquid water, and atmospheric temperature lapse rate such as

$$b = \sum_{i=1}^{N} \beta_i p_i, \tag{10.2}$$

where the cloud water predictor is turned off for other instruments. Note that the first two terms in Eq. (10.1) correspond to the background constraints for $\mathbf{x}$ and $\boldsymbol{\beta}$, the third term is the bias-corrected observation constraint, and the final term is an additional constraint (e.g., balance wind). The minimum of the cost function is found from an iterative process that computes the descent direction at the state $\mathbf{x}$ and $\boldsymbol{\beta}$.

For the radiance assimilation, the adjoint operator of Jacobian matrix or the derivative of the radiance with respect to the state vector is required during the minimization process. This is a typical three-dimensional variational data assimilation scheme implemented in the current NWP models. The tangent linear and adjoint techniques allow avoiding an explicit computation of Jacobians in the data assimilation model. The tangent linear operator of a forward model analytically computes the output perturbations corresponding to the input perturbations with a computational cost typically only about twice as much as that of the forward model.

In a data assimilation system, we are often exposed to numerous variables such as analysis field, state vector, observation, departure, or innovation. They are summarized as follows.

An analysis is the production of an accurate image of the true state of the atmosphere at a given time, represented in a model as a collection of numbers. An analysis can be useful in itself as a comprehensive and self-consistent diagnostic of the atmosphere. It can also be used as input data to another operation, notably as the initial state for a numerical weather forecast or as a data retrieval to be used as a pseudo-observation. It can provide a reference against which the quality of observations is checked. The basic objective information that can be used to perform the analysis is a collection of observed values provided by observations of the true state. If the model state is overdetermined by the observations, then the analysis reduces to an interpolation problem. In most cases, the analysis problem is underdetermined because data is sparse and only indirectly related to the model variables. In order to make it a well-posed problem, it is necessary to rely on some background information in the form of an *a priori* estimate of the model state. Physical constraints on the analysis problem can also aid in this. The background information can be climatological or a trivial state; it can also be generated from the output of a previous analysis, using some assumptions of consistency in time of the model state, such as stationarity (hypothesis of persistence) or the evolution predicted by a forecast model. In a well-behaved system, one expects that this allows the information to be accumulated in time into the model state and to propagate to all the variables of the model. This is the concept of data assimilation.

State vector in a forecast model is a collection of variables required to represent the atmospheric state of the model, which is a column matrix. How the vector components relate to the real state depends on the choice of discretization, which is mathematically equivalent to a choice of the basis. The state vector can be derived from a best set of observations, forecasts, and analysis fields.

Control variables are those components that are allowed for performing the correction to the background field. Observation vector denotes the number of observed values. To use them in the analysis procedure, it is necessary to be able to compare them with the state vector. It would be better if each degree of freedom was observed directly. In practice, there are fewer observations compared to variables in the model, and they are irregularly disposed, so that the only correct way to compare the observations with the state vector is through the use of a function from model state space to observation space called an observation operator. Departure or innovation is the discrepancy between observations and the state vector.

The process of assimilating satellite microwave sounding data in NWP systems is typical and requires robust data quality control (QC) including checks of geolocation errors, calibration uncertainties, and the effects of cloud and rain. Satellite measurements must also be corrected to be consistent with the simulations from the forward models. This is also referred to as bias correction processes.

10.3 ATMS Data Assimilation in HWRF

10.3.1 Hurricane Weather Research and Forecast (HWRF) System

The HWRF system was developed at NOAA's National Weather Service (NWS). The HWRF has a non-hydrostatic mesoscale model dynamic solver [273, 274]. The initial single-domain version of the HWRF system became an operational hurricane track and intensity guidance tool in 2007 [275]. A 3-km nesting domain was then added to the operational HWRF system for improved hurricane intensity forecasts [276–279]. Finally, a triply nested version of the HWRF system was developed, which was configured with a parent domain of 27 km horizontal resolution and about 750×750 model gridpoints, an intermediate two-way telescopic moving nesting domain at 9 km with about 238×150 gridpoints, and an innermost two-way telescopic moving nesting domain at 3 km with about 50×50 gridpoints [276]. Figure 10.1 provides an example showing the outer domain, the ghost domain, the middle nest, and the inner nest for forecasting the track and intensity change of Hurricane Beryl. The surface pressure field from the background field at 0000 UTC on May 27, 2012 within the outer domain is also shown. It is seen that the outer domain is reasonably large for capturing tropical cyclone (TC) environmental flow evolution.

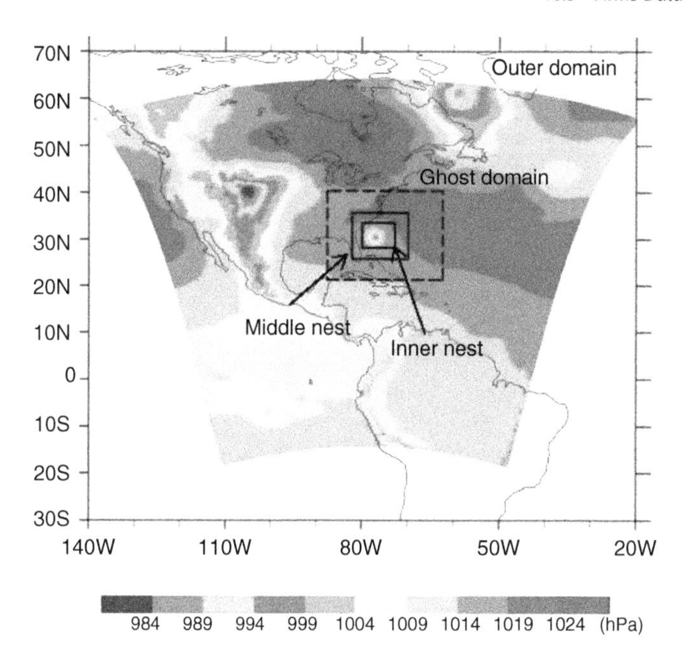

Figure 10.1 Surface pressure (shaded) from the background field at 0000 UTC 27 May 2012, for Hurricane Beryl. The outer domain, ghost domain, middle nest, and inner nest are also indicated. (Zou *et al.* 2013 [233]. Reproduced with permission of Wiley.)

The HWRF atmospheric model employs the following suite of advanced physics parameterization: Ferrier microphysics, NCEP GFS boundary layer physics, GFS SAS deep convection, and GFS shallow convection. The HWRF also includes the Geophysical Fluid Dynamics Laboratory (GFDL) GFDL surface physics, including GFDL land surface model and radiation, to account for air–sea interaction over warm water and under high wind conditions. The atmosphere component is coupled to the Princeton Ocean Model (POM) for all three domains, which employs feature-based initialization of loop current, warm and cold core eddies, and cold wake during the spin-up phase of the TCs. This version of the model also includes surface and boundary layer physics appropriate for higher resolution [280].

The triply nested 2012 version of the HWRF system is used for this study. Both the intermediate and innermost domains are centered at the initial storm location and configured to follow the projected path of the storm. The HWRF has 43 hybrid vertical levels with more than 10 model levels located below 850 hPa and a model top located at 50 hPa. The 50 hPa model top is too low for including many upper-level ATMS channels in data assimilation. Figure 10.2a shows the weighting functions for all 22 ATMS channels, the pressures of the 43 vertical levels, and the pressure differences between two adjacent vertical levels. In order to include those high-level ATMS channels with their weighting functions peaking in the stratosphere, the model top is raised to 0.5 hPa, and the model levels are increased to 61 accordingly (Figure 10.2b).

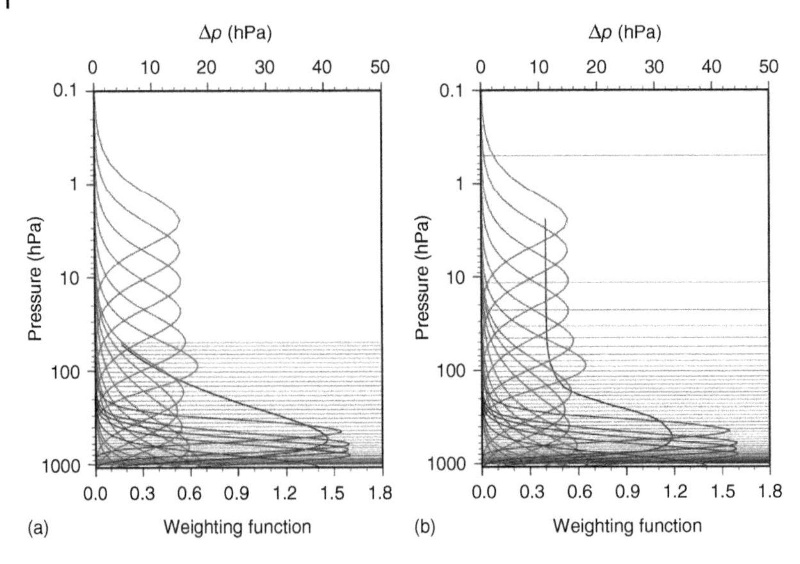

Figure 10.2 Weighting functions for ATMS channels 1–16 (light gray) and channels 17–22 (moderate gray), the pressures of the 61 vertical levels (gray horizontal line), and the pressure differences between two adjacent vertical levels (black curve) for (a) 43-level and (b) 61-level HWRF models. (Zou *et al.* 2013 [233]. Reproduced with permission of Wiley.)

The vortex initialization in the HWRF is performed at the 9-km resolution domain, with model fields in the 3-km resolution domain being downscaled from those over the 9-km domain. It consists of merging a specified bogus vortex with an environmental field extracted from the GFS analysis. Two prespecified symmetric vortices are made available: one for shallow or medium vortex, and the other for a deep vortex. A storm size correction and an intensity correction are performed to the prespecified vortex fields according to the tropical prediction center (TPC)-provided storm size and intensity data.

The merged field with a corrected vortex and the GFS analysis environment is used as the background field for data assimilation of conventional observations. Such a procedure is repeated at 6-h intervals. To eliminate the complications associated with double use of data, in this study, the 6-h HWRF forecasts are used for the environmental fields described earlier.

10.3.2 Hurricane Events in 2012

Tropical storms Beryl and Debbie and hurricanes Isaac and Sandy, which occurred in 2012, over the Atlantic Ocean, are selected for this investigation. They were the four landfall cases. The best track of each of the cases is shown in Figure 10.3.

Tropical storm Beryl developed from a tropical low on May 22, 2012, and became a subtropical storm on May 26 and a tropical storm on May 27, 2012. It

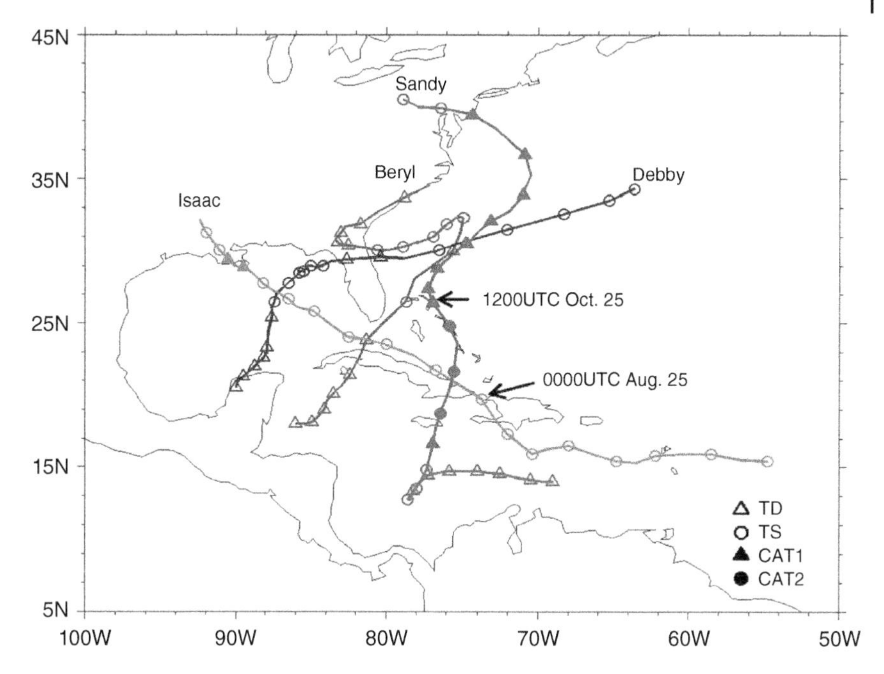

Figure 10.3 Best tracks of Hurricane Beryl from 0000 UTC 22 May to 1800 UTC 30 May Debby from 0000 UTC 21 June to 0000 UTC 30 June, Isaac from 0000 UTC 22 August to 1800 UTC 30 August, and Sandy from 1200 UTC 19 October to 1800 UTC 30 October 2012. The storm intensity is indicated at 12-h intervals at 0000 UTC and 1200 UTC. (Zou *et al.* 2013 [233]. Reproduced with permission of Wiley.)

first moved northeastward, then turned to move southwestward on May 24, and, finally, made landfall around 0410 UTC 28 May near Jacksonville Beach, Florida. After landfall, Beryl moved northeastward across northeastern Florida. Beryl was the strongest pre-season tropical storm on record. Tropical storm Debby developed from a low-pressure system in the Gulf of Mexico on June 23, 2012. Debby turned from a northward movement to an east-northward movement on June 24 and made landfall in Florida on June 26 (Figure 10.3).

Hurricane Isaac was initiated from a tropical wave in the west coast of Africa on August 21, 2012. It became a tropical storm later that day. Isaac moved westward before August 25 and moved northwestward afterward (Figure 10.3). Isaac remained as a tropical storm in the subsequent 7 days and intensified into a category-1 hurricane in the morning of August 28 before its landfall.

Hurricane Sandy developed from a tropical wave in the western Caribbean Sea on October 22, 2012. It quickly strengthened and developed into a tropical storm on the same day. Sandy became a category-1 and -2 hurricane after October 24. Sandy moved initially westward in the Caribbean Sea, northward over Bahamas, then northeastward when entering the middle latitudes, and, finally, northwestward on October 28 (Figure 10.3). It made landfall near Atlantic City on October 30. Hurricane Sandy was the largest Atlantic hurricane on record, which earned it a nickname Superstorm Sandy by the media and government agencies. It affected 24 states, resulting in severe damage particularly in New Jersey and New York. It caused an estimated damage of over \$63 billion and a casualty of at least 111 in the United States.

10.3.3 ATMS Data Quality Control

The QC for ATMS data in GSI employs six parameters associated with cloud, water vapor, and temperatures, three parameters associated with surface emissivity estimated from ATMS-observed and CRTM-simulated brightness temperatures, and one parameter related to the observation error. First, a cloud liquid water path (LWP) index (LWP^o_{index}) and a total precipitation water index (TPW^o_{index}) are calculated over ocean using the ATMS measurements at 23.8 GHz (channel 1) and 31.4 GHz (channel 2) [281]:

$$\begin{aligned}\mathrm{IWP}^o_{index} &= \mu[c_1 - \mu(c_2 - \mu c_3) + c_4 \times \log(285 - T^o_{b,1}) \\ &\quad - c_5 \times \log(285 - T^o_{b,2})],\end{aligned} \tag{10.3}$$

$$\begin{aligned}\mathrm{TPW}^o_{index} &= \mu[t_1 - \mu(t_2 - \mu t_3) - t_4 \times \log(285 - T^o_{b,1}) \\ &\quad + t_5 \times \log(285 - T^o_{b,2})],\end{aligned} \tag{10.4}$$

where $T^o_{b,i}$ ($i = 1, 2$) represent the ith ATMS channel measurements, and $\mu = \cos\theta$ with θ representing the satellite zenith angle; c_i ($i = 1, 2, \ldots, 5$) are regression coefficients for IWP whose values are set to 8.24, 2.622, 1.846, 0.754, and 2.265, respectively; and t_i are regression coefficients for TPW whose values are set to 247.92, 69.235, 44.177, 11.627, and 73.409, respectively. By replacing $T^o_{b,i}$ with $T^m_{b,i}$ ($i = 1, 2$) in Eqs. (10.3) and (10.4), IWP^m_{index} and TPW^m_{index} are also calculated.

Then, a cloud liquid water index is calculated as follows:

$$\mathrm{CLW}_{index} = \begin{cases} -0.754 \times \dfrac{T^o_{b,1} - T^m_{b,1} - \alpha_1}{285 - T^m_{b,1}} \\ \quad +2.265 \times \dfrac{T^o_{b,2} - T^m_{b,2} - \alpha_2}{285 - T^m_{b,2}} & \text{if } T^m_s > 273.15\ \mathrm{K}, \\ 0 & \text{otherwise} \end{cases} \tag{10.5}$$

where α_i is the scan-angle-dependent bias of the ith channel ($i = 1, 2$); $T^m_{b,i}$ ($i = 1, 2$) are the brightness temperature simulations of the ith channel; and T^m_s is the 2-m surface air temperature from the background fields. Two more indices, f_1 and f_2, are then calculated as follows:

$$f_1 = \begin{cases} \left(\dfrac{\mu \times \text{LWP}_{index}}{0.3}\right)^2 + \left(\dfrac{T^o_{b,5} - T^m_{b,5}}{1.8}\right)^2 & \text{ocean surface} \\ 0.36 + \left(\dfrac{T^o_{b,5} - T^m_{b,5}}{1.8}\right)^2 & \text{other surface} \end{cases}, \tag{10.6}$$

$$f_2 = \begin{cases} \left(\dfrac{\text{ds}}{10}\right)^2 + \left(\dfrac{T^o_{b,7} - T^m_{b,7}}{0.8}\right)^2 & \text{ocean surface} \\ 0.64 + \left(\dfrac{T^o_{b,7} - T^m_{b,7}}{0.8}\right)^2 & \text{other surface} \end{cases}, \tag{10.7}$$

where

$$\text{ds} = \left(2.41 - 0.098 \times \left(T^o_{b,1} - \alpha_1\right)\right) \times \left(T^o_{b,1} - T^m_{b,1}\right)$$
$$+ 0.454 \times \left(T^o_{b,2} - T^m_{b,2}\right) - \left(T^o_{b,16} - T^m_{b,16}\right).$$

Three additional surface-emissivity-related parameters are calculated for ATMS channels 1–3:

$$r_i = \frac{T^o_{b,i} - T^m_{b,i}}{\varepsilon_i}, \quad (i = 1,\ 2,\ 3), \tag{10.8}$$

where ε_i is the surface emissivity of the ith channel.

The last parameter that is employed in the GSI QC algorithm is the modified observation error e'_i, which is defined as follows:

$$e'_i = \begin{cases} \hat{e}_i \times f_{H2} \times f_{H4,i} \times f_{tropic} \times \tau_i^{top} & \text{for } i = 1-8,\ 16-22 \\ \hat{e}_i & \text{otherwise} \end{cases}, \tag{10.9}$$

where

$$\hat{e}_i = \begin{cases} 5 \times (\text{IWP}_{index} - 0.05) \times \left(e_i^{cloud} - e_i\right) & \text{if } 0.05 \le \text{IWP}_{index} < 0.25 \\ 4 \times (\text{IWP}_{index} - 0.05) \times (e_i^{cloud} - e_i) & \text{if } 0.25 \le \text{IWP}_{index} < 0.5 \\ & e_i \text{ otherwise} \end{cases}, \tag{10.10}$$

In Eqs. (10.9) and (10.10), e_i is the observation error of the ith channel (see Table 10.1); τ_i^{top} is the transmittance at the model top for the ith channel; $\text{IWP}_{index} = 0.5 \times (\text{IWP}^o_{index} + \text{IWP}^m_{index})$; e_i^{cloud} is the observation error for cloudy

Table 10.1 Prescribed ATMS observation error and maximum observation error.

Channel	Observation error (K)	Maximum observation background error (K)	Channel	Observation error (K)	Maximum observation background error (K)
1	5.0	4.5	12	0.4	1.0
2	5.0	4.5	13	0.55	1.0
3	5.0	3.0	14	0.8	2.0
4	3.0	3.0	15	3.0	4.5
5	0.55	1.0	16	5.0	4.5
6	0.3	1.0	17	2.5	2.0
7	0.3	1.0	18	2.5	2.0
8	0.3	1.0	19	2.5	2.0
9	0.3	1.0	20	2.5	2.0
10	0.3	1.0	21	2.5	2.0
11	0.35	1.0	22	2.5	2.0

radiance and is currently set to zero; and f_{tropic}, f_{H2}, and f_{H4} are defined as follows:

$$f_{tropic} = \begin{cases} 0.01 \times \varphi + 0.75 & f \ \varphi \in [25S, 25N] \\ 1 & \text{otherwise} \end{cases}, \tag{10.11}$$

$$f_{H2} = \begin{cases} \frac{2000}{H} & \text{if } H > 2000 \text{ m} \\ 1 & \text{otherwise} \end{cases}, \tag{10.12}$$

$$f_{H4,i} = \begin{cases} \frac{4000}{H} & \text{if } H > 4000 \text{ m}, \quad i = 8 \\ 1 & \text{otherwise} \end{cases}, \tag{10.13}$$

where φ is latitude and H is the terrain height.

The QC procedure in the GSI system is implemented based on the values of the aforementioned 10 parameters calculated by Eqs. (10.3), (10.4), (10.6), and (10.9), as well as the sum of the mixing ratios of cloud liquid water content and ice water content from the background field (i.e., $q_h = q^m_{liquid} + q^m_{ice}$). It consists of the following nine tests:

1) If $q_h > 0$ and $f_2 > 1$, data outside the latitudinal range [60S, 60N] are rejected for channels 1–7 and 16–22.
2) If $q_h > 0, f_2 \leq 1$, and $f_1 > 0.5$, data outside the latitudinal range [60S, 60N] are rejected for channels 1–6 and 16–22.
3) If $q_h > 0$ and $|T_{b,i}^{\ o} - T_{b,i}^{\ m}| > 3e'_i$, all data are rejected.
4) If $q_h > 0$ and either $\text{LWP}^o_{index} > 0.5$ or $\text{LWP}^m_{index} > 0.5$, data over ocean within the latitudinal range [60S, 60N] are rejected for channels 1–6 and channels 16–22.

5) If $q_h = 0$ and $f_2 > 1$, data of channels 1–7 and 16–22 are rejected.
6) If $q_h = 0$ and $f_2 \leq 1$ and $f_1 > 0.5$, data of channels 1–6 and 16–22 are rejected.
7) If $q_h = 0$ and $|T_{b,i}^{o} - T_{b,i}^{m}| > \max\{3e_i', e_{i,\max}\}$, all data are rejected, where $e_{i,\max}$ is the maximum observation error (see Table 10.1).
8) If $q_h = 0, f_2 \leq 1, f_1 \leq 0.5$, but either $r_1 > 0.05$, or $r_2 > 0.03$, or $r_3 > 0.05$, data over ocean for channels 1–6 and 16–22 are rejected.
9) All data over a mixed surface are rejected, where a mixed surface is defined as a surface with none of the ocean, land, ice, or snow cover interpolated from the background field exceeding 99%.

The ATMS channel 15 is not assimilated over both land and ocean. The first three QC tests remove the outliers under cloudy conditions when model simulations greatly deviate from observations. The fourth QC test removes the data points when either modeled (LWP^{m}_{index}) or observation-retrieved (LWP^{o}_{index}) LWP is greater than 0.5 kg/m^2. The last five QC tests remove the outliers under cloudy conditions. The fifth and seventh tests identify the outliers under clear-sky conditions when model simulations greatly deviate from observations. The sixth QC test considers not only the model and observation differences but also the sensitivity to LWP. The eighth test is used to remove the outliers associated with uncertainty in surface emissivity. The ninth test is to remove a field of view (FOV) over which any single type of surface covers less than 99% of the FOV area.

An example of data distributions retained and removed by the aforementioned QC procedure is shown in Figure 10.4. The brightness temperatures of the Advanced Very-High-Resolution Radiometer (AVHRR) channel 4 at 10.4 μm are used for showing the cloud distribution (Figure 10.4a) (O − B) values of ATMS channel 19 for those data that pass GSI QC (Figure 10.4b) and data points removed by different QC criteria (Figure 10.4c) within and around Hurricane Sandy at 0600 ± 0300 UTC 26 October 2012 are also shown. As expected, cloud is populated within and around Hurricane Sandy. However, still there are many ATMS data points distributed under clear-sky conditions within and around Hurricane Sandy. Most cloudy radiances (i.e., cross symbols in clouds as shown in Figure 10.4a–c) are successfully identified and removed by the GSI QC procedure for the ATMS data described earlier. Data points rejected by the GSI QC in this case are mostly by QC criteria 5 and 9 listed earlier.

Bias correction radiance measurements from meteorological satellites are not absolutely calibrated. Therefore, a global constant observation bias is expected with passive microwave data from a polar-orbiting satellite. For a cross-tracking satellite instrument, observations at large scan angles could be obstructed by the spacecraft radiation, which is difficult to quantify. Therefore, an angular-scan-dependent observation bias is also expected. Model simulations could also have biases. Although the limb effect of a cross-track instrument is modeled in a forward radiative transfer model, the atmospheric inhomogeneity arising from cloud and other sources within an FOV, which is not explicitly/accurately simulated in radiative transfer models, is larger at larger scan angles. In addition, the altitude of the peak weighting function increases with the scan angle, and the

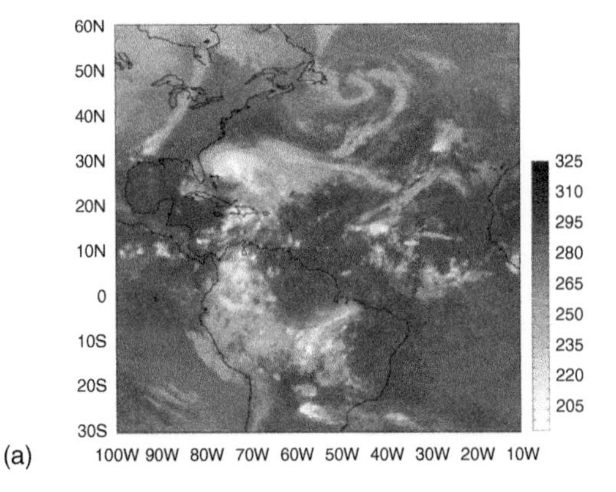

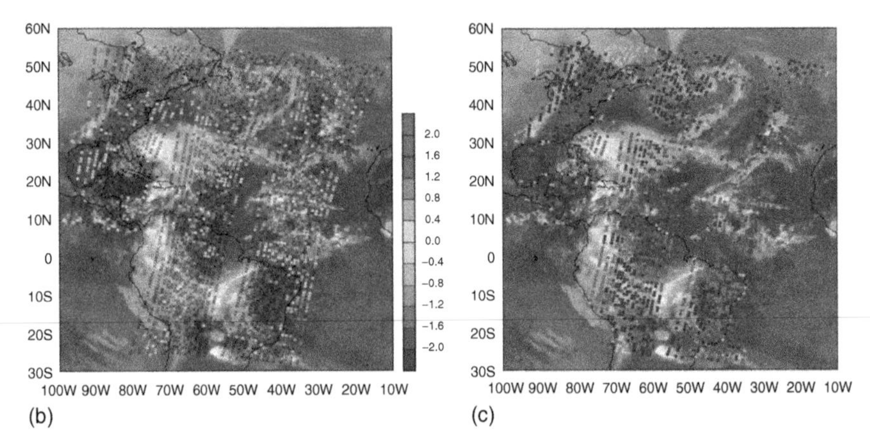

Figure 10.4 Spatial distributions of (a) brightness temperatures of AVHRR channel 4 ($T_{b,4}^{AVHRR}$), (b) (O − B) values of ATMS channel 19 for those data that pass (dots) and fail in (crosses) the GSI QC overlapped on $T_{b,4}^{AVHRR}$ (shading), and (c) data points rejected by the fifth to ninth QC criteria at 0600 UTC 26 October 2012. (Zou *et al.* 2013 [233]. Reproduced with permission of Wiley.)

atmospheric inhomogeneity within an FOV varies with altitude. The atmospheric radiative transfer models are more accurate near nadir than at large scan angles. Therefore, an angular-scan-dependent model bias is expected.

On the other hand, all data assimilation systems are developed under the assumption that observation and forward model errors are unbiased. Therefore, any bias related to satellite instruments and forward models must be removed in satellite data assimilation. Therefore, it is important for the ATMS data biases to be properly quantified and removed prior to assimilation of ATMS data [147, 152, 233].

The GSI bias correction consists of the following three parts: (i) scan biases, which are calculated as mean differences between observations and model simulations for each scan position; (ii) residual biases that depend on air mass distributions and geographical locations [282]; and (iii) residual biases that vary with time on both diurnal and seasonal scales [283]. Specifically, the biases for the ith channel and the jth FOV of ATMS data, $b_{i,j}$ ($i = 1, \ldots, 22$; $j = 1, \ldots, 96$), consist of a static scan bias term ($b_{i,j}^{(1)}$), an air-mass-dependent bias term ($b_{i,j}^{(2)}$), and an adaptive scan bias term ($b_{i,j}^{(3)}$) written as follows:

$$\begin{aligned} b_{i,j} &= b_{i,j}^{(1)} + b_{i,j}^{(2)} + b_{i,j}^{(3)} \\ &\equiv \alpha_{i,j} + \sum_{l=1}^{5} \omega_l \times p_{i,j,l} + \sum_{m=1}^{M} \gamma_{i,j,m} (\beta_{i,j})^m, \end{aligned} \tag{10.14}$$

where $\alpha_{i,j}$ denotes the static scan biases, which are calculated as mean differences between observations and model simulations for each scan position; ω_l ($l = 1, \ldots, 5$) are the five bias correction coefficients for the five air mass predictors $p_{i,j,l}$; and $\beta_{i,j}$ and $\gamma_{i,j,m}$ ($m = 1, \ldots, M$) are the scan angle and the polynomial coefficients up to the order of M, respectively. The adaptive scan bias term ($b_{i,j}^{(3)}$) is not applied for ATMS bias correction.

The five air mass predictors are defined as follows:

$$p_{i,j,1} = 1, \tag{10.15a}$$

$$p_{i,j,2} = \left(\frac{1}{\cos\theta} - 1\right)^2, \tag{10.15b}$$

$$p_{i,j,3} = \begin{cases} (\cos\theta)^2 \times \mathrm{CLW}_{index}, & \text{if } q_{liquid}^m + q_{ice}^m = 0 \\ 0, & \text{otherwise} \end{cases}, \tag{10.15c}$$

$$p_{i,j,4} = (\Gamma_{i,j}^{\tau} - \overline{\Gamma_i^{\tau}})^2, \tag{10.15d}$$

$$p_{i,j,5} = \Gamma_{i,j}^{\tau} - \overline{\Gamma_i^{\tau}}, \tag{10.15e}$$

where $\Gamma_{i,j}^{\tau}$ is the lapse rate of transmittance of the ith channel and the jth FOV, and $\overline{\Gamma_i^{\tau}}$ is the mean lapse rate of transmittance of the ith channel for all FOVs. Values of $\alpha_{i,j}$ ($i = 1, \ldots, 22; j = 1, \ldots, 96$), $\overline{\Gamma_i^{\tau}}$ ($i = 1, \ldots, 22$) and ω_l ($l = 1, \ldots, 5$) are fixed and provided in the GSI input file.

The lapse rate of transmittance that appears in Eqs. (10.15d) and (10.15e) is calculated by the following expression:

$$\Gamma_{i,j}^{\tau} = -\sum_{k=1}^{K} \frac{\tau_{i,k+1} - \tau_{i,k}}{T_{j,k+1} - T_{j,k-1}}, \tag{10.16}$$

where the subscript k indicates the kth vertical level, K is the total number of model levels, $\tau_{i,k+1}$ is the atmospheric transmittance of the ith channel integrated from the kth vertical level to the top of the atmosphere, and T_k is the atmospheric temperature at the kth vertical level.

The biases of ATMS data $b_{i,j}$ described earlier are subtracted from the term $T^o_{b,i} - T^m_{b,i}$, which appears in the cost function of the 3D-Var method in the GSI system.

10.3.4 Comparison between (O − B) and (O − A) Statistics

Figure 10.5 shows the data count distributions as a function of the differences between ATMS data and background (O − B) or the differences between ATMS data and analysis (O − A) at the 8th, 13th, 18th, … , 93th ATMS FOVs for three tropospheric ATMS channels 6–8 and three stratospheric channels 9–11. The angle-dependent biases and standard deviations are also plotted in Figure 10.5. It is seen that the (O − B) data spread is much broader than that of (O − A) for all ranges of (O − B) or (O − A) brightness temperature values. There is a scan-angle-dependent bias such as channels 7 and 11 in (O − B) distributions. The biases and standard deviations are also reduced at all scan angles. The (O − A) biases remain constant for all scan angles.

Values of the mean and standard deviation of (O − B) and (O − A) at nadir for all ATMS channels assimilated in the experiment SAT+ATMS for Hurricane Isaac are presented in Figure 10.6. The biases are reduced for all channels except for channels 13 and 18. The standard deviations of (O − A) are significantly reduced for all channels except for channel 14. This confirms a better fit of NWP model fields to ATMS observations resulting from data assimilation.

10.3.5 Impact of ATMS Data on Forecasting Track and Intensity

Impacts of ATMS data assimilation on hurricane track and intensity forecasts are examined in this section. First, we show daily variations of an added value of ATMS data to conventional data on Hurricane Isaac's track and intensity forecasts (Figures 10.7–10.10). Figure 10.7 shows the forecast tracks of Hurricane Isaac with model forecasts initialized at 0000 UTC and 1200 UTC during August 23–29, 2012 for CTRL1 and CTRL1+ATMS. Forecast results initialized at 0600 UTC and 1800 UTC during the same time period from August 23 to 29, 2012, are similar and not included in Figure 10.7 for clarity. The forecast tracks have an overall eastward bias compared to the observed track. Impacts of ATMS data assimilation on track forecasts were seen for the forecast started at 0000 UTC on August 25 and became more significant on August 26 and afterward. While the center of the observed Hurricane Isaac moved over the Gulf of Mexico, the forecast tracks at 0000 UTC on 25, 0000 UTC, and 1200 UTC on August 26 were over the coastal land in the CTRL experiment. Such an error in forecast track will introduce a significant error in the intensity forecasts.

Daily impacts of ATMS data assimilation on both track and intensity forecasts for Hurricane Isaac are shown in Figure 10.8. Specifically, an average of four 5-day forecasts initialized at 0000 UTC, 0600 UTC, 1200 UTC, and 1800 UTC on each

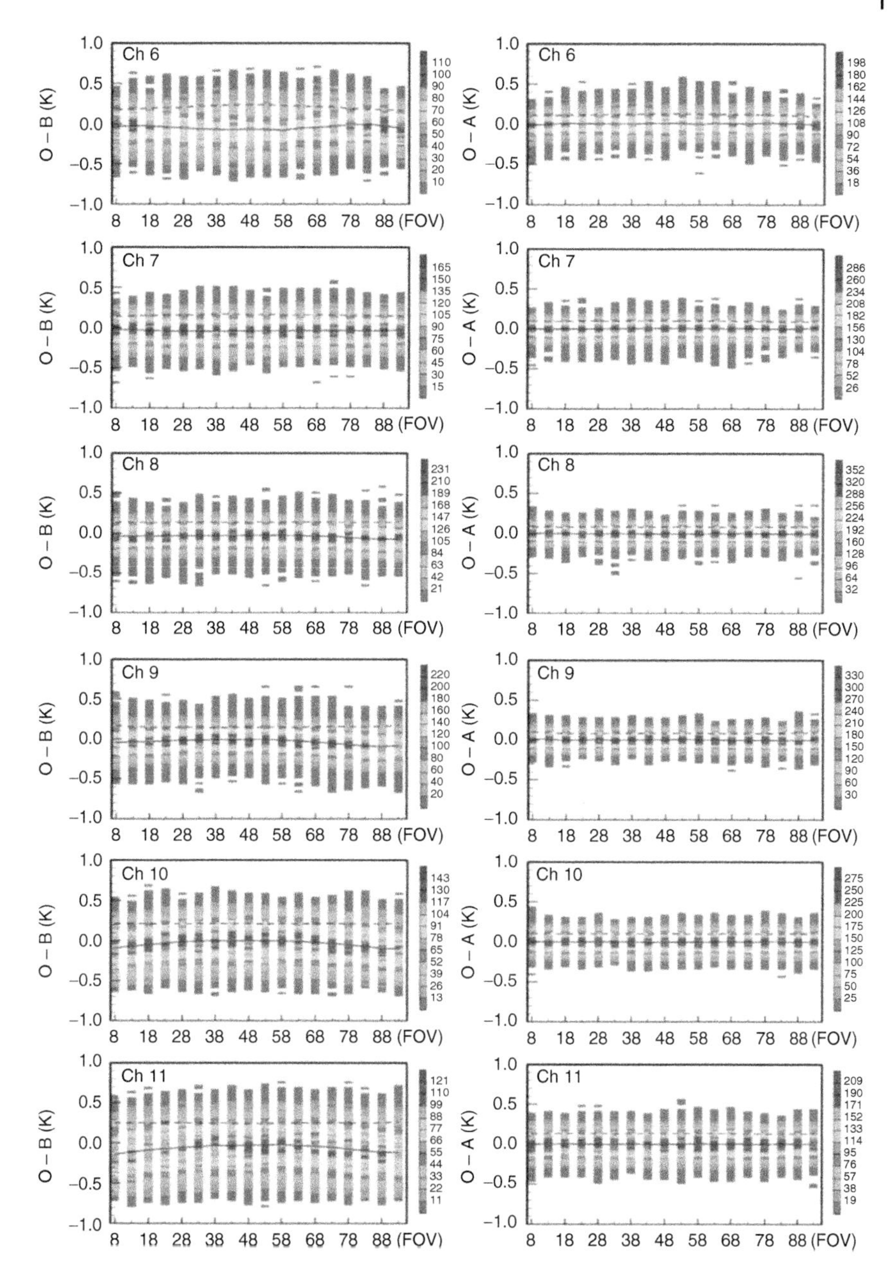

Figure 10.5 Data counts calculated at an interval of 0.025 K (color shading) as a function of FOV and the difference between observations and model simulations calculated from the background fields (left panels) and the analysis fields (right panels) for three tropospheric ATMS channels 6–8 in the experiment CTRL2+ATMS for Hurricane Isaac. The angle-dependent biases and standard deviations are indicated in solid and dashed curves, respectively. (Zou *et al.* 2013 [233]. Reproduced with permission of Wiley.)

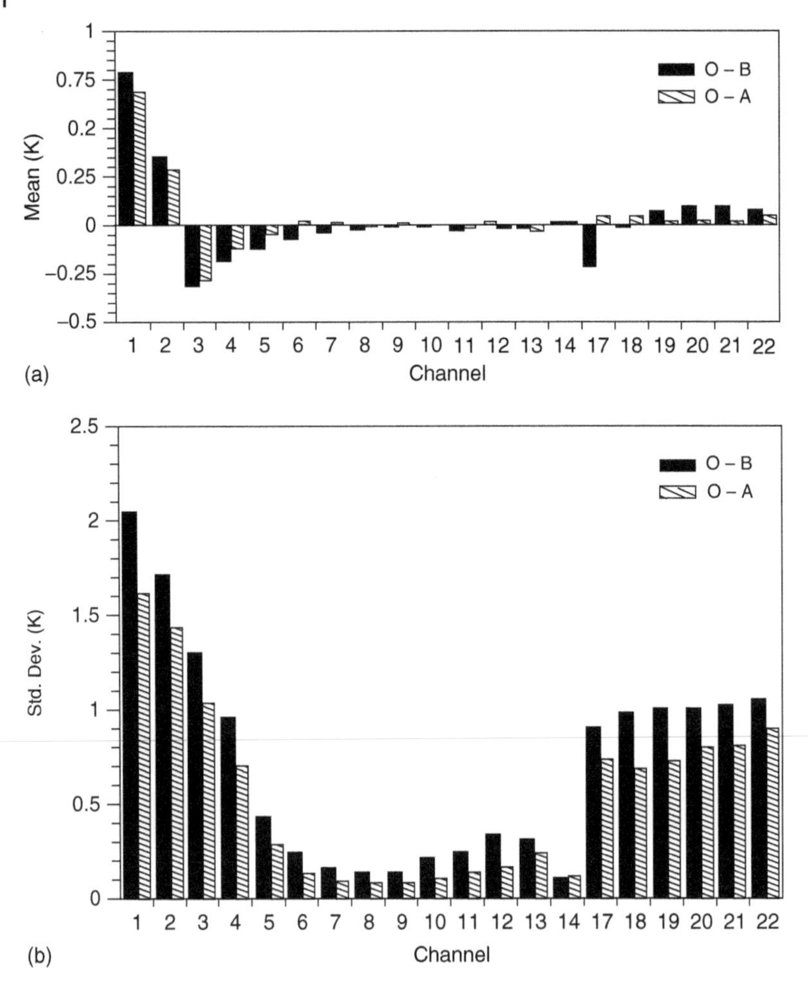

Figure 10.6 (a) Mean and (b) standard deviation of O − B (solid bar) and O − A (dashed bar) at nadir (FOV 48) from the experiment CTRL2+ATMS for Hurricane Isaac. (Zou *et al.* 2013 [233]. Reproduced with permission of Wiley.)

day from August 22 to 27, 2012, is calculated as a function of forecast lead time in both the CTRL1 and CTRL1+ATMS experiments. The track errors are relatively small on the first 3 days from August 22 to 23, 2012, in both experiments when Hurricane Isaac was in the open ocean. A rapid increase in track error with forecasting time occurred on August 25 and 26 in the CTRL1 experiment when Hurricane Isaac moved into Porto Rico. The track errors exceed 400 and 600 nm when the forecast time reaches 96 h in the CTRL1 experiment. When ATMS data are added for data assimilation, the track errors remain around and below 200 nm even during the entire 5-day forecast period of August 25–27, 2012. In other words, ATMS data assimilation had a marginal impact on Hurricane

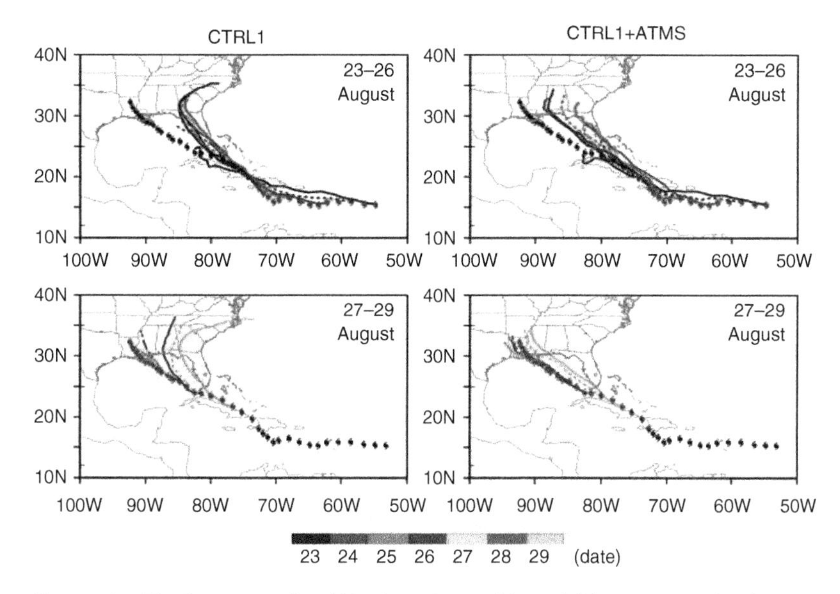

Figure 10.7 The forecast tracks of Hurricane Isaac with model forecasts initialized at 0000 UTC (solid) and 1200 UTC (dotted) during August 23–26 (top panels) and August 27–29 2012 (bottom panels) for CTRL1 (left panels) and CTRL1+ATMS (right panels). The observed track is indicated by hurricane symbols. (Zou *et al.* 2013 [233]. Reproduced with permission of Wiley.)

Isaac's track forecasts when the CTRL1 experiment had a reasonably good track forecast in the first 3 days. It had a significant positive impact on Hurricane Isaac's track forecasts when the CTRL1 experiment did not perform well. Impacts of ATMS data assimilation on intensity forecasts for Hurricane Isaac were most significant on August 25 and 26, 2012. Improvements of ATMS data to intensity forecasts occurred at the same time when track forecasts were improved for Hurricane Isaac.

The maximum wind speeds and the minimum sea level pressure (SLP) from all the 5-day forecasts, four times a day for August 23–29, 2012, for Hurricane Isaac, in CTRL1, CTRL2, CTRL1+ATMS, and CTRL2+ATMS experiments are presented in Figures 10.9 and 10.10, respectively. The HWRF forecasts without satellite data assimilation tend to produce a stronger Sandy than observed. Such a bias in intensity forecasts is reduced after satellite data are assimilated. Assimilating ATMS data reduces the spread of both the maximum wind speed and the minimum SLP from different 5-day forecasts throughout the time period of October 23–31.

The track forecast of Hurricane Sandy was a well-known challenge in 2012. Impacts of ATMS data assimilation on Hurricane Sandy's track forecasts are shown in Figures 10.11 and 10.12. Figure 10.11 is a spaghetti map showing the observed and model-predicted tracks of the 5-day forecasts initialized at 0000 TUC, 0600 UTC, 1200 UTC, and 1800 UTC from October 23 to 29, 2012, by the four experiments CTRL1, CTRL2, CTRL1+ATMS, and CTRL2+ATMS. The

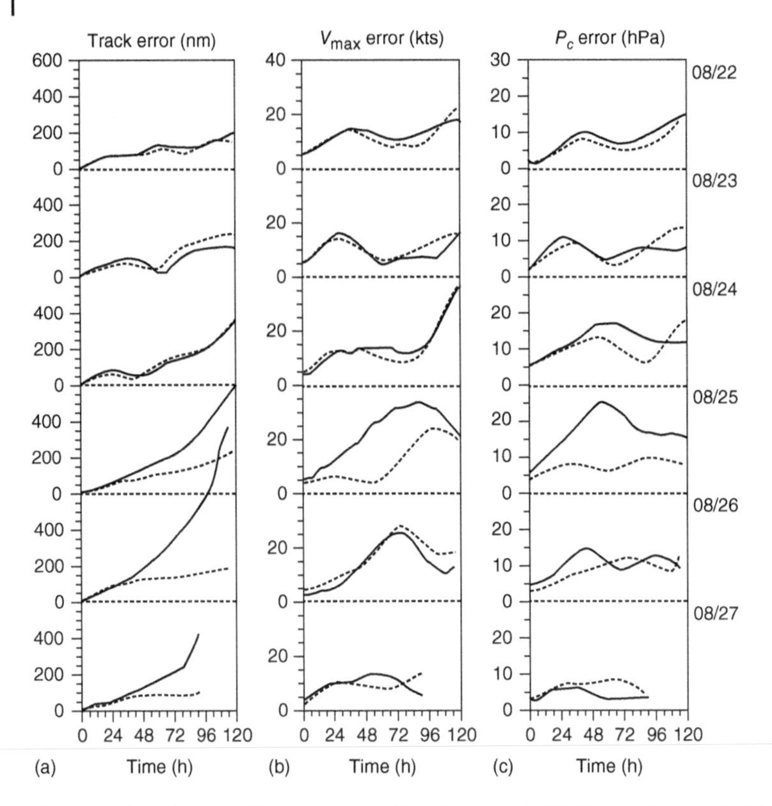

Figure 10.8 Daily mean forecast errors of hurricane track (a), the maximum wind (b), and the central SLP (c) for hurricanes Isaac from the experiments CTRL1 (solid) and CTRL1+ATMS (dashed) from August 22 to 27, 2012. (Zou *et al.* 2013 [233]. Reproduced with permission of Wiley.)

forecast tracks from the CTRL1 experiment have a systematic northeastward bias. Such a track bias is significantly reduced for forecasts initialized after 1200 UTC 26 October 2012. The forecast tracks from the CTRL2+ATMS experiment followed the observed track very closely.

In order to show the forecast track differences between CTRL2 and CTRL2+ATMS more clearly, the 5-day forecast tracks of Hurricane Sandy with the HWRF model forecasts initialized at 0000 UTC and 1200 UTC during the first 4 days (October 23–26) and the later 3 days (October 27–29) are separately presented in Figure 10.12. It is seen that the forecast tracks by the CTRL2 experiment before October 25 all moved northeastward while the observed track turned from its northeastward to northwestward moving direction. Assimilation of ATMS observations results in a much improved track prediction. The track of the CTRL2+ATMS forecasts followed the observed track when the forecast model is initialized as early as 0000 UTC 25 October. The CTRL2+ATMS experiment produced a reasonably good track forecast one day earlier than the CTRL2

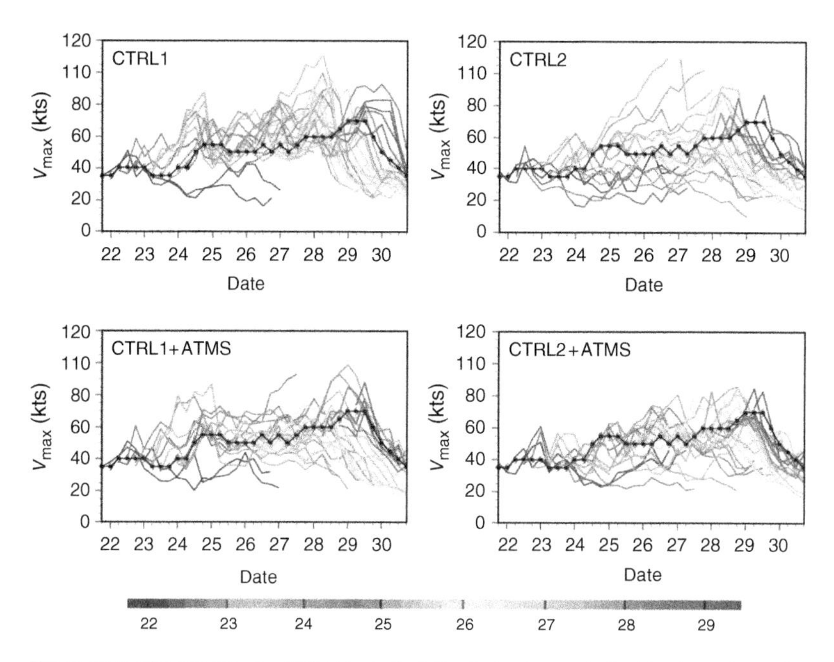

Figure 10.9 The maximum wind speed of all the 5-day forecasts for Hurricane Isaac from CTRL1, CTRL1+ATMS, CTRL2, and CTRL2+ATMS. (Zou *et al.* 2013 [233]. Reproduced with permission of Wiley.)

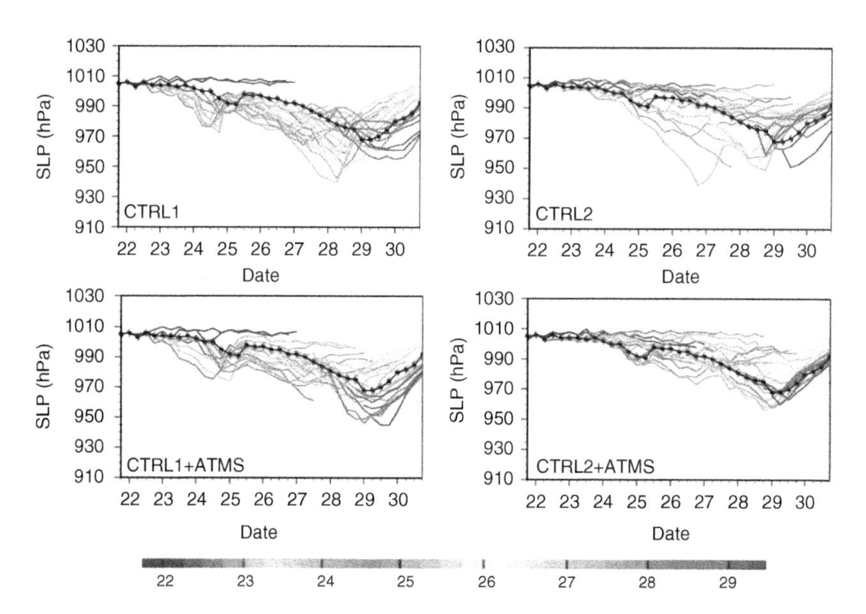

Figure 10.10 Same as Figure 10.9 except for the minimum central SLP. (Zou *et al.* 2013 [233]. Reproduced with permission of Wiley.)

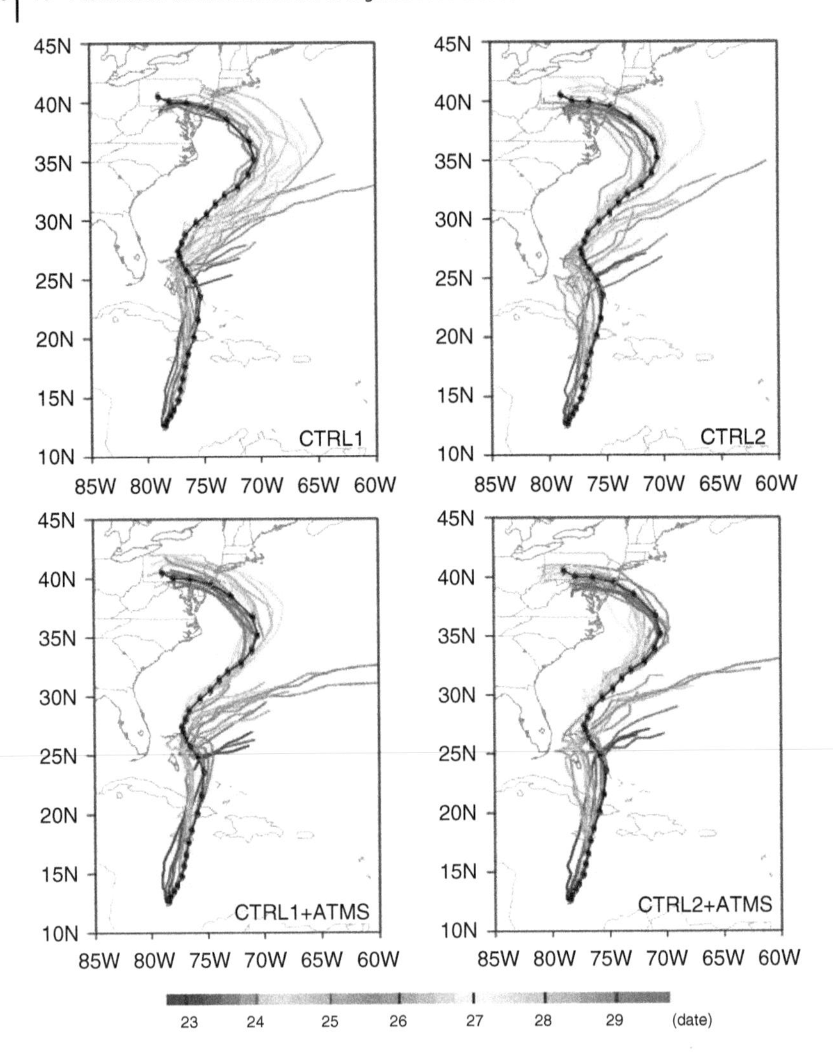

Figure 10.11 Five-day forecast tracks of Hurricane Sandy by the four experiments CTRL1, CTRL1+ATMS, CTRL2, and CTRL2+ATMS with HWRF initialized from October 23 to 29, 2012 at 6-h intervals. The NHC best track is shown in black. (Zou *et al.* 2013 [233]. Reproduced with permission of Wiley.)

experiment. The forecast tracks of both CTRL2 and CTRL2+ATMS experiments made the right turn from their northeastward movement to a northwestward movement for all the forecasts initialized during October 27–29. However, the landfall position from the CTRL2+ATMS forecasts initialized on October 27 is more precise than that in the CTRL2 experiment.

The largest difference in the track forecasts of Hurricane Sandy with and without ATMS data assimilation occurred in the forecasts initialized at 1200 UTC 26 October 2012. It is found that the improvement on track forecasts by ATMS

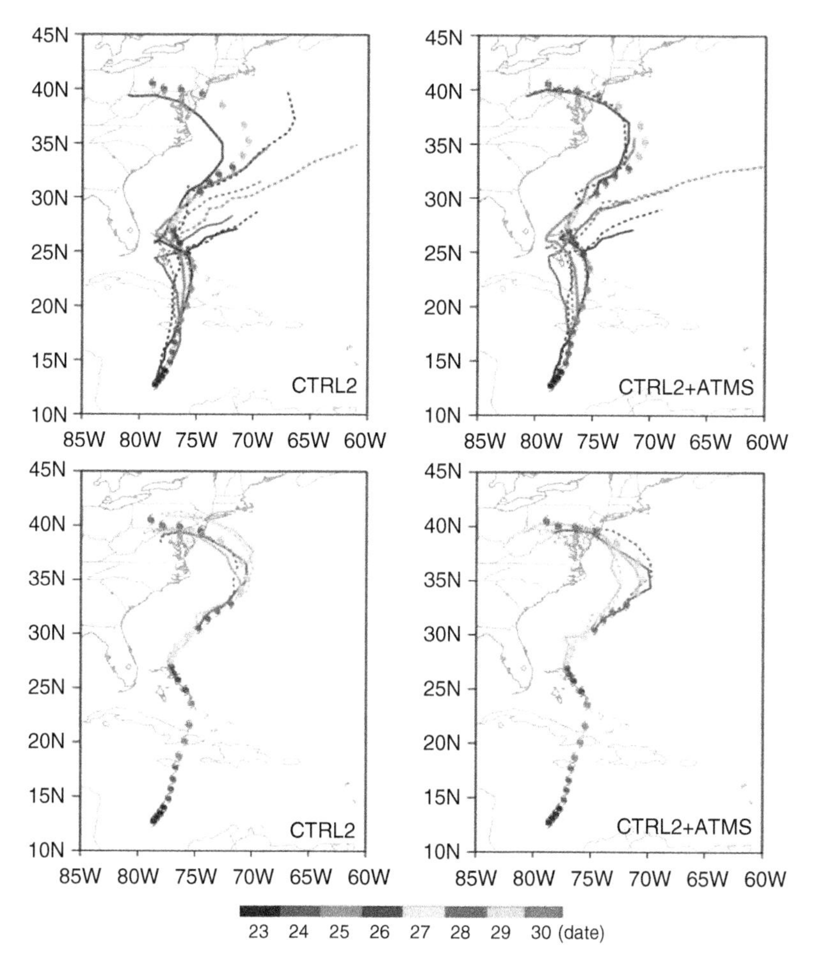

Figure 10.12 Tracks of Hurricane Sandy with the HWRF model forecasts initialized at 0000 UTC (solid) and 1200 UTC (dotted) during October 23–29 2012 for CTRL2 (left panels) and CTRL2+ATMS (right panels). The observed locations of Hurricane Sandy in different days are indicated by colored hurricane symbols. (Zou *et al.* 2013 [233]. Reproduced with permission of Wiley.)

data arises from a more realistic development of an Ω-shaped ridge in the middle latitudes when Sandy moved into it. Figure 10.13 shows the potential vorticity and wind distributions at 200 hPa at 1200 UTC 28 October, 1200 UTC 29 October, and 0000 UTC 30 October 2012, from the NCEP GFS analysis. The position of Hurricane Sandy is indicated. Based on the large-scale environmental flow pattern shown in Figure 10.13, it is pointed out that the track of Hurricane Sandy changed from a northeastward movement to a northwestward movement due to the fact that it moved into a mid-latitude trough or the west-side neck of an Ω-shaped ridge downstream of the trough. The cyclonic flow of the trough, or the anticyclonic flow of the ridge, seems to significantly contribute

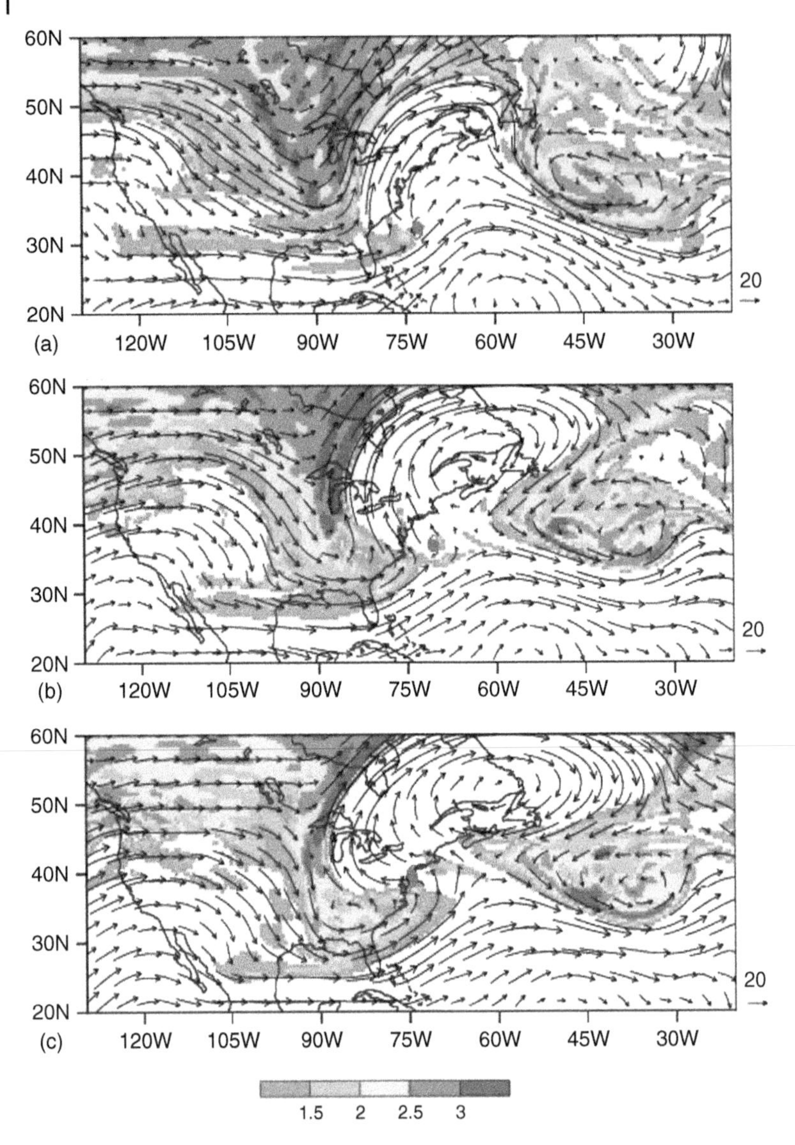

Figure 10.13 Potential vorticity (shading) and wind vector (black arrow) at 200 hPa at (a) 1200 UTC 28 October, (b) 1200 UTC 29 October, and (c) 0000 UTC 30 October from the NCEP GFS analysis for Hurricane Sandy. Purple hurricane symbol indicates the center location of Hurricane Sandy. (Zou *et al.* 2013 [233]. Reproduced with permission of Wiley.)

to the northwestward movement of Hurricane Sandy. As seen in Figure 10.13c, Hurricane Sandy made landfall at 0000 UTC 30 October 2012.

Improvements of Hurricane Sandy's track forecasts thus depend on how well these large-scale features in the middle latitudes are forecast by HWRF. The 48-, 72-, and 84-h forecasts from the two forecast experiments initialized at 1200 UTC

26 October 2012, for CTRL2 and CTRL2+ATMS are presented in Figures 10.14 and 10.15, respectively. It is mentioned that the 48-, 72-, and 84-h forecasts shown in Figures 10.14 and 10.15 are valid at 1200 UTC 28 October 1200 UTC 29 October, and 0000 UTC 30 October 2012. Compared with Figure 10.13, it is concluded that the large-scale flow patterns are better predicted by the CTRL2+ATMS experiment. The model-predicted Ω-shaped ridge downstream north of Sandy is too weak in the CTRL2 experiment. In other words, the ATMS data assimilation positively contributes to the prediction of Hurricane Sandy's environmental flow.

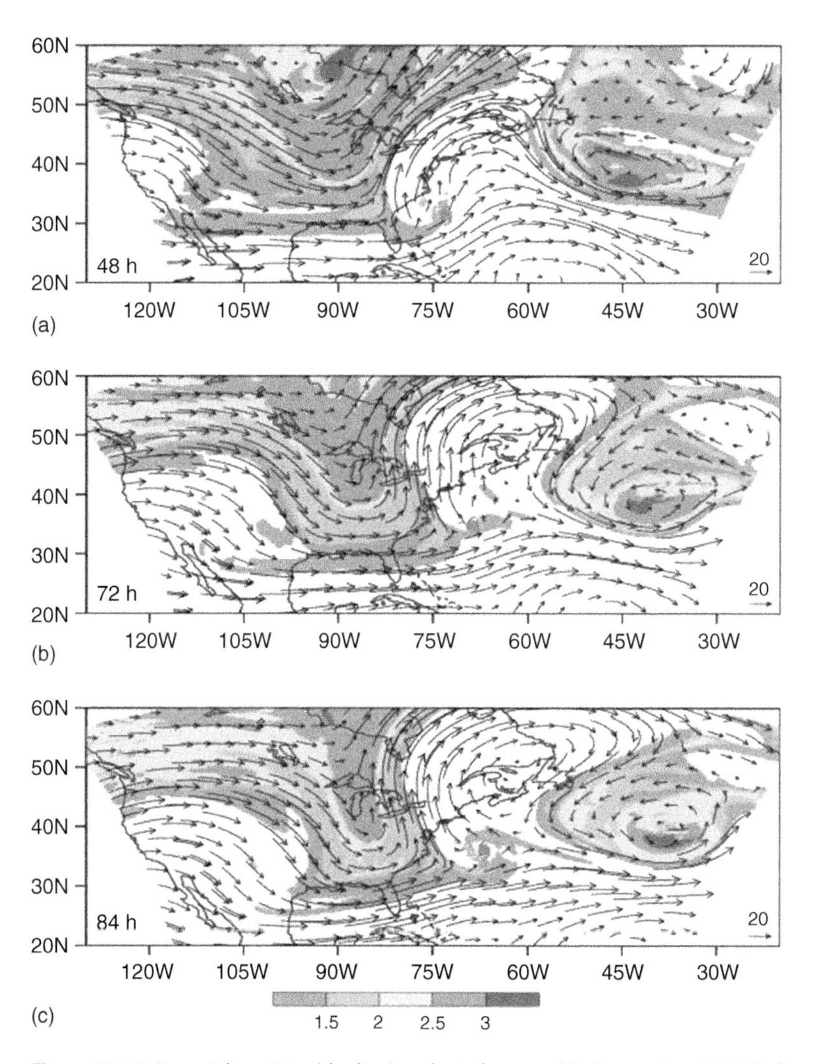

Figure 10.14 Potential vorticity (shading) and wind vector (black arrow) at 200 hPa from (a) 48 h, (b) 72 h, and (c) 84 h the CTRL2 forecast initialized at 1200 UTC 26 October 2012. Purple hurricane symbol indicates Sandy's center location predicted by the CTRL2 experiment. (Zou *et al.* 2013 [233]. Reproduced with permission of Wiley.)

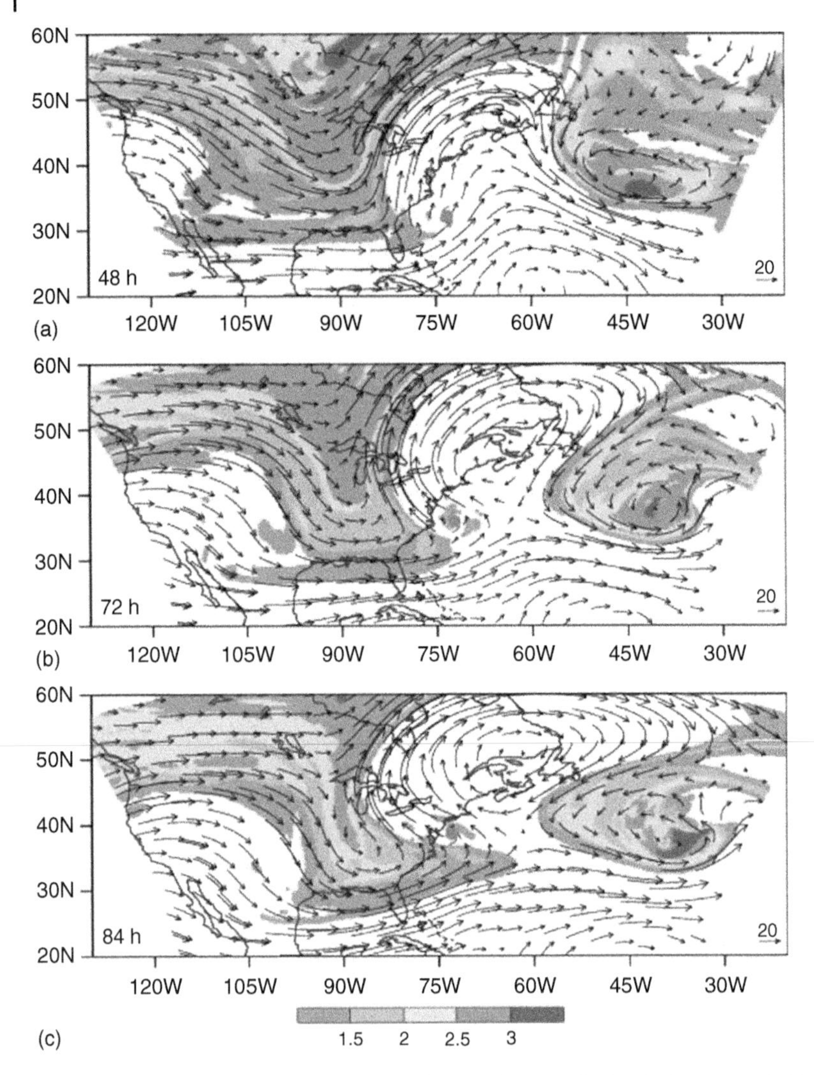

Figure 10.15 Same as Figure 10.14, except for experiment CTRL2+ATMS. (Zou *et al.* 2013 [233]. Reproduced with permission of Wiley.)

10.4 SSMIS Data Assimilation

10.4.1 SSMIS Instrument

The US Defense Meteorology Satellite Program (DMSP) F-16 satellite was launched successfully on October 18, 2003. On board F-16, the Special Sensor Microwave Imager Sounder (SSMIS) measures the Earth's radiation at 24

Table 10.2 Characteristics of SSMIS channels.

Channel	Center frequency (GHz)	3-dB width (MHz)	Frequency stability (MHz)	Polarization	NEDT (K)	Sampling interval (km)
1	50.3	380	10	*v*	0.34	37.5
2	52.8	389	10	*h*	0.32	37.5
3	53.596	380	10	*h*	0.33	37.5
4	54.4	383	10	*h*	0.33	37.5
5	55.5	391	10	*h*	0.34	37.5
6	57.29	330	10	RCP	0.41	37.5
7	59.4	239	10	RCP	0.40	37.5
8	150	1642(2)	200	*h*	0.89	12.5
9	183.31 ± 6.6	1526(2)	200	*h*	0.97	12.5
10	183.31 ± 3	1019(2)	200	*h*	0.67	12.5
11	183.31 ± 1	513(2)	200	*h*	0.81	12.5
12	19.35	355	75	*h*	0.33	25
13	19.35	357	75	*v*	0.31	25
14	22.235	401	75	*v*	0.43	25
15	37	1616	75	*h*	0.25	25
16	37	1545	75	*v*	0.20	25
17	91.655	1418(2)	100	*v*	0.33	12.5
18	91.655	1411(2)	100	*h*	0.32	12.5
19	63.283248 ± 0.285271	1.35(2)	0.08	RCP	2.7	75
20	60.792668 ± 0.357892	1.35(2)	0.08	RCP	2.7	75
21	$60.792668 \pm 0.357892 \pm 0.002$	1.3(4)	0.08	RCP	1.9	75
22	$60.792668 \pm 0.357892 \pm 0.0055$	2.6(4)	0.12	RCP	1.3	75
23	$60.792668 \pm 0.357892 \pm 0.016$	7.35(4)	0.34	RCP	0.8	75
24	$60.792668 \pm 0.357892 \pm 0.050$	26.5(4)	0.84	RCP	0.9	37.5

channels (see Table 10.2) from 19 to 183 GHz and scans conically at an Earth incidence angle of 53° (see Figure 10.16). This instrument can, in principle, provide improved atmospheric temperature and water vapor sounding under all weather conditions after imaging and sounding channels attain the same viewing geometry. A simultaneous retrieval of surface and atmospheric parameters is developed to produce operational SSMIS products [255]. As shown in Figure 10.17, the upper atmospheric sounding is extended to 100 km altitude through the use of Zeeman absorption lines (discussed in Chapter 2). Compared with the previous

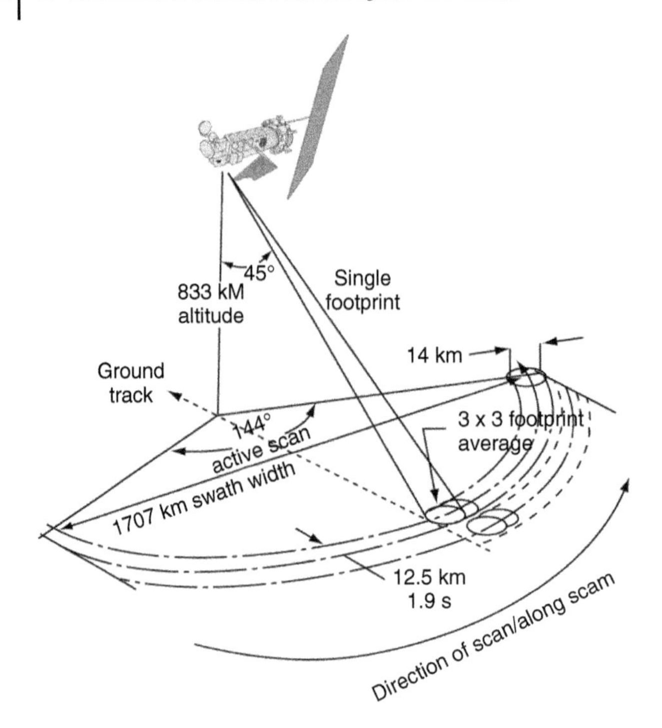

Figure 10.16 SSMIS scan geometry [289].

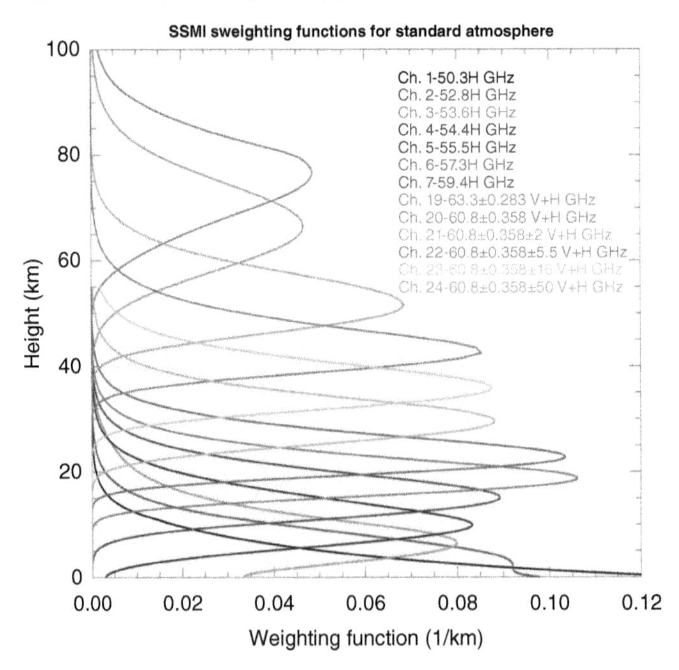

Figure 10.17 SSMIS weighting functions for a standard atmosphere [289].

SSM/I instrument on board F-8 to F-15 satellites, the SSMIS scan swath increases to 1700 km with much less orbit gap from the 833-km spacecraft altitude.

SSMIS brightness temperatures from the Earth scene are derived from a satellite microwave radiometer through a series of instrument calibrations. First, a raw receiver voltage count is converted to an antenna-beam-averaged temperature through a so-called calibration equation, which requires the measurements from warm and cold calibration targets. This antenna temperature is then converted to the main-beam-referenced brightness temperature where the side-lobe contribution is removed. For SSMIS polarization channels, an additional correction is made for a spillover effect from the cross-polarization leakage. Thus, any remaining error in the antenna brightness temperature (TDR) is translated into sensor brightness temperature (SDR) during the calibration process. It has been noted that several anomalies are shown in SSMIS TDR data for all channels, which result primarily from the thermal emission of the main reflector as well as the calibration target count perturbation [284]. Similar anomalies are also observed in SSMIS SDR data. These anomalies can severely downgrade the applications of SSMIS measurements in various environmental applications. Figure 10.18 displays the biases of SSMIS observations to the simulations at 55.5 GHz, where the simulations are calculated from temperature and water vapor profiles analyzed in the GFS analysis of NCEP GFS data assimilation system (GDAS). Normally, the (O − B) should be fairly uniform for the microwave sounding channels on NOAA satellites. Here, the bias varies with the latitude by an order of several degrees kelvin.

The root cause of these anomalies has been investigated by the SSMIS instrument calibration team since the instrument was launched and is mainly the

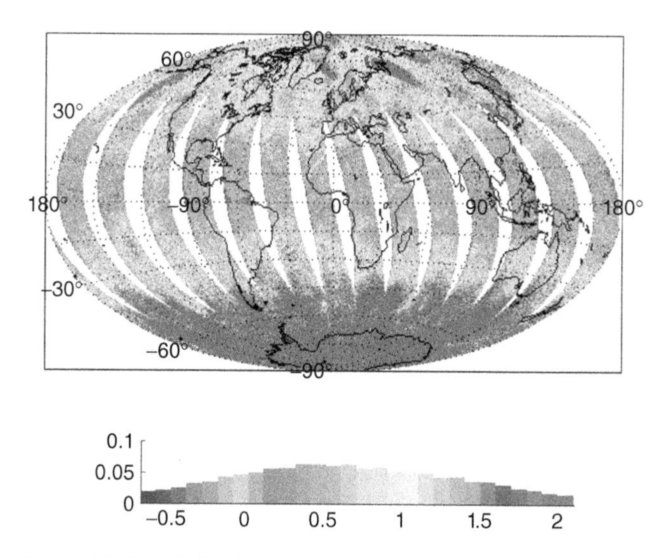

Figure 10.18 F-16 SSMIS O − B biases on ascending node at channel 5 (55.5 GHz) on February 15, 2012.

emission of the main reflector and the direct solar and stray light contamination on two calibration targets [284]. SSMIS antenna and calibration subsystem consists of a main reflector, six corrugated feedhorns, a warm calibration load, and a subreflector for viewing cold space. In each scan, the radiation from the Earth's atmosphere is focused via the main reflector on to the feedhorns and then to the receiver subsystem. At the end of the scan, the feedhorns also pass beneath a stationary warm load and cold space reflector, providing periodic calibration of the measured Earth-viewing radiances at the input to the feedhorns. The receiver subsystem accepts the energies from the six feedhorns and performs amplification, filtering, and additional frequency multiplexing to output 24 discrete frequency bands to the signal processing subsystem. Then, the received scene count of the Earth-viewing radiance is converted to antenna temperature (scene temperature) by a calibration equation, which is associated with the temperatures and counts of two calibration targets. Thus, the emission from the main reflector and the solar heating of the calibration targets can result in the anomalies in the SSMIS antenna temperatures.

A methodology is developed to correct the SSMIS anomalies at the TDR level, which includes predicting the reflector face temperature and detecting and removing the warm target calibration count anomalies [285]. Essentially, the total energy in terms of antenna brightness temperature received at the feeds is

$$T'_A = T_A + \varepsilon_R(T_R - T_A), \tag{10.17}$$

where T_A is the antenna brightness temperature corresponding to the Earth scene temperature, and T_R the reflector face temperature. If the reflector emissivity and face temperature are known, T_A can be calculated. From the NRL antenna model, the emissivity of the main reflector is estimated as 0.012, 0.016, 0.020, 0.025, and 0.035 at 19.35, 37, 60, 91.65, and 183 GHz, respectively, for vertically polarized channel at an incident angle of 20° [284]. The reflector face temperature can be estimated from antenna arm temperatures [286, 287].

The locations of the SSMIS warm target anomalies are detected through the fast Fourier transform (FFT) analysis of one-orbit of calibration target information (warm count and platinum resistance thermometer (PRT) count) [287]. From a linear algorithm, the anomaly in antenna brightness temperature can be expressed as

$$\Delta T_A = -\frac{T_A - T_C}{C_W - C_C}\Delta C_W - \frac{T_W - T_A}{C_W - C_C}\Delta C_C + \frac{C_S - C_C}{C_W - C_C}\Delta T_W, \tag{10.18}$$

where ΔC_W, ΔC_C, ΔT_W are jumps in warm and cold counts, PRT temperature, respectively. Thus, for solar heating, the warm target count anomaly is positive and depresses the antenna temperature, whereas the PRT temperature anomaly is also positive and increases the antenna temperature. From our analyses, ΔT_A from ΔC_W varies from −0.5 to −1.5 K at most of channels. Among the three anomalies (ΔC_W, ΔC_C, ΔT_W), ΔC_W dominates and depresses the SSMIS antenna temperatures at almost all channels.

The antenna emission and the calibration target anomaly are important error sources to produce the anomalous SSMIS radiances over most of the global areas. The antenna emission occurs at all channels over all global areas, but it becomes significant (1 – 3 K) near the North Pole, where the solar heating of the main reflector reaches a maximum. Among the calibration target anomalies, the warm load count anomaly is the most noticeable and can alone depress the SSMIS temperature by about 1 K. The warm load count anomaly occurs only over several latitudinal zones, but it affects almost all channels. More importantly, the location and magnitude of the emission and calibration target anomaly vary with season.

10.4.2 SSMIS Data Quality Control

The SSMIS instrument has nine channels measuring the thermal emission from the Earth's surface and atmosphere at 19.35, 22.235, 37, and 91 GHz with both horizontal and vertical polarizations, except for the 22.35 GHz channel which is vertically polarized. In the cases of clouds over oceans, each of the four channels responds differently to cloud liquid water with different sizes of the cloud droplets. The 19.35 GHz channel provides the most direct measurement of liquid water for rain-bearing clouds; the 37 GHz channel is more sensitive to the nonprecipitating clouds with small droplet sizes; and the 85 GHz channel has higher sensitivity to the low-lying thin stratus clouds, which have the smallest amount of liquid water and are strongly affected by the scattering from precipitation of ice particles. To measure the large dynamic range of LWP, Weng and Grody [149] developed a composite LWP algorithm for SSM/I as

$$\mathrm{LWP} = \begin{cases} \mathrm{LWP}^{19V}, & \mathrm{LWP}^{19V} > 0.70\,\mathrm{mm}, \\ \mathrm{LWP}^{37V}, & \mathrm{LWP}^{37V} > 0.28\,\mathrm{mm\ or\ WVP} \geq 30\,\mathrm{mm}, \\ \mathrm{LWP}^{85H} & \text{otherwise}, \end{cases} \tag{10.19}$$

where WVP is the vertically integrated water vapor, which is calculated as

$$\begin{aligned} \mathrm{WVP} = {} & 232.89 - 0.1486 \times TB_{19h} - 0.3695 \times TB_{37h} \\ & - (1.8291 - 0.006193 \times TB_{22v}) \times TB_{22v}. \end{aligned} \tag{10.20}$$

The composite LWP is the vertically integrated liquid water path, and each item on the RHS of Eq. (10.19) is derived by a regression equation using measurements of two SSM/I channels as follows:

$$\mathrm{LWP}^{freq} = a_0[\ln(290 - T_a^{freq}) - a_1 - a_2 \ln(290 - T_a^{22V})], \tag{10.21}$$

where the superscript "*freq*" represents 19*v*, 37*v*, and 85*h*, respectively, and T_a^{freq} is the antenna temperature at a specific SSM/I frequency. The coefficients a_0,

Table 10.3 SSM/I coefficients in LWP algorithm.

Frequency (GHz)	a_0	a_1	a_2
19v	−3.20	2.80	0.420
37v	−1.66	2.90	0.349
85h	−0.44	−1.60	−1.354

Table 10.4 Remapping coefficients for linearly remapping the F16 SSMIS antenna temperatures at SSM/I-like channels to F15 SSM/I antenna temperatures.

Frequency (GHz)	α	β
19h	0.00424	1.00027
19v	−2.03627	1.00623
22v	−2.52875	0.99642
37h	0.80170	0.99139
37v	−3.86053	1.00550
91v	−7.43913	1.03121
91h	1.53650	0.99317

a_1, and a_2 are listed in Table 10.3 [149]. As seen in Eq. (10.19), this algorithm is developed for SSM/I based on known antenna temperatures. However, the brightness temperatures are provided to the operational model HWRF. Therefore, the brightness temperatures need to be first converted to the corresponding antenna temperatures before they can be used in Eq. (10.19).

The composite LWP retrieval algorithm described earlier was designed for SSM/I. To apply it to SSMIS (on board F16–F19) in the current GSI system, the imager channels 12–18 of SSMIS need to be remapped to SSM/I-like channels based on the following equation:

$$T_{SSM/I}^{freq} = \alpha T_{SSMIS}^{freq} + \beta, \tag{10.22}$$

where $T_{SSM/I}^{freq}$ and T_{SSMIS}^{freq} are the antenna temperatures of SSM/I-like channel and SSMIS channel (*freq*), respectively. The remapping coefficients α and β used in the current GSI system are provided by Yan and Weng [287] and listed in Table 10.4.

10.4.3
SSMIS Bias Correction

Biases in the GSI system consist of three parts. Specifically, for the bias of the ith channel of SSMIS ($b_{i,j}$), they are (i) a static scan bias term ($b_{i,j}^{(1)}$), (ii) an air-mass-dependent bias term ($b_{i,j}^{(2)}$), and (iii) an adaptive scan bias term ($b_{i,j}^{(3)}$). In the current

HWRF GSI system, only the first two terms are applied to SSMIS bias correction, shown as follows:

$$b_{i,j} = b_{i,j}^{(1)} + b_{i,j}^{(2)} = \alpha_{i,j} + \sum_{k=1}^{5} \omega_k \times p_{i,j,k}, \tag{10.23}$$

where $\alpha_{i,j}$ is the static scan bias of the ith channel at jth scan position, and ω_k ($k = 1, \ldots, 5$) are the five bias correction coefficients for the five air mass predictors $p_{i,j,k}$, respectively. Values of $\alpha_{i,j}$ ($i = 1, \ldots, 24$; $j = 1, \ldots, 60$) and ω_k are predefined. The five air mass predictors are defined as the same expressions by Eqs. (10.15) and (10.16).

The described SSMIS biases are subtracted from the term $T_{b,i}^o - T_{b,i}^m$ in the ith channel, which appears in the cost function of the 3D-Var method in the GSI system. In the default configuration of SSMIS bias correction, the predefined parameters are all set to be zero:

$$\alpha_{i,j=0} (i = 1, \ldots, 24; j = 1, \ldots, 60), \tag{10.24a}$$

$$\omega_k = 0 (k = 1, \ldots, 5), \tag{10.24b}$$

$$\overline{\Gamma_i^\tau} = 0 (i = 1, \ldots, 24). \tag{10.24c}$$

To determine the data affected by scattering effect in SSMI/S channels 9–11, the parameter r_i, is defined in the current GSI system:

$$r_i = f_i^{(1)} - f_i^{(2)} T_{b,17}^s + f_i^{(3)} T_{b,8}^s, \tag{10.25}$$

where $i = 9$, 10, or 11. $T_{b,8}^s$ and $T_{b,17}^s$ are the simulated brightness temperatures of channel 8 and 17, respectively. The coefficients $f_i^{(1)}$, $f_i^{(2)}$, and $f_i^{(3)}$ are listed in Table 10.5.

The QC procedures in the GSI system are based on the values of LWP and WVP and implemented over ocean and land separately. In general, there are eight steps in the current QC procedures:

1) In the ith channel, if the absolute brightness temperature difference of O – B is greater than (or equal to) 3.5 K, that is, $|T_{b,i}^o - T_{b,i}^m| \geq 3.5$ K, data from that channel will be rejected.

Table 10.5 Coefficients $f_i^{(1)}$, $f_i^{(2)}$, and $f_i^{(3)}$ for detecting scattering effected data.

i	$f_i^{(1)}$	$f_i^{(2)}$	$f_i^{(3)}$
9	271.252327	0.485934	0.473806
10	272.280341	0.413688	0.361549
11	278.824902	0.400882	0.270510

Table 10.6 SSMIS channel characteristics and the related LWP* (kg/m²) for cloud check.

Channel	Center frequency (GHz)	Polarization	LWP* (kg/m²)
1	50.3	V	0.10
2	52.8	v	0.20
3	53.596	v	0.60
4	54.4	v	2.00
5	55.5	v	2.00
6	57.29	RCP	2.00
7	59.4	RCP	2.00
8	150	h	0.10
9	183.31 ± 6.6	h	0.10
10	183.31 ± 3	h	0.10
11	183.31 ± 1	h	0.10
12	19.35	h	0.20
13	19.35	v	0.20
14	22.235	v	0.20
15	37	h	0.20
16	37	v	0.20
17	91.655	v	0.10
18	91.655	h	0.10
19	63.283248 ± 0.285271	RCP	10.00
20	60.792668 ± 0.357892	RCP	10.00
21	$60.792668 \pm 0.357892 \pm 0.002$	RCP	10.00
22	$60.792668 \pm 0.357892 \pm 0.0055$	RCP	10.00
23	$60.792668 \pm 0.357892 \pm 0.016$	RCP	10.00
24	$60.792668 \pm 0.357892 \pm 0.05$	RCP	10.00

RCP, right-hand circular polarization.

2) If $r_i - T^o_{b,i} - T^m_{b,i} - T^s_{b,i} > 2\,\mathrm{K}$ (=9, 10, or 11), data from the ith channel will be rejected.
3) If the LWP retrieval described in Eqs. (10.19)–(10.21) fails, or the retrieved TPW is less than zero, data obtained over that particular part of ocean will be rejected.
4) If the retrieved LWP > LWP* over sea, the data will be rejected. The LWP* is used as a threshold listed in Table 10.6.
5) If cloud is detected while $|T^o_{b,ch2} - T^m_{b,ch2} - b_{ch2,j}| \geq 1.5\,\mathrm{K}$ not over sea, data of channels 1, 2, and 12–16 are rejected.
6) Data from channels 1–3 and 8–18 data will be rejected if the data is over a mixed surface, which is a surface other than a sole sea, ice, land, or snow surface.
7) Data from channel 9 data will be rejected if the surface elevation is higher than 2 km.
8) Data from channels 3 and 10 will be rejected if the surface elevation is higher than 4 km.

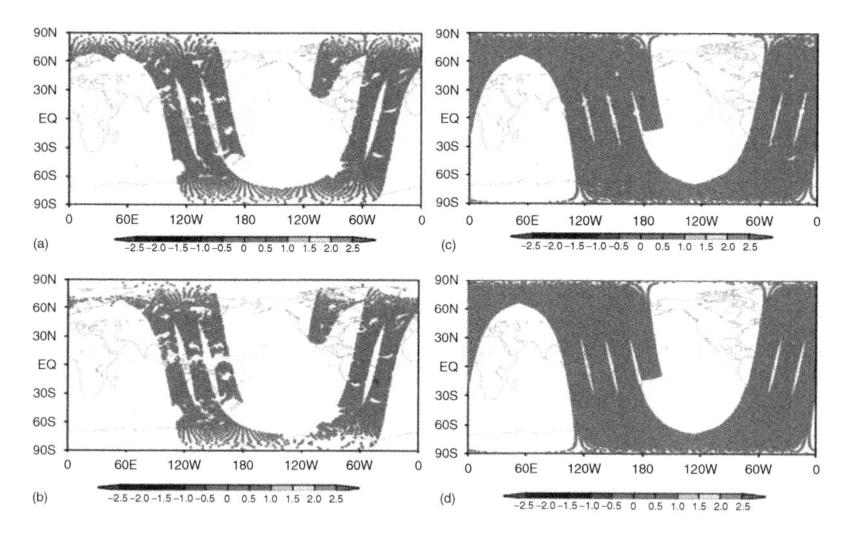

Figure 10.19 Distribution of brightness temperature differences on August 5, 2008, for F-16 UPP at (a) 54.4 GHz and (b) 55.5 GHz, and for MetOp-A AMSU-A at (c) 54.4 GHz and (d) 55.5 GHz, which are calculated using the data passing the current GFS quality control test after the GFS bias correction [270] is applied. The acronym sdt represents the standard deviation of the brightness temperature difference. (Yan and Weng 2012 [267]. Reproduced with permission of American Meteorological Society.)

Figure 10.19a,b displays distributions of brightness temperature differences (DTB) on August 5, 2008, for F-16 UPP at 54.4 and 55.5 GHz, respectively. For comparison, the biases at these two frequencies in the data from the AMSU-A on board the Meteorological Operational Satellite Programme (MetOp) satellite are also shown in Figure 10.19c,d, respectively. It is seen that the biases from the SSMIS data are dependent on the latitude and orbit node and vary substantially from month to month. Note that the results shown in the figures are generated after applying the original GFS bias correction (BC) algorithm [270]. This implies that the aforementioned regional biases in F-16 SSMIS data cannot be simply removed by GFS BC algorithm. An additional BC algorithm is required for further improvements.

For SSMIS data distributed from DMSP ground processing, the biases are high in some regions but do not change significantly on a weekly basis. Thus, the weekly mean O − B biases are first generated according to the satellite ascending/descending orbit and latitudes (θ). The mean biases are referred to as $\Delta T_B^{Cal}(\theta, \text{node})$. At a given time and location, the brightness temperature is corrected as

$$T_B^{Cal} = T_B^{obs} - \Delta T_B^{Cal}(\theta, \text{node}), \tag{10.26}$$

where T_B^{Obs} is the original UPP brightness temperatures. The O − B bias, $\Delta T_B^{Cal}(\theta, \text{node})$, is derived from longitudinal mean O − B biases attained in the previous week. Since some of the lower atmospheric sounding (LAS) channels (e.g., 50.3 and 52.8 GHz) are affected by clouds and/or surface emissivity, a QC procedure must be developed. For the LAS channel at 52.8 GHz, which is affected by raining clouds, a threshold approach is used to detect cloud contamination, depending on the surface type. This channel is also sensitive to surface emissivity, especially for high-elevation terrain. Thus, $\Delta T_B^{Cal}(\theta, \text{node})$ is derived from all the data over the land where the surface pressure is greater than 700 mb.

Figure 10.20a,b displays the variations in the daily averaged biases at channels 4 (54.4 GHz) and 5 (55.5 GHz) from August 5 to 11, 2011. The main features of the O − B biases vary slowly from day to day, including the locations and magnitudes of the maximum and minimum biases and the magnitude of the longitudinal average of the O − B biases with latitude. The slow change in ΔT_B with time is related to

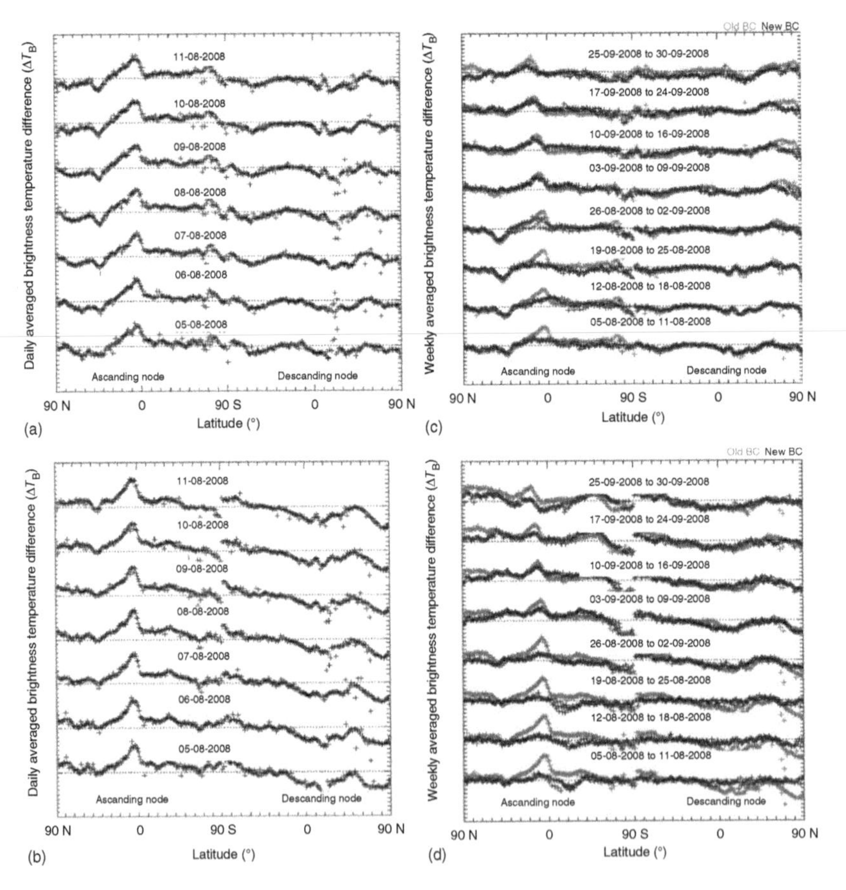

Figure 10.20 Daily average of longitudinally averaged biases from August 5 to 11, 2011, at (a) 54.4 GHz and (b) 55.5 GHz. Weekly mean of longitudinally averaged bias from August 5 to September 30, 2011, at (c) 54.4 GHz and (d) 55.5 GHz. (Yan and Weng 2012 [267]. Reproduced with permission of American Meteorological Society.)

the pattern of the original calibration anomaly in F-16 [284, 285]. Figure 10.20c,d displays the time series of the weekly averaged ΔT_B (in moderate gray color) at the same channels for a few weeks from August 5 to September 30, 2011. Although the regionally dependent biases exist at each channel during 2 months, their locations and magnitude have a small change from week to week. Therefore, the weekly averaged $\Delta T_B^{Cal}(\theta, \text{node})$ is used to correct the residual biases for each day. Using the correction in Eq. (10.26) to the original SSMIS brightness temperature data, the time series of the longitudinal mean, ΔT_B, at 54.4 and 55.5 GHz is replotted in Figure 10.20c,d, and the bias is more uniform than the original data. After this analysis is applied to other LAS channels except for 50.3 GHz channel, the bias distribution becomes more uniform across all the latitudes for all the channels. Note that the 50.3 GHz channel is not analyzed here because of a large uncertainty in the simulated brightness temperature caused by the uncertainty in surface emissivity.

10.4.4 Impacts from SSMIS and AMSU-A Data Assimilation

Since the frequencies of the SSMIS LAS channels are similar to those of many of the AMSU-A channels, it was expected that the SSMIS LAS data would produce similar effects as AMSU-A when they are used in NWP systems. In addition, the SSMIS is a conically scanning instrument and has a constant viewing angle, so the bias should be independent of the scan position. Using simulated data, Rosenkranz *et al.* [288] showed that the retrieval accuracy from a conically scanning instrument is better than that from cross-track scanning data. However, assimilation of real satellite data involves a number of other issues, such as bias characterizations and corrections, and QC criteria, which are very different from the uses of simulated data. Therefore, it is necessary to compare the impacts on forecast skill from conically (e.g., SSMIS LAS) and cross-track (e.g., AMSU-A) scanning data in our NWP model. In the control run (Crntl exp), only conventional data (radiosondes, buoy data) are assimilated. In AMSUA runs (AMSUA Exp1 and AMSU-A Exp2), conventional data and AMSU-A data from NOAA-18 and MetOp-A, respectively, are assimilated. In SSMIS runs (SSMIS Expl and SSMIS Exp2), SSMIS UPP (Unified Preprocessing Package) LAS data without and with the new BC are assimilated.

In SSMIS experiments, the LAS data from 52.8 GHz (channel 2) to 59.4 GHz (channel 7) are assimilated with and without the new BC correction. For the AMSU-A experiments, similar data from channel 4 (52.8 GHz) to channel 9 (57.3 GHz) are assimilated. The 50.3 GHz channel in both the AMSU-A and SSMIS datasets is not used in the experiment since the SSMIS UPP data at this channel have a large uncertainty.

Figure 10.21a,b shows the ACCs at 500 mb geopotential height for Northern Hemisphere (NH) and Southern Hemisphere (SH), respectively, for a 2-month period from August 1–30 to September 30, 2008. Both SSMIS LAS and AMSU-A

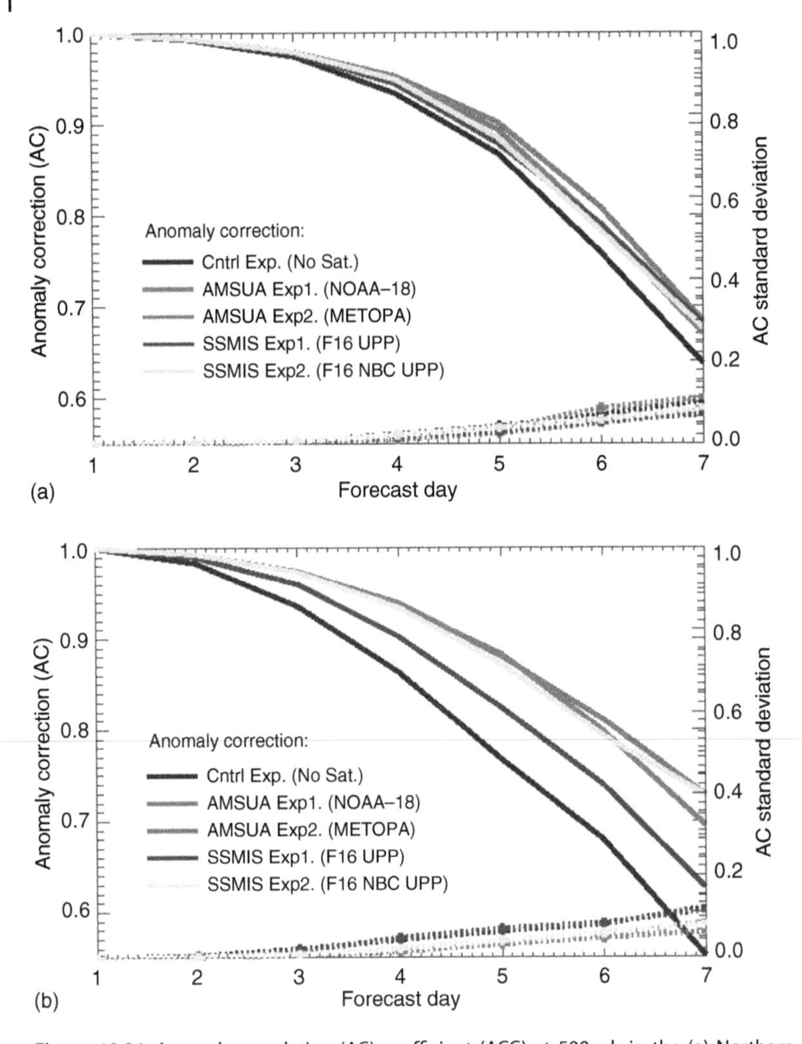

Figure 10.21 Anomaly correlation (AC) coefficient (ACC) at 500 mb in the (a) Northern Hemisphere and (b) Southern Hemisphere for one control and four experimental runs, which cover the period from August 1 to September 30, 2008. (Yan and Weng 2012 [267]. Reproduced with permission of American Meteorological Society.)

have positive impacts on the global medium-range forecasts in both hemispheres. The impact of the satellite data, including the SSMIS LAS data, in NH is smaller than that in SH. With the new bias correction algorithm, the impact from assimilating the SSMIS LAS data is comparable to that of the AMSU-A data from NOAA-18 and MetOp-A over both NH and SH, while the impact of the SSMIS data without the new BC displays a smaller impact on the forecast skill compared to the AMSU-A in SH. This demonstrates that the impact of the data can be reduced if the bias of the satellite data is geographically dependent. To further infer the relevant importance of the anomaly correlation, the standard deviations of the ACCs are also shown in Figure 10.21. Generally, the ACC standard deviation increases with forecast length in all experiments. Among the five experiments, the ACC standard deviations of the three experiments (NOAA-18 AMSU-A, MetOp-A AMSU-A, and F-16 UPP LAS with the new

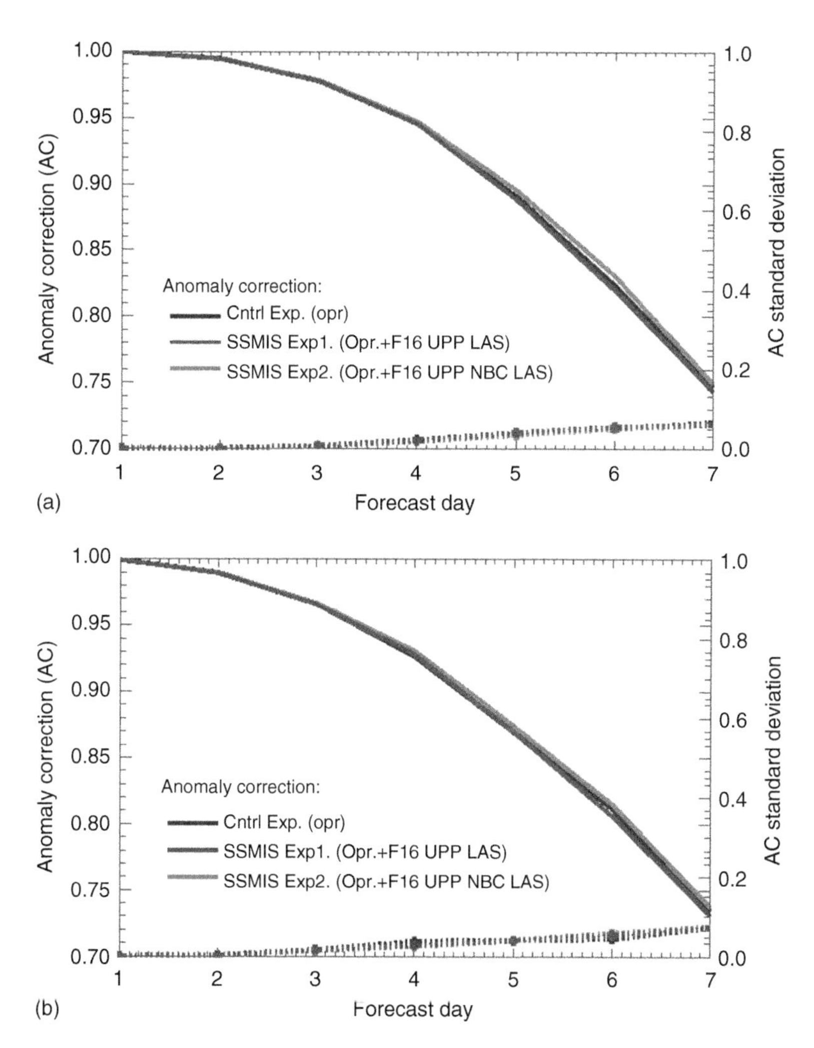

Figure 10.22 Anomaly correlation coefficient in SH at (a) 500 mb and (b) 1000 mb for one control and two experimental runs for the period from August 1 to September 15, 2008. (Yan and Weng 2012 [267]. Reproduced with permission of American Meteorological Society.)

BC) are smaller than those of the control experiment and the SSMIS experiment without the new BC. This suggests that the new BC improves the impact of the SSMIS LAS data on forecasts.

10.4.5 Impact of SSMIS LAS Data on GFS Operational Forecasts

It is also important to assess the impact of the SSMIS LAS data on the GFS operational forecasts. GFS/GSI has assimilated the data from conventional and many satellite data sources (e.g., HIRS, AIRS, and IASI) and microwave sounders (AMSU-A and MHS), Geostationary Operational Environmental Satellite (GOES) sounder radiances, SSM/I ocean surface wind speeds, Moderate Resolution Imaging Spectroradiometer (MODIS) winds, and so on. For this purpose, a new control run and two experimental runs are designed. The control run uses all the GFS operational data during the period from August 1 to September 15, 2008, without SSMIS data. Two SSMIS experimental runs are conducted by adding the SSMIS UPP LAS data with and without the new BC. Note that the SSMIS LAS data only includes channels from 52.8 GHz (channel 2) to 59.4 GHz (channel 7). Numerical results show that in NH, the impact of SSMIS LAS data on the forecast skill at both 500 and 1000 mb is neutral. In SH, the SSMIS LAS data with the new BC algorithm produce a slightly positive impact on the forecast skill at both 500 and 1000 mb (Figure 10.22a,b). The limited impact of the SSMIS LAS data here is due primarily to the use of many other satellite data sources in the control run. This is also demonstrated by the fact that the AC standard deviations in the SSMIS experiments are similar to those in the control experiment.

10.5 Summary and Conclusions

ATMS data are added to conventional and other satellite data streams and assimilated in HWRF forecast fields through GSI. Specifically, the added values of ATMS radiances to conventional data for improved TCs over the Atlantic Ocean are compared with a conventional-data-only experiment and with an experiment in which the conventional data and satellite data were from the other three types of instruments (i.e., AMSU-A, AIRS, and HIRS). It is found that ATMS radiance data assimilation in the HWRF system positively contributes to both the track and intensity. The improvements brought by the ATMS data assimilation are more significant when the benchmark HWRF forecasts, without incorporating ATMS data, deviate more from the best track data.

Assimilation of the SSMIS LAS data into the GFS is developed and turned into operation. A new algorithm is used for removing the geographically dependent biases in the SSMIS LAS data. It is shown that the new bias correction improves assimilation of LAS data with positive impacts on the forecast skills. The impact of SSMIS data is similar to that of the AMSU-A data from either NOAA-18 or MetOp-A after the regionally dependent biases are removed in the LAS data. When the control run uses all the GFS operational data, the F-16 SSMIS LAS data produces some neutral or slightly positive impacts on the forecast skill. This new bias correction methodology can also be applied to other instruments that have calibration anomaly similar to that of SSMIS.

11 Applications of Microwave Data in Climate Studies

11.1 Introduction

The Microwave Sounding Unit (MSU) and the Advanced Microwave Sounding Unit-A (AMSU-A) on board the National Oceanic and Atmospheric Administration (NOAA) polar-orbiting satellites measure the atmospheric radiation from 23 to 89 GHz. Under a clear-sky atmospheric condition, the radiative energy primarily comes from the oxygen emission. Since the oxygen concentration is nearly uniformly distributed through the Earth's atmosphere and is stable with time, MSU and AMSU-A are unique satellite instruments for remotely sounding the atmospheric temperature. The MSU instruments on board Tiros-N, NOAA-6 to NOAA-14 have four channels and provided data from 1979 to 2006. Each of the four channels provides measurements of a weighted average of radiation emitted from a particular layer of the atmosphere at a specified frequency. The relative contributions to the total measured radiance from different levels of the atmosphere are quantified by the so-called weighting function (WF), which is channel dependent. The measured radiation is most sensitive to the atmospheric temperature at the altitude where WF reaches the maximum value. The AMSU-A instruments on board NOAA-15 to NOAA-19 have 15 channels, among which four channels (i.e., AMSU-A channels 3, 5, 7, and 9) are similar, but not identical, to the four MSU channels in frequency (50.30, 53.74, 54.96 and 57.95 GHz). The other 11 AMSU-A channels sample more atmospheric layers compared to MSU. By combining MSU and the MSU-like AMSU-A channels, a long-term series of global satellite microwave sounding data of more than 30 years is available for climate study related to atmospheric temperature changes [290, 291].

The atmospheric temperature trends derived from the MSU and AMSU-A on board the NOAA polar-orbiting satellites have been a subject of debate. Pioneering investigations by Spencer and Christy [292, 293] and their follow-on work at the University of Alabama at Huntsville (UAH) [160, 294, 295] showed nearly no warming trends for the mid-tropospheric temperature time series derived from the MSU channel 2 (53.74 GHz) and AMSU-A channel 5 (53.71 GHz) observations (called T2). However, the Remote Sensing Systems (RSS) [296, 297] and University of Maryland [159] and NOAA/NESDIS/Center

Passive Microwave Remote Sensing of the Earth: For Meteorological Applications, First Edition. Fuzhong Weng.

for Satellite Applications and Research (STAR) [290] groups obtained a small warming trend from the same satellite observations. The most recent analysis of different data sets shows a global ocean mean T2 trend of 0.080 ± 0.103 K/decade for UAH, 0.135 ± 0.113 K/decade for RSS, 0.22 ± 0.07 K/decade for University of Maryland (UMD), and 0.200 ± 0.067 K/decade for STAR for the time period from 1987 to 2006 [290]. In the past, the analysis of the MSU and AMSU-A decadal trend focused on the following areas: (i) diurnal adjustment of various instruments; (ii) correction of Earth incidence angle; and (iii) intercalibration of co-orbiting satellite [297]. Qin *et al.* [298] pointed out a nonlinear behavior of the decadal climate trends of most AMSU-A channels using data from NOAA-15 satellite over the time period from October 26, 1998 to August 7, 2010.

Earlier a concern was also raised on how much of the MSU signal arises from nonoxygen emission. For example, for MSU channel 2, Spencer *et al.* [299] predicted a small influence of contamination from precipitation-sized ice in deep convection, cloud water, water vapor, and surface emissivity. It was concluded that the largest effects come from precipitation-sized ice in deep convection, causing a brightness temperature depression of up to several degrees Celsius. Therefore, the MSU data have been filtered to remove this particular contamination. Spencer [300] also developed a technique for calculating the anomalous temperature increase in MSU channel 1 (50.3 GHz) and then calibrating it to a rainfall rate. Prabhakara *et al.* [301, 302] suggested that a substantial hydrometeor effect exists in the MSU records, which was criticized by Spencer *et al.* [303] who argued that the residual hydrometeor contamination effects were greatly overestimated by Prabhakara *et al.* [301, 302].

In this chapter, we first discuss the climate trend uncertainty derived from the observations related to the instrument noise and the data record length. To produce a robust climate data record (CDR) from satellites, we illustrate that the longer data record can be generated after all the instruments from the same category are cross-calibrated with sensor-to-sensor biases. In understanding the impacts of both cloud and precipitation on trending the atmospheric temperature, tropospheric and low-stratospheric temperature trends are generated and compared with and without precipitation contributions. We also aim at directly deriving the atmospheric temperature at different pressure levels from all four MSU and MSU-like AMSU channels so that the climatology of the atmospheric temperature at specific pressure levels could be deduced globally.

11.2 Climate Trend Theory

Global warming is a well-known phenomenon. However, large uncertainties exist in the quantitative estimate of global climate trend of the atmosphere. Here, we briefly describe a simple statistical method – the linear regression method – for climate trend detection using observations and point out a few factors controlling the precision of such an estimate.

Given a time series of data $\{x_i^{obs} = x^{obs}(t_i), \quad i = 1, 2, \ldots, N\}$, where x represents a measured variable (such as annual global mean near-surface atmospheric temperature), t_i represents the ith measurement in time, and N is the total number of measurements in the time series. Assume that the mean value of the variable x for the N measurements has been removed from the data, that is, $\overrightarrow{x_i^{obs}} = 0$.

Firstly, we can express the observed time series as follows: $x^{obs} = x^{true} + \varepsilon$, where x^{true} represents the truth, and ε is the observation error whose variance is denoted as σ^2_{obs}. Secondly, the variable x is modeled by a linear function of time: $x^{model} = a(t - \bar{t})$, where a is the regression coefficient (e.g., the climate trend to be determined) and $\bar{t}$ represents the average year. Thirdly, the true value of the variable x is expressed as $x^{true} = x^{model} + e$, where e is the nonlinear term representing the natural variability whose variance is denoted as σ^2_{nc}. We may write these expressions in the following matrix form:

$$\mathbf{x}^{true} = \mathbf{A}a + \mathbf{e}, \tag{11.1}$$

$$\mathbf{x}^{true} = \begin{pmatrix} x_1^{true} \\ x_2^{true} \\ \vdots \\ x_N^{true} \end{pmatrix}, \mathbf{e} = \begin{pmatrix} e_1 \\ e_2 \\ \vdots \\ e_N \end{pmatrix}, \mathbf{A} = \begin{pmatrix} t_1 - \bar{t} \\ t_2 - \bar{t} \\ \vdots \\ t_N - \bar{t} \end{pmatrix}.$$

The linear regression coefficient a is obtained by a least-squares fit, which minimizes the differences between the observations and the linear regression model:

$$\sigma^2(a) = (\mathbf{x}^{obs} - \mathbf{A}a)^T(\mathbf{x}^{obs} - \mathbf{A}a) \tag{11.2}$$

Assuming that there is no temporal error correlation for both observations and regression model and the observation and model errors are independent, one may obtain the minimum solution of $a^* = \min_a \sigma^2(a)$ by setting the first derivative to zero, $\partial\sigma^2(a)/\partial a = 0$, which gives the following expression for trend detection:

$$a^* = \frac{\sum_{i=1}^{N} \left(x_i^{obs} - \overline{x^{obs}}\right)(t_i - \bar{t})}{\sum_{i=1}^{N} (t_i - \bar{t})^2} = \frac{\sum_{i=1}^{N} x_i^{obs}(t_i - \bar{t})}{\frac{N^3 - N}{12}}. \tag{11.3}$$

Equation (11.3) is used for estimating the trend from the data. It is pointed out that in the derivation of Eq. (11.3), we used the following equality for equating the denominators in Eq. (11.3) (notice $\bar{t} = (N + 1)/2$):

$$\sum_{i=1}^{N} (t_i - \bar{t})^2 = \sum_{i=1}^{N} \left(t_i^2 + \bar{t}^2 - 2t_i\bar{t}\right) = \sum_{i=1}^{N} t_i^2 + N\left(\frac{N+1}{2}\right)^2 - 2\frac{N(N+1)}{2}\frac{N+1}{2}$$
$$= \frac{N(N+1)(2N+1)}{6} - N\left(\frac{N(N+1)}{2}\right)^2 = \frac{N^3 - N}{12}.$$

By substituting Eq. (11.3) into Eq. (11.2), one may obtain the precision for the trend detection using Eq. (11.3), which is equal to

$$\sigma_{trend}^2 \equiv \sigma^2(a^*) = \frac{12\left(\sigma_{obs}^2 + \sigma_{nv}^2\right)}{N^3 - N}, \tag{11.4}$$

where σ_{obs}^2 is the observation error variance and σ_{nv}^2 represents the natural variability. Based on Eq. (11.4), it is seen that the precision of the trend deduced from the data depends on the observation error (σ_{obs}^2), the length of data (N), and the natural variability (σ_{nv}^2) of the variable whose trend is under investigation. The larger the observation error and the natural variability are, the longer the required data record for an accurate estimate of climate trend.

An example is provided in Figures 11.1 and 11.2 to show the sensitivity of climate trend calculated from data to measurement precision. Firstly, three monthly mean temperature time series are generated over a 300-year period by adding three different random noises (with $\sigma_{obs} = 0.1,\ 0.3,$ and 1 K) to the same climate trend of 0.2 K/decade (i.e., the truth). The natural variability is assumed to be $\sigma_{obs} = 0.1$ K. The trends calculated by Eq. (11.2) from these three time series with varying length of data are presented in Figure 11.1. It is seen that the true trend of 0.2 K/decade could be deduced from a shorter time series when the observation error is smaller ($\sigma_{obs} = 0.1 - 0.3$ K). When the observation error is increased to 1 K, a much longer time series of data is required to deduce the decadal trend. The precision for the trend estimate (i.e., σ_{trend} in Eq. (11.4)) is shown in Figure 11.2. It is indicated that the measurement precision has a significant impact on climate

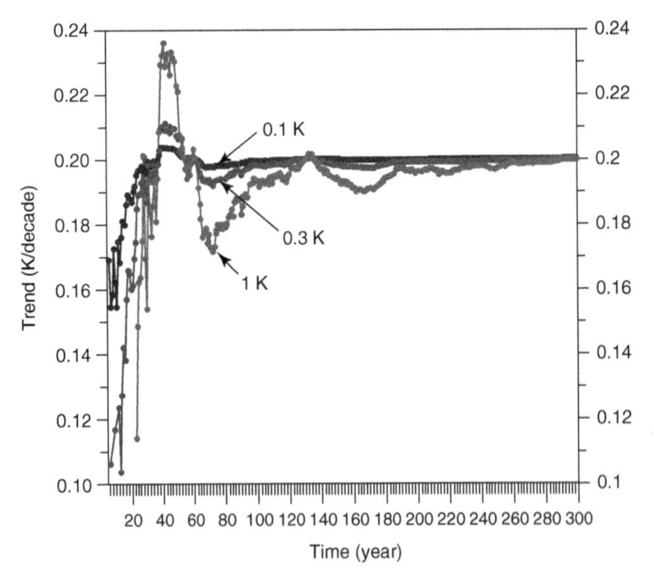

Figure 11.1 Climate trend calculated from different lengths of time series with three different observation error variances: 0.1 K, 0.3 K, and 1 K.

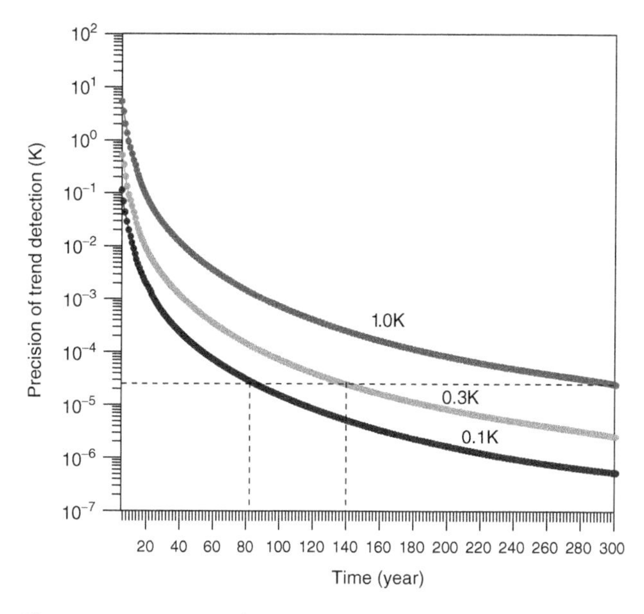

Figure 11.2 Variations of σ_{trend} with respect to data length for the trends shown in Figure 11.1.

trend detection. As a consequence, the required data length increases from 82 years when σ_{obs} = 0.1 K to 140 years when σ_{obs} = 0.3 K and to 300 years when σ_{obs} = 1 K, for detecting a climate trend on the order of 0.2 K/decade with the same precision of about $\sigma_{trend} = 1.5 \times 10^{-5}$ K.

It is thus concluded that providing measurement precision along with data is extremely important for climate trend detection. When obtaining and applying data for climate study, it is important to also obtain the measurement bias and precision from data providers so that the data length required for reliable climate trend detection can be estimated.

11.3
A Long-Term Climate Data Record from SSM/I

The history of environmental satellite measurements now spans several decades, which is relatively sufficient to form a CDR for climate-related studies. Based on the 2004 National Research Council's (NRC) definition, a CDR is a time series of measurements of sufficient length, consistency, and continuity to determine the climate variability and change. The Special Sensor Microwave Imager (SSM/I) series provides one of the longest time series of the satellite microwave measurements from July 1987 to the present. These observations are being followed with a similar sensor, the Special Sensor Microwave Imager/Sounder (SSMIS). The long record of consistent measurements from multiple similar sensors is extremely important in generating CDRs for climate change research and analysis. However, the long-term multiple SSM/I measurements are not accurate enough to be directly applied in climate-related studies. The uncertainty caused by simply stitching multiple SSM/I data together as a CDR arises from instrument offsets, instrument degradation, signal interference, satellite orbital drift, missing data, and so on. In addition, when these Defense Meteorology Satellite Program (DMSP) instruments were originally designed, the instrument calibration was mainly focused on their weather and environment applications, and their long-term performance stability has not been thoroughly assessed to date. Therefore, different SSM/I sensors have to be carefully calibrated to a reference satellite or a stable reference system in order to produce a consistent and high-quality CDRs for climate analysis and reanalysis.

Intersensor calibration between SSM/I and SSMIS was explored and geared toward climate applications [304, 305]. The SSM/I anomaly due to the antenna field-of-view intrusion by the spacecraft and the glare suppression system was successfully corrected (e.g., [306]). The SSM/I measurements are calibrated with respect to the radiative transfer simulations over oceans [307], and the well-calibrated SSM/I data reduces the discrepancy between the observed precipitation trend and the climate model prediction [304]. Recently, the observational anomalies of the first SSMIS on board the F16 satellite were investigated and found to be caused by solar illumination on the SSMIS warm calibration target and antenna reflector emission [287, 308].

Several approaches have been commonly used for satellite intersensor calibrations, including the following: (i) intercomparison between satellite and ground-based measurements; (ii) comparison with clear-sky radiative transfer model simulations; (iii) analysis of two overlapping sensor measurements at nearly simultaneous temporal and spatial locations; and (iv) matching up the statistical properties of two sensor measurements at selected spatial scales. The simultaneous nadir overpass (SNO) technique was developed by Cao *et al.* [309] for infrared sensor cross-calibration and has been widely used for MSU and AMSU temperature retrievals [162]. A simultaneous conical overpass (SCO) technique was developed for conically scanning instruments, and the preliminary results show that the SCO calibration scheme can effectively remove the biases between SSM/I or SSMIS sensors [287, 305].

11.3.1 Simultaneous Conical Overpassing (SCO) Method

For DMSP satellites, the intersensor calibration can be performed at either antenna temperature (T_a) or brightness temperature (T_b). The data record of satellite T_b measurements is called the sensor data record (SDR). T_a is the effective blackbody temperature of the radiance on the feedhorn, while T_b is the calibrated effective blackbody temperature of the radiance on the antenna reflector. The conversion of T_a to T_b is to conduct the antenna pattern correction (CPA) in correcting the incomplete radiometric coupling between the reflector and feedhorn and the cross-polarization coupling between the channels and side-lobe

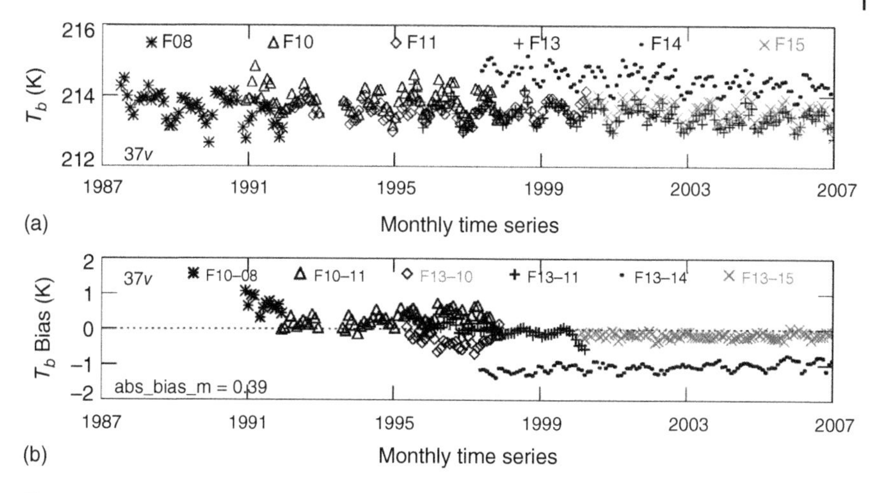

Figure 11.3 (a) shows the time series of oceanic rain-free monthly T_b (K) at the SSM/I 37v GHz during the time period of 1987–2006; and (b) presents the time series of the SSM/I intersensor bias of oceanic rain-free monthly T_b at the 37v GHz during the time period of 1987–2006 for any overlapped sensors. The mean absolute bias (K) against F13 is shown at the lower-left corner. (Yang *et al.* 2011 [305]. Reproduced with permission of American Meteorological Society.)

contamination [306]. Here, we present a methodology for intercalibrating the SSM/I SDR. The impacts of this scheme on the SSM/I-derived total precipitable water (TPW) are demonstrated.

The SSM/I measurements are available at five frequency channels with polarization, except at the water vapor channel, that is, 19.35 (v, h), 22.235 (v), 37 (v, h), and 85.5 (v, h) GHz. Figure 11.3 displays the time series of rain-free monthly mean brightness temperature (T_b) at 37v GHz from valid SSM/I measurements over the 60S–60N oceanic areas. It is obvious that SSM/I instruments provide a continuous measurement averaged over the oceanic region since July 1987; however, the brightness temperature trends from different sensors are quite different. The T_b biases of all overlapped SSM/I sensors shown in the bottom panel indicate that the bias varies with time and different SCO pairs by as much as ±1.2 K, while their mean absolute intersensor T_b bias against the F13 is 0.39 K. This large bias among different sensors demonstrates that the SSM/I SDRs and their derived environmental data records (EDRs) from the existing calibrations may not be suitable for climate studies. Thus, the calibration efforts for all SSM/I sensors must be conducted in order to generate unbiased and high-quality CDRs for climate change analysis and trend studies.

The SSM/I measurements during their overlapping periods provide an alternative way to check their consistency and to select a reference frame for the SSM/I intersensor calibrations. Figure 11.4 shows the local equatorial crossing times of the SSM/I sensors at their ascending nodes. It is evident that F13 and F14 have more overlapping time periods with other SSM/I sensors. F13 has the longest data

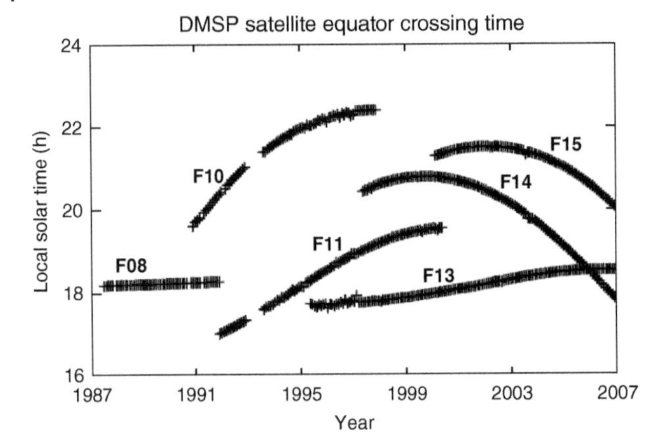

Figure 11.4 The local equatorial crossing times (h) of available DMSP satellites with SSM/I instruments on board for ascending node, except F08 on descending mode due to its 12 h out of phase with others. (Yang *et al.* 2011 [305]. Reproduced with permission of American Meteorological Society.)

record and the smallest change of equatorial crossing time from the satellite orbit drift and is selected as a reference satellite.

The SSM/I measurements from a pair of the DMSP satellites are matched when they are simultaneously overpassing a local area, typically at high latitudes. These measurements are called the SCO pairs. They are supposed to be identical if the sensors with the incident and azimuth angles are all well calibrated. Figure 11.5 shows the locations of the SCO pairs in the Northern and Southern hemispheres. However, a bias between two different SSM/I sensors normally exists due to many factors such as instrument calibration, instrument degradation, sources of interference to signals, satellite orbital drift, and incidence and azimuth angles. The bias between the SCO pairs should be removed in order to generate consistent SDRs, EDRs, and CDRs. We do not explicitly correct any possible error due to different sensor incident and azimuth angles although this bias should be very small

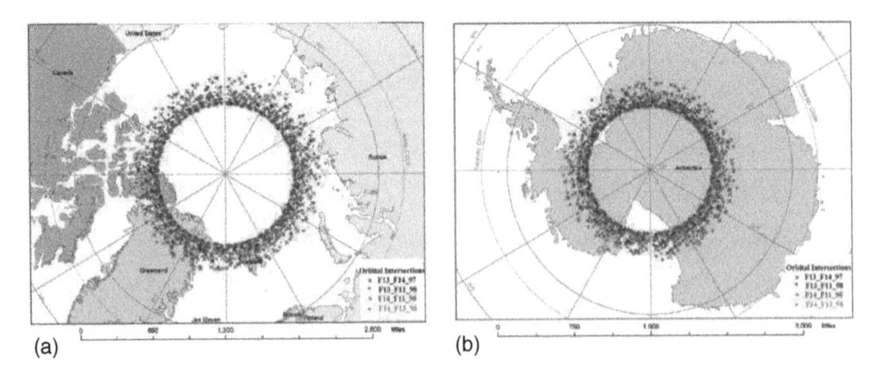

Figure 11.5 The locations of the selected SSM/I SCO pairs using F13 as a reference satellite near the (a) North Pole region and (b) the South Pole region. (Yang *et al.* 2011 [305]. Reproduced with permission of American Meteorological Society.)

due to the SCO pair selection procedure. The data quality control of these SCO pixels is one of the key procedures in the calibration process. A double-difference technique (DDT) is applied for SSM/I data sets if there is no direct interception between F13 and another satellite (S_A), that is, a third satellite (S_B), which has good overlapping with both F13 and S_A, will be used as a transfer radiometer to connect F13 and S_A. Thus, F14 is applied as the transfer radiometer between F13 and F15 for SCO over water surface.

11.3.2
Bias Characterization of Specific SSM/I Instrument

The SSM/I scan-angle-dependent bias was previously reported by Colton and Poe [306] to be due to the antenna field-of-view intrusion by the SSM/I spacecraft near the beginning of the scan and the glare suppression system near the end of the scan. The spacecraft intrusion and the glare suppression system have the least impact on SSM/I at its scan central position, so a measurement at this position can be regarded as an accurate reference. This bias is sensor dependent and varies at different channels and satellite orbit orientations and must be removed prior to the SSM/I SDR intersensor bias corrections.

As an example, Figure 11.6 shows the scan-angle-dependent bias at 37 GHz averaged from all available SSM/I measurements of oceanic rain-free pixels

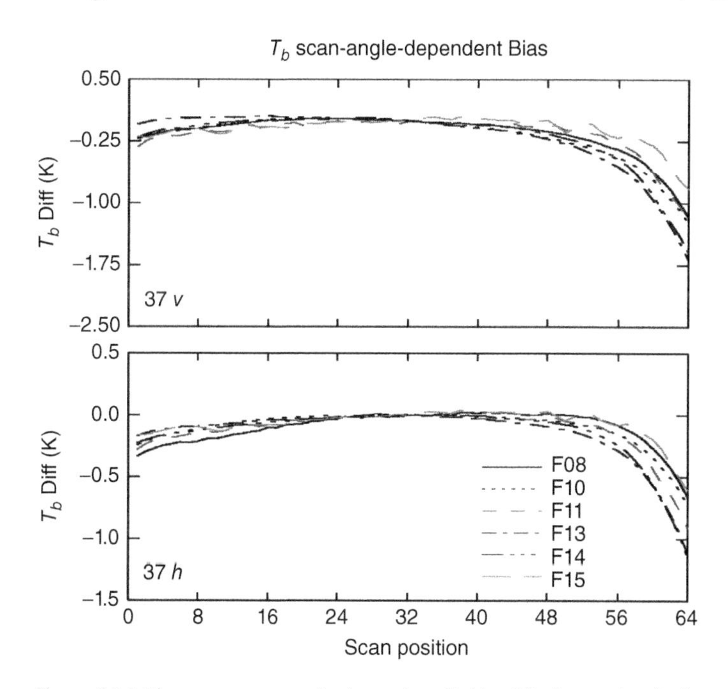

Figure 11.6 The mean scan-angle-dependent T_b bias (K) of oceanic rain-free pixels within 60S–60N against the scan central position for all SSM/I sensors at the 37 GHz vertical (a) and horizontal (b) polarization. (Yang *et al.* 2011 [305]. Reproduced with permission of American Meteorological Society.)

within 60S–60N at the low-frequency channel pixel locations against its central scan position. It is evident that the scan-angle-dependent bias patterns are similar for each SSM/I sensor, but with certain differences. The bias reaches −1.75 K near the end of the scan, but only −0.3 K near the start of the scan. The bias patterns are also similar for other SSM/I low-frequency channels. Similar bias patterns with a maximum of −2.5 K and a minimum of −0.5 K are seen at SSM/I high-frequency channels. In addition, there are also prominent bias differences for ascending and descending nodes. During the calibration processes, the five-scan position weighted average bias curve is applied so that the possible noise associated with the scan positions is minimized. A linear interpolation scheme is used to estimate the bias at any pixel position of a SSM/I high-frequency channel that is not collocated with the position of the low-frequency channel. Finally, the seasonal variability of the bias patterns is analyzed. The results indicate a small seasonal change of the scan-angle-dependent bias patterns, but this seasonal variation is not considered in this study.

11.3.3 RADCAL Beacon Interference with F15 SSM/I

RADCAL, a system of instruments on the DMSP F-15 spacecraft, consists of redundant C-band transponders with unique antennas, Doppler transmitters operating at 150 and 400 MHz, and a deployable antenna. The purpose of the RADCAL C-band transponder/antennas is to provide a signal source for ground-based C-band radar interrogation and tracking, while the primary purpose of the Doppler transmitters and antenna is to determine the satellite position for comparison with the radar data. A secondary purpose of the Doppler systems is to support the Coherent Electromagnetic Tomography (CERTO) experiment. RADCAL has been operational since August 14, 2006, and considerably affects the F15 SSM/I and Special Sensor Microwave Temperature-2 (SSMT-2) sensor data. In particular, the 150 MHz beacon produces considerable increases in the brightness temperatures at 22 GHz [310].

Since the RADCAL beacon interference on F15 22v GHz channel is steady, the correction algorithm is applied to remove the interference. Figure 11.7 exhibits the SSM/I F15 T_a error at 22v GHz due to the RADCAL beacon interference on August 30, 2006, as a function of the scan position for the ascending and descending nodes, respectively [285, 311]. The errors are based on the mean differences between the global SSM/I observations and the radiative transfer model simulations under oceanic cloud-free conditions. A polynomial function is applied to fit the error curves. Then, this error is subtracted from the raw T_a at 22v GHz.

$$T_a(\text{err}) = a_0 + a_1 \times X + a_2 \times X^2 + a_3 \times X^3, \tag{11.5}$$

$$T_a(\text{cal}) = T_a(\text{obs}) - T_a(\text{err}). \tag{11.6}$$

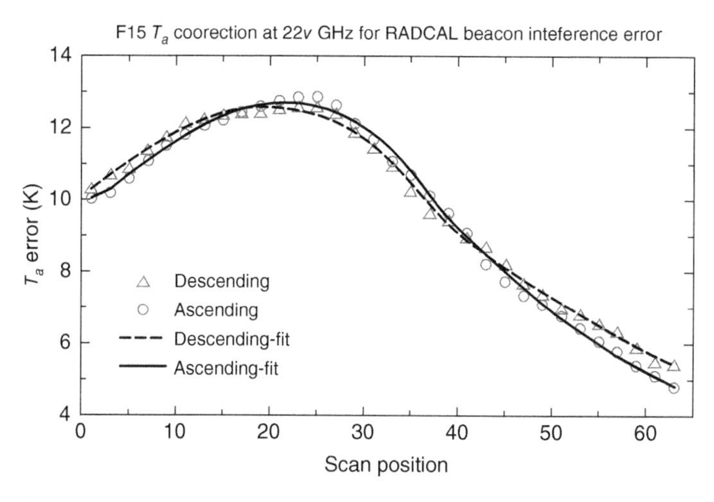

Figure 11.7 F15 antenna temperature (T_a) error (K) at the SSM/I 22.235v GHz resulting from the RADCAL beacon interference starting on August 30, 2006, as a function of the scan positions at low-frequency channels for ascending (open circle) and descending (open triangle) node, respectively. The solid and dashed lines are their related fitting curves. T_a error is the mean difference between SSM/I observations and radiative transfer model simulations for oceanic cloud-free conditions. (Yang *et al.* 2011 [305]. Reproduced with permission of American Meteorological Society.)

where X is the scan position. The coefficients of a_0, a_1, a_2, and a_3 are given in Table 11.1. The fitted lines are overlapped, as shown in Figure 11.7, to clearly show the reasonable fitting applied for this study. This correction of the F15 RADCAL beacon interference is applied before the scan angle bias correction.

Table 11.1 The T_a correction coefficients used in Eq. (11.5) for the RADCAL beacon interference error at F15 SSM/I 22v GHz.

Satellite status	Coefficients	Scan position			
		1	2–37	38–62	63–64
Ascending node	a_0	10.01136	9.607	2.2655	4.78003
	a_1	0	2.2651×10^{-1}	-3.7596×10^{-1}	0
	a_2	0	-1.2794×10^{-3}	1.5069×10^{-4}	0
	a_3	0	-1.2039×10^{-4}	2.1233×10^{-5}	0
Descending node	a_0	10.28697	9.989	28.191	5.40614
	a_1	0	2.286×10^{-1}	-8.4842×10^{-1}	0
	a_2	0	$-3.38.4\times10^{-3}$	1.21016×10^{-2}	0
	a_3	0	-7.877×10^{-5}	-6.8437×10^{-5}	0
Unknown node	a_0	10.14	9.798	28.578	5.05
	a_1	0	2.2756×10^{-1}	-8.52×10^{-1}	0
	a_2	0	-2.3314×10^{-3}	1.1749×10^{-2}	0
	a_3	0	-9.9581×10^{-5}	-6.4806×10^{-5}	0

However, Poe *et al.* [310] indicate an increased RADCAL interference during Earth's shadow, resulting in approximately 1.3 K increase in the vicinity of the maximum interference at beam position 25. This effect is apparent for the descending orbits, which enter the Earth's shadow for part of the year due to the Earth's elliptical orbit and leads to a clear impact on EDRs, especially TPW during Earth's shadow. This Earth shadow effect on TPW is small at global or tropical ocean averages, while it could create a notably artificial spatial variation at regional scales, especially when the time of Earth shadow increased considerably during the summer of 2007. In addition, the RADCAL interference also affects the F15 85.5 GHz channels at a lesser degree. Since only 4-month F15 data sets affected by the RADCAL interference were used in this study, impact on the 85 GHz channels should not substantially change the results from this study. Therefore, we do not discuss the influence of the interference, during Earth's shadow and its impact on the 85.5 GHz channels in this chapter.

11.3.4 SSM/I Intersensor Bias Correction

Based on the assumption that simultaneous measurements at a location from two different sensors of the same design should be highly correlated. If one sensor is regarded as a reference, the other can be calibrated to this reference. The skill of the SCO technique requires minimization of the measurement differences caused by noninstrumental factors. Thus, the SCO differences between two different sensors are primarily due to instrumental errors, which should be removed during the postlaunch calibration processes.

Many experiments with different SCO constraints are conducted for an optimal result. All possible SCO pairs are first quality-controlled for the same orbital node (e.g., ascending or descending) and similar pixel positions. A spatial distance (Δd) of 3 km between the SCO pair is used to ensure that at least the footprints of two SSM/I instruments are overlapped by 75–90%. A reasonable time difference (Δt) between two sensors is required to lead to a reliable analysis. Figure 11.8 displays the mean bias, standard deviation, and the SCO pair samples against the time difference criterion at 22v GHz between F13 and other SSM/I sensors over water surface. It is evident that the bias and standard deviation vary slightly with different time criterion, except the F13/F11 bias increases dramatically when the time difference is less than 2 min due to the increase in uncertainties with limited samples. Similar results exist for other channels. Over land (figure omitted), the SCO bias difference is very small with the different time criterion (0.5, 1, and 2 min). However, the bias increases considerably when the time difference is greater than 5 min. Thus, in general, the 2 min criterion is used, and the SCO samples are enough for the analysis of the intersensor bias correction. In addition, the samples under inhomogeneous background conditions are eliminated by applying the standard deviation (σ) of nine neighboring pixels surrounding a candidate SCO pair. The SCO pair is taken as a good one if σ is less than 2 K for a regular surface. Similar features to the water surface are found for SCO pairs over ice and

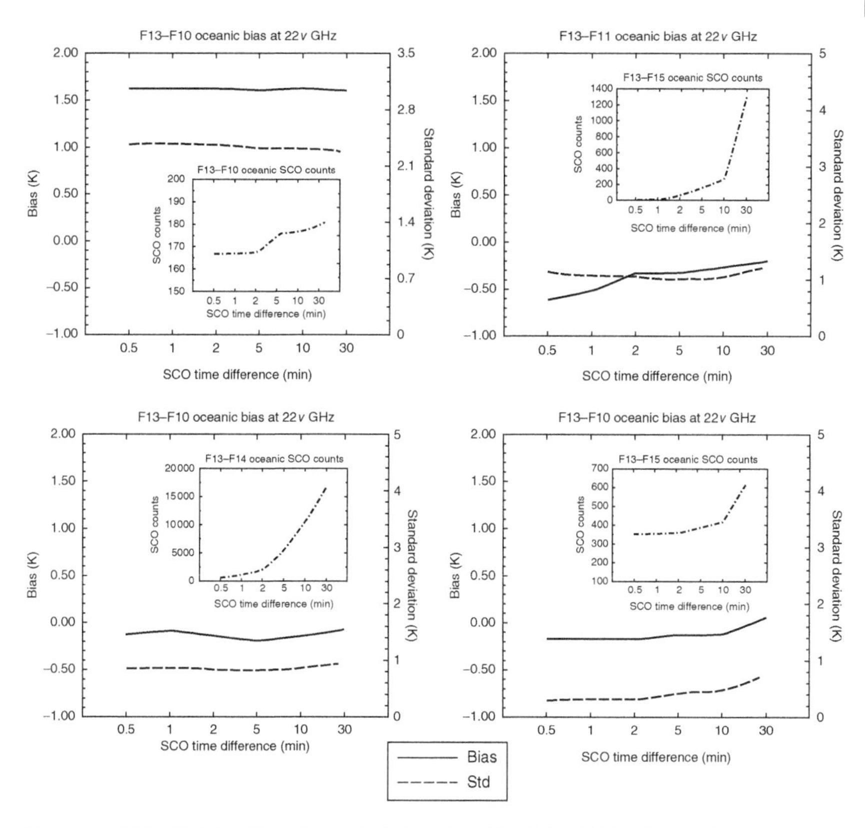

Figure 11.8 The T_b bias (K) and standard deviation (K) of the SSM/I SCO pixels over ocean between F13 and other DMSP satellites at the 22v GHz channel as a function of the SCO time difference. The inset plot shows the number of the SCO samples as a function of the SCO time difference. (Yang *et al.* 2011 [305]. Reproduced with permission of American Meteorological Society.)

coast surface types, except that σ of less than 5 K is used for coastal areas. Finally, the absolute T_b difference ($|\Delta T_b|$) of an SCO pair should be less than 10 K. Only about 1% (0.06%) of oceanic (continental) SCO pairs were excluded from this criterion.

After careful analysis of the SCO pairs using F13 as a reference satellite, the criteria of collecting high-quality SCO pairs are set up as follows:

$$\Delta d \leq 3\,\text{km}, \Delta t \leq 2\,\text{min}, \sigma \leq 2\,\text{K (5 K for coast case), and } |\Delta T_b| \leq 10\,\text{K}$$

In addition, the same orbital nodes and similar scan positions (less than three-scan position difference) are also required. The bias distribution with the SCO time differences is inspected to eliminate any potentially large bias caused by small samples of the SCO pairs that were not well distributed.

All SCO pixels are sorted into four categories based on the surface type (i.e., water, land, ice, and coast) to avoid the contamination caused by mixing these

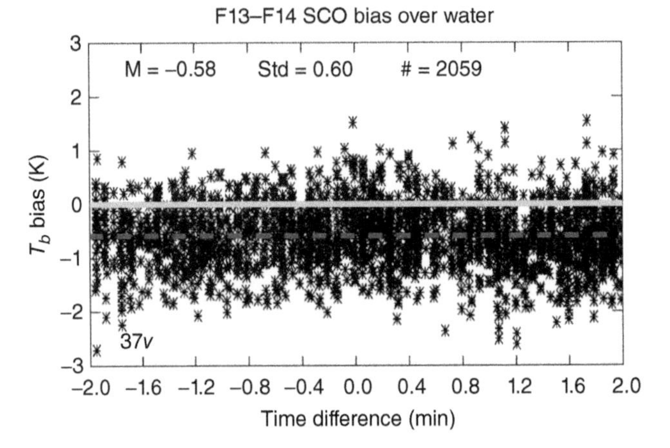

Figure 11.9 The F13 and F14 SCO T_b bias (K) distribution over water surface as a function of the SCO time difference for the 37*v* GHz channel. The heavy dark gray dash line denotes the mean bias. The light gray line denotes the zero bias. M, Std, and # denote the statistical mean, standard deviation, and total sample of the SCO pixels, respectively. (Yang *et al.* 2011 [305]. Reproduced with permission of American Meteorological Society.)

surface types and carefully analyzed accordingly. As an example, Figure 11.9 presents the SCO T_b bias $[T_b(\text{F13}) - T_b(\text{F14})]$ at 37*v* GHz against their interception time difference over the water background. It is evident that 2059 SCO pixels are well distributed around the exact interception time ($\Delta t = 0$) between F13 and F14. The statistical mean bias is −0.58 K with a standard deviation of 0.60 K. Since it is difficult to find many exactly simultaneous measurements between any two SSM/I sensors, the statistical mean bias of the SCO pixels with nearly simultaneous measurements is a reliable choice for obtaining the bias between these two sensors. Similar processes are conducted for all other SSM/I channels to estimate the intersensor bias coefficients between F13 and other SSM/I satellites.

Due to the limited SCO pixels between F13 and F15 over water surface, a DDT approach is utilized to estimate T_b bias, that is, using F14 (which has better overlaps with both F13 and F15) as a transfer radiometer to connect them so that subtraction of T_b bias between F15 and F14 from that between F13 and F14 results in the T_b $\text{Bias}_{\text{F13}-\text{F15}}$ because the F14 effect is literally cancelled out from the double difference ($\text{Bias}_{\text{F13}-\text{F15}} = \text{Bias}_{\text{F13}-\text{F14}} - \text{Bias}_{\text{F15}-\text{F14}}$). Note that the DDT can be applied only when there are insufficient good-quality SCO pixels between two SSM/I sensors so that a third sensor must be used as a transfer radiometer. Finally, no bias correction is applied with F08 because there are no reliable SCO pixels between F08 and F13 and no sufficient F08–F10 matchups to apply a DDT. The final intersensor bias correction coefficients and standard deviations for F10, F11, F13, F14, and F15 using F13 as the reference satellite are listed in Table 11.2.

Table 11.2 The calibration coefficients of the SSM/I intersensor bias (K) derived from the SCO technique using F13 SSM/I as a reference radiometer for surface types of water, land, ice and cost.

Satellite	Surface type	19v		19h		22v		37v		37h		85v		85h	
		Bias	σ	Bias	σ	Bias	σ	Bias	σ	Bias	σ	Bias	σ	Bias	σ
F10	W	0.22	2.84	−0.27	2.04	1.62	2.37	−0.03	2.05	−0.05	4.26	−0.71	3.01	−1.12	4.04
	L	0.18	1.30	0.65	1.34	1.82	1.79	0.49	1.20	1.10	1.50	0.02	2.42	0.05	2.76
	I	0.08	1.05	0.14	1.77	1.39	1.03	0.20	0.85	0.58	1.97	−0.01	1.13	0.06	3.37
	C	−0.47	1.75	0.18	2.38	1.40	1.35	0.22	1.05	1.18	1.92	−0.11	2.65	0.41	2.29
F11	W	−0.32	0.76	0.0	0.88	−0.33	1.05	0.35	0.65	0.89	1.17	−0.38	1.26	−0.53	1.63
	L	0.04	1.04	−0.16	1.09	0.23	1.10	0.40	0.90	0.34	1.03	−0.54	1.70	−1.14	1.58
	I	0.31	0.88	−0.03	1.01	0.45	0.93	0.81	0.94	0.79	1.15	−0.22	2.03	−0.99	2.06
	C	0.02	1.18	−0.03	1.28	0.19	1.06	0.89	0.91	0.79	1.61	0.0	2.27	−0.80	2.36
F14	W	−0.16	0.68	0.28	0.60	−0.14	0.83	−0.58	0.60	0.34	0.84	−0.11	1.06	0.20	2.11
	L	0.0	0.76	0.12	0.76	0.09	0.89	−0.42	0.63	−0.09	0.72	0.02	1.31	0.20	1.27
	I	0.02	0.25	0.25	0.85	0.13	0.91	−0.39	0.65	−0.04	0.81	0.17	1.36	0.34	1.41
	C	0.29	0.83	0.44	1.23	0.41	0.96	−0.11	0.76	0.23	1.24	0.13	2.05	0.55	2.25
F15	W	0.77	0.65	−0.14	0.69	0.26	0.82	0.11	0.55	0.21	0.77	0.55	0.95	0.05	1.74
	L	0.41	1.05	0.13	0.98	0.59	1.47	0.14	0.80	−0.08	0.84	0.12	1.13	−0.21	1.21
	I	0.59	0.84	0.11	0.75	0.83	1.07	0.26	0.66	0.10	0.70	0.21	1.14	−0.11	1.30
	C	0.56	1.11	−0.07	1.72	0.24	1.21	0.45	0.93	0.28	1.29	0.65	2.33	0.44	2.32

σ is the standard deviation (K).
Criterion: $|\Delta t| \leq 2$ min; $\Delta d \leq 3$ km; $\sigma \leq 2$ K; $|\Delta T_b| \leq 10$ K.
W, water; L, land; I, Ice; C, Coast.
Coast, $\sigma \leq 5$ K; F13_F15 (water) bias is the combined F13_F14 and F14_F15 SCO biases.

11.3.5
Impact of Cross-Calibration on SSM/I SDR

With the continuously increasing time span of environmental satellite measurements, the satellite remote-sensing-derived EDRs are becoming more important in climate-related studies. Since the climate trends of meteorological parameters are small compared to the natural variability in the system, improvements in satellite remote-sensing data sets are important for use in quantifying the climate trends. After implementation of the SDR intersensor calibration procedure, the trends from all SSM/I SDR time series are now more consistent. Figure 11.10 shows the time series of SSM/I oceanic rain-free monthly mean intersensor calibrated T_b and their intersensor bias at the 37v GHz channel. When compared to Figure 11.13, it is apparent that the new time series has a considerably improved consistency among these SSM/I sensors, with the intersensor bias being reduced dramatically. The variation of the intersensor biases with time and different overlapping SSM/I sensors is only around ±0.5 K, indicating a 58% decrease in the bias (±1.2 K) after intersensor calibration. We use the mean absolute bias to summarize how calibration offsets affect the consistency of the monthly T_b time series among

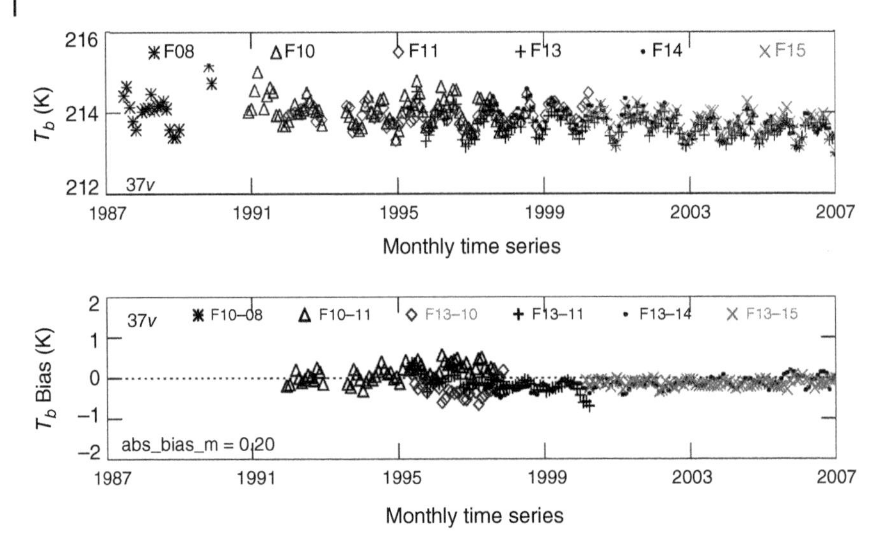

Figure 11.10 Similar to Figure 11.3, except for the SDR intersensor calibrated T_b (K) at 37 GHz. (Yang *et al.* 2011 [305]. Reproduced with permission of American Meteorological Society.)

different sensors. This parameter is the average absolute monthly T_b biases, which highlights the spread, as opposed to the grand-average offsets captured by a sample average of the monthly biases. The mean absolute bias after calibration is only 0.20 K, indicating a bias deduction of 49%. Similar results are found for other SSM/I channels. Therefore, these results demonstrate that the newly developed SSM/I SDR intersensor calibration scheme is very useful and can substantially improve the consistency of the SSM/I SDR time series with dramatically reduced intersensor biases that have a considerably smaller temporal variation.

The improved consistency of the SSM/I SDRs should lead to a more reliable SDR trend analysis. Thus, the trend analysis is only based on SSM/I measurements from the time period 1990–2006. Figure 11.11 presents the time series of the SSM/I oceanic rain-free monthly T_b before and after intersensor SDR calibration at the 37*v* GHz channel. The SSM/I T_b time series before calibration presents a standard deviation (σ) of 0.49 K and a linear trend of −0.12 K/decade at 2.5% significance level. After calibration, the monthly T_b time series has the same mean value, but much smaller σ (0.32) and larger trend (−0.32 K/decade) at 0.1% significance level. Similar analyses are conducted for other channels, and the results are summarized in Table 11.3. It is obvious that SDR intersensor calibration using F13 as the reference satellite does not substantially change the averaged monthly T_b but leads to a more consistent time series among multiple SSM/I sensors with a mean reduced σ of 21.4% for all channels. The most important impact of this calibration is its role in changing the T_b trend considerably at every SSM/I channel, that is, the trend is reduced by 37.1%, 72.2%, 74.1%, and 77.3% for channels 19*v*, 37*h*, 85*v*, and 85*h* GHz, respectively, while increasing by a larger percentage at 19*h*, 22*v*, and 37*v* GHz because the original trend was small.

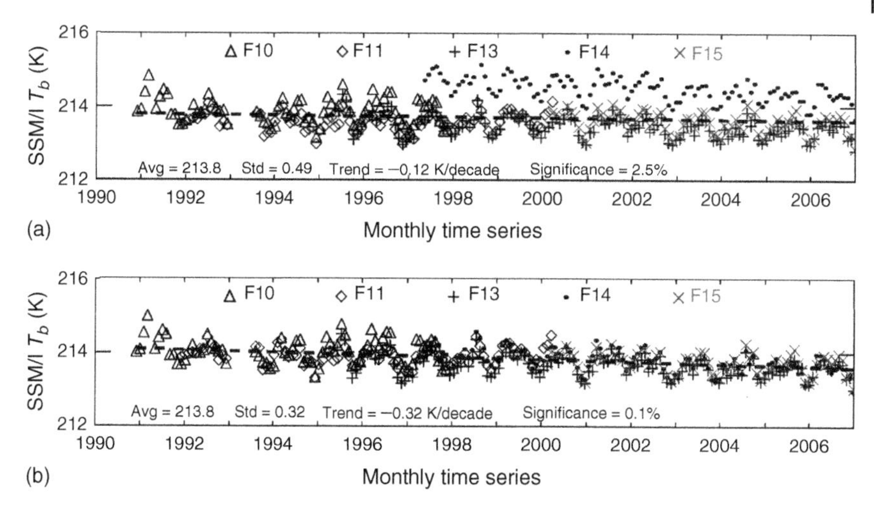

Figure 11.11 Intercomparison of the SSM/I oceanic rain-free monthly mean time series of T_b (K) at 37 GHz (a) before and (b) after SDR intersensor calibration. The heavy dash line denotes the linear fitting curve based on the least absolute deviation method. The key stats of mean (K), standard deviation (K), trend (K/decade), and *t* test significance (%) are listed at the bottom of each panel. (Yang *et al.* 2011 [305]. Reproduced with permission of American Meteorological Society.)

11.3.6
Impacts of SSM/I Intersensor Calibration on TPW

The TPW product is a key variable retrieved from SSM/I measurements [312]. Although there are discrepancies between different TPW and precipitation algorithms, the oceanic TPW and precipitation estimates based on satellite passive microwave measurements are considered as robust retrievals [313, 314]. Therefore, they are used to demonstrate the intersensor calibration impact on the SSM/I-based EDRs. Figure 11.12 presents the monthly TPW intersensor bias over the global ocean and the tropical ocean between any two overlapped SSM/I sensors of F10, F11, F13, F14, and F15 before and after the intersensor SDR calibration. The relatively large TPW intersensor biases before calibration are obvious, especially between F10 and F11 and between F10 and F13 that have large biases of −1.5 mm over the global ocean and −3.0 mm bias over the tropical ocean. The averaged absolute TPW intersensor bias is 0.358 and 0.264 mm for the global and tropical oceans, respectively. After the intersensor calibration, the amplitude of the associated TPW intersensor biases is only about ±0.10 mm over the global ocean and ±0.50 mm over the tropical ocean, showing a dramatic bias reduction from before calibration. By the same token, the mean absolute TPW bias is only about 0.089 and 0.209 mm, respectively. Thus, SSM/I intersensor calibration has a large positive impact on the TPW retrievals with a resultant decrease in the mean absolute intersensor bias by 75% over the global ocean and 20% over the tropical ocean.

Table 11.3 Comparison of mean (M, unit: K), standard deviation (σ, unit: K), and linear trend (Trd, unit: K/decade) of the SSM/I oceanic rain-free monthly T_b time series before and after SDR intersensor calibration for F10, F11, F13, F14, and F15 and their associated percentage change.

	19*v*			19*h*			22*v*			37*v*			37*h*			85*v*			85*h*		
	M	σ	Trd	M	σ	Trd	M	σ	Trd	M	σ	Trd	M	σ	Trd	M	σ	Trd	M	σ	Trd
Before calibration	196.2	0.58	−0.70	131.1	0.84	−0.02	221.6	0.84	−0.02	213.8	0.49	−0.12	154.9	0.50	0.36	256.8	0.62	−0.58	224.3	0.91	−0.66
After calibration	196.3	0.44	−0.44	131.0	0.54	0.03	221.8	0.70	−0.31	213.8	0.32	−0.32	155.1	0.45	0.09	256.8	0.53	−0.15	224.1	0.78	−0.05
Change percentage (%)	0.05	−24.1	−37.1	−0.08	−35.7	250	0.09	−16.7	1450	0	−34.7	166.7	0.13	−10.0	−72.2	0	−14.5	−74.1	−0.09	−14.3	−77.3

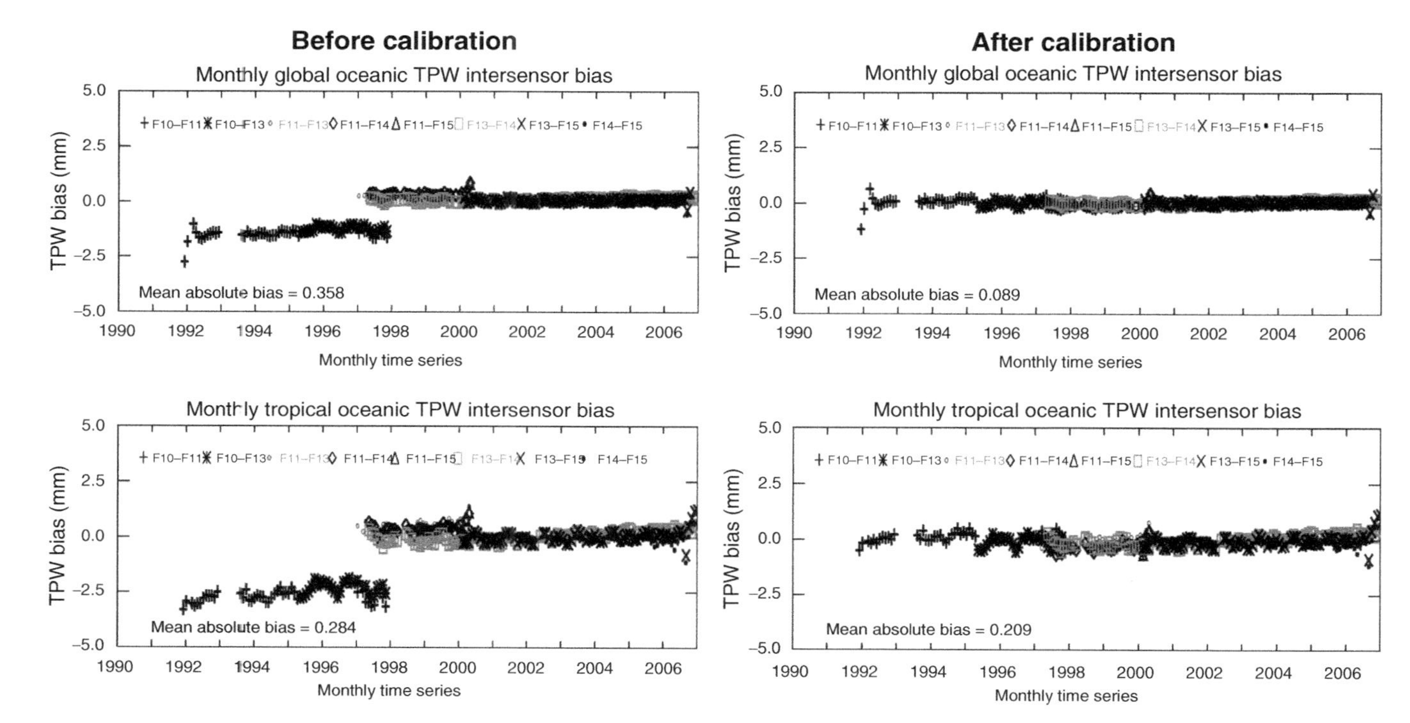

Figure 11.12 Time series of monthly TPW intersensor bias (mm) for any overlapped SSM/I sensors of F10, F11, F13, F14, and F15. Different SSM/I pairs are marked by different symbols. The left and right panels are for before and after SDR intersensor calibration, respectively. The top panels are for the global ocean, while the bottom panels are for the tropical ocean (20S–20N). The averaged absolute TPW intersensor bias (mm) is shown in the bottom-left corner of each panel. (Yang *et al.* 2011 [305]. Reproduced with permission of American Meteorological Society.)

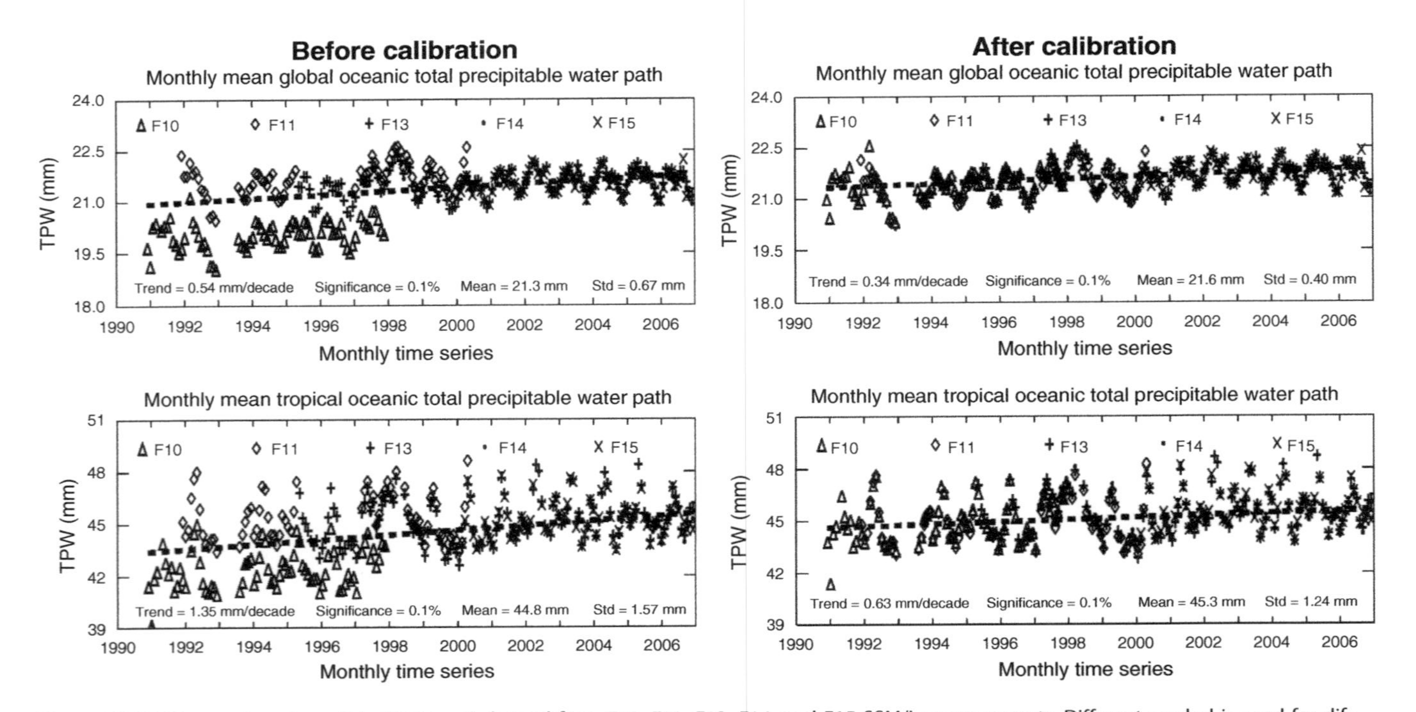

Figure 11.13 Time series of monthly TPW (mm) derived from F10, F11, F13, F14, and F15 SSM/I measurements. Different symbol is used for different SSM/I satellite. The left and right panels are for before and after SDR intersensor calibration, respectively. The top panel is for the global ocean, while the bottom panel is for the tropical ocean. The overlapped heavy dash line denotes the linear fitting curve based on the least absolute deviation method. The key stats of trend (mm/decade), *t* test significance (%), mean TPW (mm), and standard deviation (mm) are listed at the bottom of each panel. (Yang *et al.* 2011 [305]. Reproduced with permission of American Meteorological Society.)

We have demonstrated that the newly developed SSM/I SDR intersensor calibration scheme improves the consistency of the multisensor SSM/I measurements and associated EDRs. Figure 11.13 presents the TPW trend analysis over both global and tropical oceans before and after the SDR intersensor calibration for F10, F11, F13, F14, and F15. The key statistics are also shown in each panel. It is apparent that the improved TPW consistency among different SSM/I sensors results in a reliable trend analysis. Prior to the intersensor calibration, the mean TPW, standard deviation, and trend are at 21.30, 0.67, 0.54 mm/decade, respectively, for the global ocean and at 44.80, 1.57, 1.35 mm/decade, respectively, for the tropical ocean. The corresponding values after the intersensor calibration are only about 21.60, 0.40, 0.34 mm/decade over the global ocean and 45.30, 1.24, and 0.63 mm/decade over the tropical ocean. The trend is at 0.1% significance level for all cases. The mean TPW is only increased by 1% with the calibration; however, the TPW standard deviation and trend are decreased by 40% and 38%, respectively, over the global ocean and 21% and 54%, respectively, over the tropical ocean. The impacts of the TPW mean absolute intersensor bias and trend from the SDR calibration are summarized in Table 11.4. Therefore, this study illustrates the importance of the SSM/I intersensor calibration in TPW climate-related studies. Although uncertainties especially at detailed horizontal distributions in TPW retrievals based on passive microwave measurements exist, different TPW algorithms are generally in agreement with each other (Sohn and Smith, 2003). The TPW retrieval error should be smaller compared with the error associated with the uncalibrated T_b. The 0.34 mm/decade (or 1.59% per decade) of the global oceanic TPW trend in this study agrees well with the analysis by Trenberth *et al.* (2005) at 0.40 ± 0.09 mm/decade (or 1.3 ± 0.3% per decade) over the 1988–2003 period and the results obtained by Wentz *et al.* (2007) at 0.354 ± 0.114 mm/decade (or 1.2 ± 0.4% per decade) over the 1987–2006 period. Our results also indicate that the TPW trend is 1.39% per decade for the tropical ocean within the 20° latitude zonal belts.

Table 11.4 Intercomparison summary of TPW mean absolute intersensor bias (mm) and trend (% per decade) before and after the SSM/I intersensor calibration and their associated percentage change.

		Mean absolute intersensor bias			Trend		
		Before calibration	After calibration	Change (%)	Before calibration (% per decade)	After calibration (% per decade)	Change (%)
TPW (mm)	Global ocean	0.358	0.089	−75.1	2.54	1.57	−38.2
	Tropic ocean	0.264	0.209	−20.8	3.01	1.39	−53.8

11.4 A Long-Term Climate Data Record from MSU/AMSU

Previous studies based on MSU on board the NOAA polar-orbiting satellites show that the tropospheric temperature warms up within a range of 0.08–0.21 K/decade [160, 162, 290, 295]. The MSU intersensor calibration also has a prominent impact on the temperature climate trend not only on increasing the confidence of estimating the climate trend but also on raising the consistency of temperature trends at the surface and in the troposphere [161, 290]. The most recent study indicates a trend of 0.2 K/decade for the global tropospheric temperature [290].

The first MSU was launched on board the first NOAA Tiros-N satellite in 1978 and performed the measurements at four frequencies (50.3, 53.74, 54.96, and 57.95 GHz). Channels 1 and 3 measure the microwave emitted and scattered radiation at the quasi-vertical polarization at the nadir, whereas channels 2 and 4 correspond to that of the quasi-horizontal polarization. The MSU antenna system was designed with a nominal beam width of 7.5° at the half-power points, which results in a cross-track resolution of 105 km near the nadir. The MSU scans across the track within ±47.4° from the nadir and produces a scan swath of 2400 km. Beam positions 1 and 11 are the extreme scan positions of the Earth views, each separated by 9.47°, while beam position 6 is in the nadir direction. The radiation from the nadir position arises from the atmosphere in the vertical direction, which is ideal for weather and climate applications. Onboard calibration using blackbody and cold space observations is performed once every 25.6 s for each scan line. The main MSU characteristics are provided in Table 11.5.

The AMSU-A has been operational since 1998 and is flown on board NOAA-15 to 19 and Metop-A and -B satellites. Similarly to MSU, AMSU-A is mainly designed to vertically probe the atmosphere under nearly all-weather conditions (except for heavy precipitation). It contains 15 channels quantifying the thermal radiation at microwave frequencies ranging from 23.8 to 89.0 GHz (see Table 11.6). The AMSU-A has an instantaneous field of view of 3.3° and scans ±48.7° from the nadir with 15 different viewing angles at both sides. Atmospheric temperature profiles are primarily based on the measurements obtained at channels near 50–60 GHz. In particular, the AMSU-A sounding channels (3–14) respond to the thermal radiation at various altitudes, whereas channels 1 and 2 are primarily

Table 11.5 MSU channel characteristics and noise.

Channel number	Center frequency (GHz)	Number of passbands	Bandwidth (MHz)	Center frequency stability (MHz)	NEΔT (K)
1	50.30	1	220	10	0.30
2	53.74	2	220	10	0.30
3	54.96	1	220	10	0.30
4	57.95	1	220	0.5	0.30

Table 11.6 AMSU channel characteristics and noise.

Channel number	Center frequency (GHz)	Number of passbands	Bandwidth (MHz)	Center frequency stability (MHz)	NEΔT (K)
1	23.80	1	251	10	0.30
2	31.40	1	161	10	0.30
3	50.30	1	161	10	0.40
4	52.80	1	380	5	0.25
5	53.59 ± 0.115	2	168	5	0.25
6	54.40	1	380	5	0.25
7	54.94	1	380	10	0.25
‘8	55.50	1	310	0.5	0.25
9	$57.29 = f_0$	1	310	0.5	0.25
10	$f_0 \pm 0.217$	2	76	0.5	0.40
11	$f_0 \pm 0.322 \pm 0.048$	4	34	0.5	0.40
12	$f_0 \pm 0.322 \pm 0.022$	4	15	0.5	0.60
13	$f_0 \pm 0.322 \pm 0.010$	4	8	0.5	0.80
14	$f_0 \pm 0.322 \pm 0.004$	4	3	0.5	1.20
15	89.00	1	2000	50	0.50

designed for obtaining the information on surface and cloud properties. Since the satellite provides a nominal spatial resolution of 48 km at its nadir, the temperature perturbations from synoptic scale to large mesoscale can be depicted reasonably well.

Figure 11.14 displays the WFs for four MSU and MSU-like AMSU channels at a 0° local zenith angle. The MSU channel 2 has two bands located at both sides of the center frequency and AMSU channel 5 has only one band. The center frequency absorption is covered in AMSU channel 5, but not in MSU channel 2. Such a difference is reflected in the WFs shown in Figure 11.14. The WF of the MSU channel 4 is slightly broader and peaks slightly higher than that of the AMSU-A channel 9 due to slight differences in the center frequency and bandwidth for these two corresponding channels. Similar but smaller differences exist between MSU channel 3 and AMSU channel 7. The WF differences between MSU channel 1 and AMSU channel 3 are the smallest.

In history, all MSU and AMSU instruments from Tiros-N, NOAA-6 to NOAA-19 were designed for day-to-day operational uses in weather forecasting. The requirements on satellite data calibration for climate studies are different from weather forecasting applications. Issues such as variable calibration accuracy (or bias) associated with each satellite instrument and accuracy (bias) changes with respect to time due to the satellite orbital drift must be resolved since they may be mistakenly interpreted as climate influences.

The MSU instruments performed sounding measurements using four channels. Channel 3 contains substantial errors in the NOAA-6 and NOAA-9 instruments and thus is only valid from 1987 onward [297]. Channel 1 (50.3 GHz) is sensitive to the lowest 2 or 3 km of the atmosphere. Data from this channel is heavily

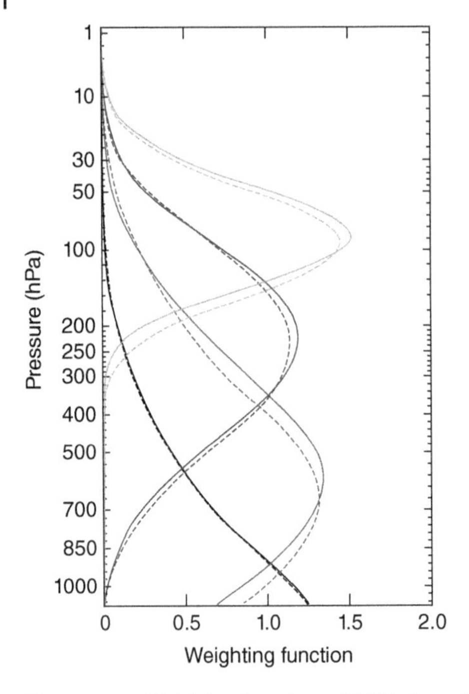

Figure 11.14 Weighting functions of MSU channels 1–4 (solid) and AMSU-A channels 3, 5, 7, and 9 (dash) for the US standard atmosphere.

contaminated by emissions from the surface and atmospheric water and ice and is of limited utility for tropospheric temperature studies [315]. Nash and Forrester [316] reported possible radiometric calibration bias existing for NOAA-9 in channels 1 and 2 when comparing NOAA-9 with NOAA-6. It was found that there is a spurious trend in the differences for AMSU channels 5–9 between the measurements performed by the NOAA-15 and NOAA-16 satellites, and they argued that NOAA-16 is the source of these trends [297]. Channels 1 and 2 on NOAA-9 ceased operation in early 1987 after only 102 days of simultaneous observations with NOAA-10.

Merging multiyear satellite data from different MSU instruments requires careful adjustments of the observations to account for drifts caused by orbital decay and changes in local observing time and determination of intersatellite offsets and errors caused by changes in the temperature of the calibration sources. NOAA/STAR has recently released its level-1c intercalibrated 30+ year (1979–2011) MSU/AMSU-A observations [317]. The instrument nonlinearity is updated using SNO data. Diurnal drift errors, incident angle errors, warm target temperature correction, and residual intersatellite biases are accounted for. Figure 11.15 provides MSU data periods on board NOAA's earlier eight polar-orbiting satellites (from Tiros-N, NOAA-6 to NOAA-14) and the AMSU-A data period on NOAA-15, which are used in this study.

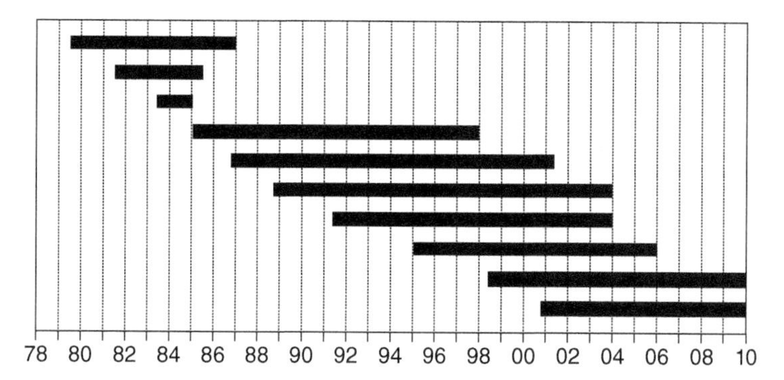

Figure 11.15 MSU data period on board NOAA's earlier eight polar-orbiting satellites from NOAA-6 to NOAA-14 and AMSU-A data period on NOAA-15.

11.4.1
Impacts of Clouds and Precipitation on AMSU-A Trends

NOAA-15 was launched on May 13, 1998, in a polar orbit with an inclination angle of 98.7°, a local equator crossing time (LECT) of about 7:30 pm in its ascending node at its launch time, and an altitude of 833 km above the Earth. Since the noises in the AMSU-A sounding channels are about 0.2–0.3 K, the data record from single AMSU-A on board from 1998 to date is long enough for climate trend study. Recently, the trend uncertainty from the natural variability of clouds and precipitation was investigated [318] and is briefly highlighted in this section.

11.4.2
Emission and Scattering Effect on AMSU-A

The emission and scattering of clouds and precipitation modulate the brightness temperatures from both MSU and AMSU-A. In order to assess the effects of the radiation from clouds and precipitation on the atmospheric temperature trend derived from microwave temperature sounding instruments (e.g., MSU and AMSU-A), cloud-affected brightness temperature measurements must be identified. Fortunately, the information from AMSU-A window channels 1 and 2 can be used for retrieving the atmospheric cloud liquid water path (LWP) using the algorithm developed by [211]. For this reason, temperature trends for AMSU-A channels 3, 5, 7, and 9 from NOAA-15 are investigated in this study. Here, the Community Radiative Transfer Model (CRTM) is first used for simulating the responses of AMSU-A brightness temperatures to the clouds characterized by rain rate [29, 238]. For calculations of cloud and aerosol absorption and scattering, lookup tables of the optical properties of six cloud and eight aerosol types are included in the cloud/aerosol optical property module. The fast doubling–adding method is used to solve the multistream radiative transfer equation.

An idealized cloud distribution similar to that found in Liu and Curry [210] is assumed for the sensitivity study. Namely, a 0.8-km nonprecipitating cloud layer with an LWP of 0.5 kg/m^2 is added to the atmosphere below the freezing level, and the precipitation layer is set below the cloud. Each specific rain rate is assumed for the entire precipitation layer. A schematic diagram of the cloud and precipitation system is illustrated in Figure 11.16. Note that an atmospheric profile including temperature and water vapor is used as input to CRTM, and the surface emissivity is set to 0.5, which is close to that of the ocean surface.

For a given atmospheric temperature condition, the emission and scattering of clouds and precipitation could increase or decrease the brightness temperatures

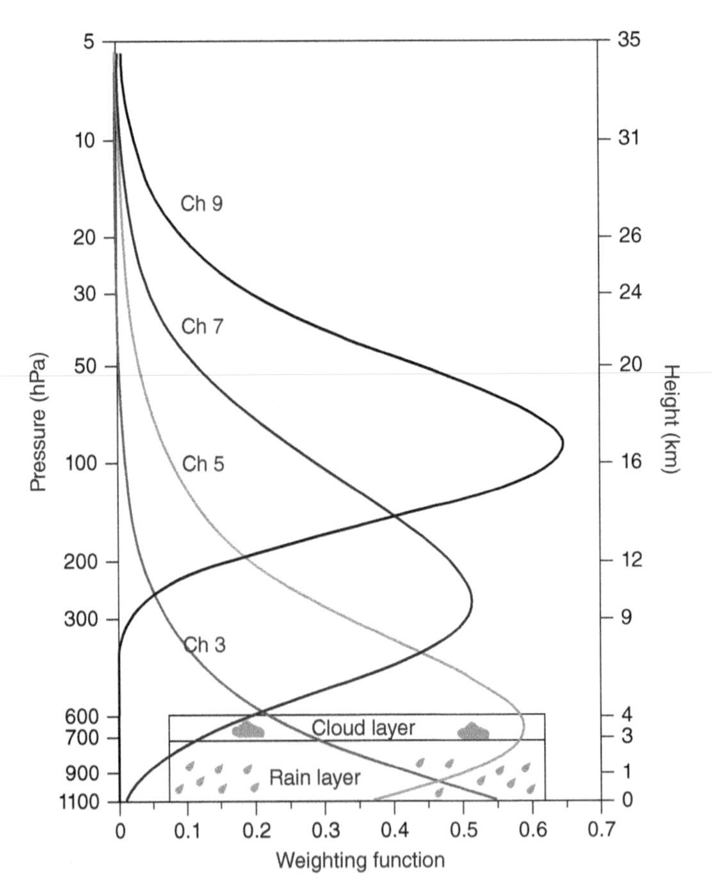

Figure 11.16 Weighting function of AMSU-A channel 3, channel 5, channel 7, and channel 9 from NOAA-15 calculated by the CRTM using the US standard atmosphere profile, which is overlapped with the schematic illustration of a stratiform cloud with rainfall consisting of a 0.8-km-deep nonprecipitating cloud layer located below the freezing level with liquid water path of 0.5 kg/m^2 and the raindrops below the nonprecipitating cloud layer with the rainfall rates unchanged vertically. Emissivity is set to 0.5. (Weng *et al.* 2014 [318]. Reproduced with permission of Springer.)

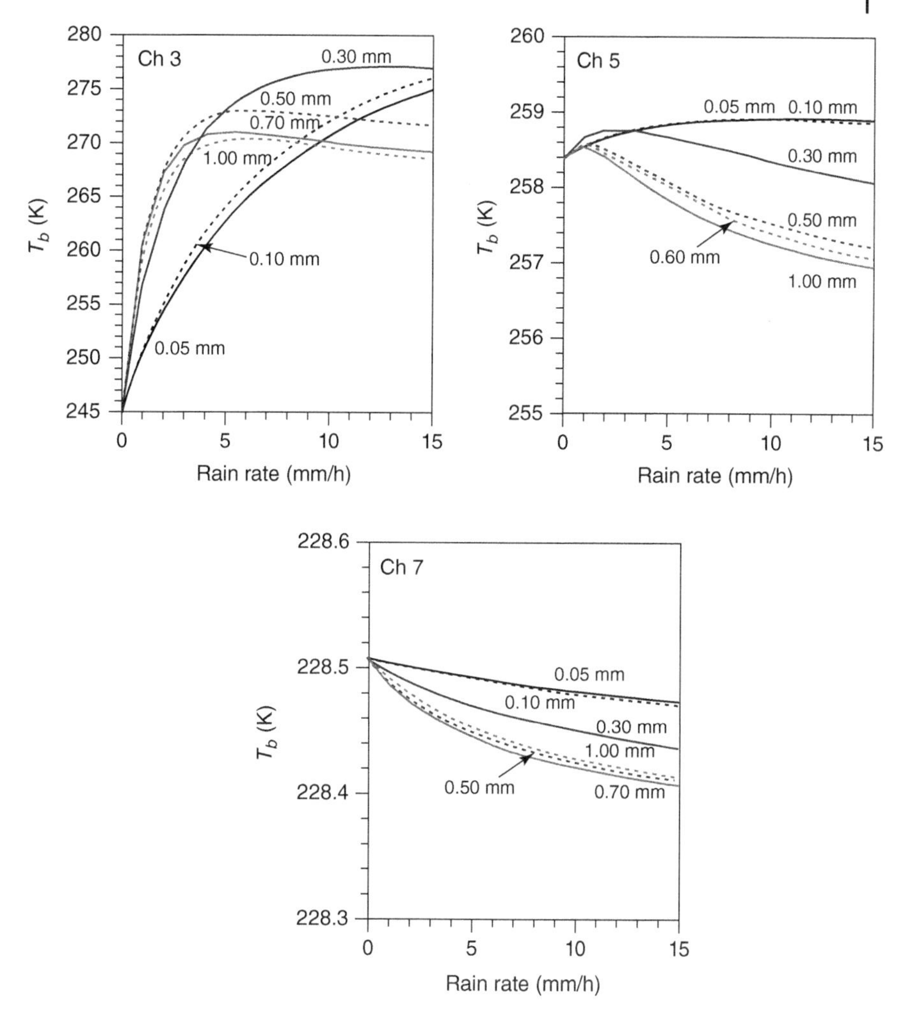

Figure 11.17 Variations of brightness temperature of AMSU-A channels 3, 5, 7, and 9 with respect to rainfall rate with effective diameters of cloud droplets of 0.05 mm, 0.1 mm, 0.3 mm, 0.5 mm, and 0.7 mm and 1.0 mm. The emissivity is set to 0.5. (Weng *et al.* 2014 [318]. Reproduced with permission of Springer.)

measured by AMSU-A. As an example, Figure 11.17 presents variations in the model-simulated AMSU-A brightness temperatures at the satellite nadir position versus rain rate with the mean raindrop size as a parameter. The droplet radius varies from 0.05 to 1.0 mm for the clouds as shown in Figure 11.16. Since the assumed cloud is distributed below 4 km, the higher the channels, the smaller the impact that the clouds have on the brightness temperature. It is seen that the brightness temperature at AMSU-A channel 3 increases as the rain rate increases due to the cloud emission when the mean radius of raindrops is less than 0.1 mm

(Figure 11.17a). When the rain droplets are less than 0.1 mm, the size parameter (i.e., $2\pi/\lambda$, where λ is the wavelength) of clouds and raindrops all fall into the Rayleigh scattering regime where the absorption dominates over scattering [319]. In addition, the absorption coefficient is proportional to the liquid water content or rain rate [211]. Note that the brightness temperature at AMSU-A channel 3 in clear-sky atmosphere is much lower due to the lower surface emissivity over ocean, and thus, an increase in cloud and raindrop absorption results in an increase in brightness temperature. When the raindrop size increases to 0.3 mm, the brightness temperature first rapidly increases with the rain rate and then remains nearly constant (e.g., saturated) as the rain rate further increases. For the larger raindrops whose radius exceeds 0.5 mm, the brightness temperature increases more rapidly with rain rate compared to smaller raindrops when the rain rate is low; the brightness temperature slightly decreases as the rain rate further increases (Figure 11.17b). Except for very small rain rate, the brightness temperature at channel 5 decreases as rain rate increases, which is due to increasing precipitation scattering (Figure 11.17b). A decrease in brightness temperature occurs when the radiation from other atmospheric layers is scattered out of the instrument field of view. At AMSU channel 7 (Figure 11.17c), cloud and precipitation have very small effects on brightness temperature since their WFs peak mostly above the clouds, and therefore, the brightness temperature decreases with rain rate for all raindrop sizes. For AMSU-A channel 9 (figure omitted), clouds and precipitation have negligible effects on brightness temperature since this channel has its WF completely above the cloud and precipitation layer (see Figure 11.16), and the radiation arises from the oxygen absorption and emission in the atmosphere.

Based on the results in Figures 11.16 and 11.17, it is concluded that the larger the raindrop size is, the larger the scattering and emission effects of clouds on the brightness temperatures are. The cloud emission dominates when the rain rate is low and the raindrop size is small. In reality, clouds are of different types and are located at different altitudes. Only the integrated effect of clouds on the temperature trend can be estimated from real data.

11.4.3 AMSU-A Brightness Temperature Trend

Since the launch of NOAA-15 satellite on May 13, 1998, the AMSU data has been used in NOAA operational applications for over 13 years. The cloud algorithm developed by Weng *et al.* [211] for NOAA-15 AMSU-A data over ocean obtains an estimate of the cloud impact on the possible trend using AMSU-A data. The two AMSU-A window channels 1 (23.8 GHz) and 2 (31.4 GHz) can be used for a physical retrieval of cloud LWP over oceans [211]. Figure 11.18a provides a distribution of an annual data count with LWP being less than 0.5 kg/m^2 within each of $5° \times 5°$ grid boxes over the global ocean for NOAA-15 AMSU-A data in 2008. The data in Figure 11.20a are further separated into two groups: one for clear-sky conditions (Figure 11.18b) and the other for cloudy (Figure 11.18c) conditions, where

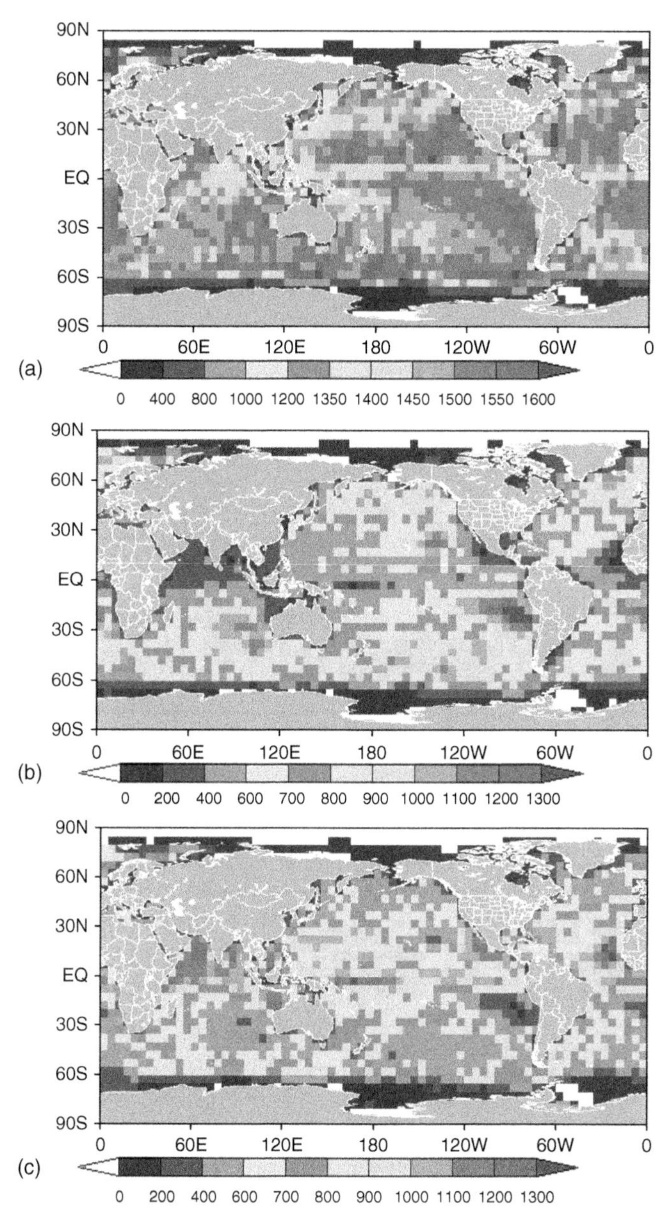

Figure 11.18 Global distributions of data counts within 5° × 5° grid boxes for NOAA 15 AMSU-A FOVs 15 and 16 in 2008 with (a) LWP <0.5 kg/m^2, (b) $0.01 \leq$ LWP < 0.5 kg/m^2, and (c) LWP <0.01 kg/m^2. (Weng *et al.* 2014 [318]. Reproduced with permission of Springer.)

the clear-sky and cloudy data are identified by LWP being less than 0.01 kg/m^2 and between 0.01 and 0.5 kg/m^2, respectively. More cloudy data with LWP values between 0.01 and 0.5 kg/m^2 are found in the east Pacific, east Atlantic, and east Indian oceans in the Southern Hemisphere compared to the middle latitudes (Figure 11.18b). In the Northern Hemisphere, more cloudy data are found in the tropical warm pool and mid-latitude storm track regions than in other oceanic regions (Figure 11.18b). The tropical area has more clear-sky areas compared to the middle latitudes (Figure 11.18c). The LWP threshold of 0.5 kg/m^2 is used for identifying the precipitation.

In the following trend analysis, a threshold of LWP being greater than 0.01 kg/m^2 is used as an indicator of cloudy AMSU data. Figure 11.19 shows a latitudinal variation of the averaged daily counts of the total and clear-sky AMSU-A nadir FOVs from NOAA-15 satellite over global oceans. In general, about 30–50% and 20–30% of oceanic observations are cloud-affected in the middle to high latitudes and the low latitudes, respectively.

The temperature trends derived from AMSU-A channels 3, 5, 7, and 9 with and without the cloudy microwave measurements are shown in Figure 11.20. The global temperature trends calculated from AMSU-A data between 60S and 60N for the window channel 3, the mid-tropospheric channel 5, the upper-tropospheric channel 7, and the upper-stratospheric and low-stratospheric channel 9 are 0.04, 0, −0.01, and −0.16 K/decade, respectively, under all-weather conditions (Figure 11.20a). When the cloud effect is eliminated, the global temperature trend for both channels 3 and 5 is increased to 0.03 and 0.07 K/decade, respectively, leading to more than 43% decrease in the warming trend. The brightness temperatures at AMSUA channels 3 and 5 are sensitive to the emission and scattering from clouds and precipitation. A decrease in the

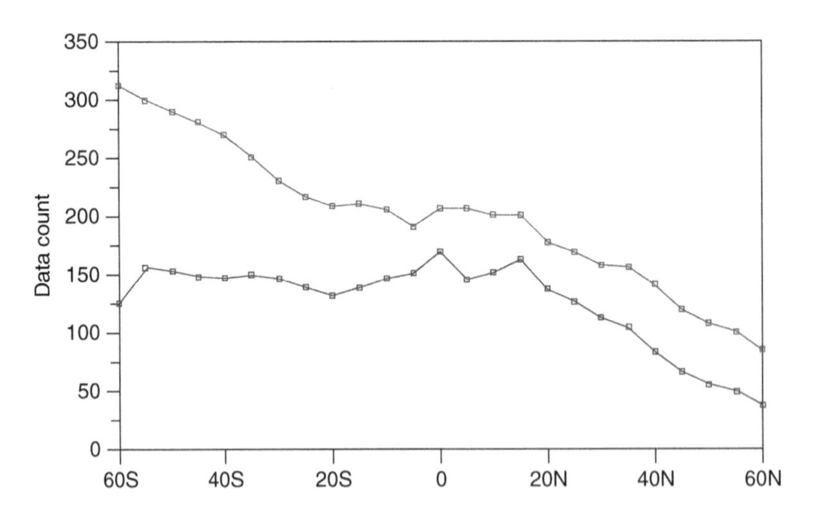

Figure 11.19 Averaged daily count of AMSU-A nadir data in each of 5° latitudinal bands averaged from August 1, 1999 to June 30, 2012, under clear-sky and all-weather conditions over ocean. (Weng *et al.* 2014 [318]. Reproduced with permission of Springer.)

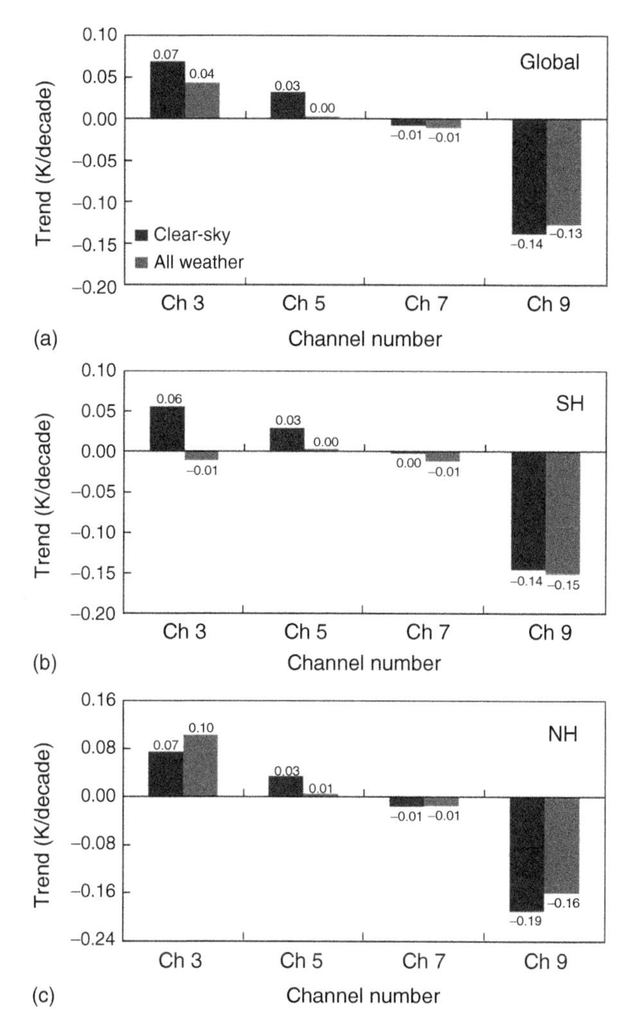

Figure 11.20 Decadal temperature trends between (a) 60S–60N, (b) 0S–60S, and (c) 0N–60N under clear-sky (left bars) and all-weather (right bars) conditions. (Weng *et al.* 2014 [318]. Reproduced with permission of Springer.)

brightness temperature can be associated with cloud and precipitation scattering, rather than with the physical temperature in the lower and middle troposphere; therefore, the trends from microwave sounding data could be misleading if the brightness temperatures from all-weather conditions are averaged as a representation of the atmospheric physical temperature.

The global impacts of clouds on temperature trends in the upper troposphere and low stratosphere are negligible. If we separate the Northern Hemisphere from the Southern Hemisphere, impacts of clouds on temperature trends become more significant compared to the global average. In the Southern Hemisphere, the warming trends for both window channel 3 and the mid-tropospheric channel

5 are 0.06 and 0.03 K/decade, respectively, which decrease to nearly zero when clouds are present (Figure 11.20b). The temperature cooling trends for channels 7 and 8 are slightly increased by the clouds (Figure 11.20b) in the Southern Hemisphere. Cloud impact on the temperature trend for channel 5 in the Northern Hemisphere is similar to that in the Southern Hemisphere, that is, cloud reduces the warming trend. Cloud impact on channel 7 is also negligible in the Northern Hemisphere. However, an opposite effect of cloud on temperature trend is found in the Northern Hemisphere near the surface and in the upper troposphere and low stratosphere: clouds cause an overestimate of the temperature warm trend of the mid-tropospheric channel 5 and an underestimate of temperature cooling trend of the upper-tropospheric and low-stratospheric channel 9. The global ocean mean trends for channel 5 within the time period from October 26, 1998 to August 7, 2010 with or without cloud-affected data obtained in this study are much smaller than those obtained by UAH, UMD, and STAR for the time period from 1987 to 2006 [298]. Further investigation is needed to clarify if the different trends can be caused by the data record length.

11.5 Atmospheric Temperature Trend from 1DVar Retrieval

11.5.1 Climate Applications of 1DVar

As discussed in Chapters 10 and 11, the best estimate of atmospheric vector can be derived by minimizing the cost function

$$J(\mathbf{x}) = \frac{1}{2}(\mathbf{x} - \mathbf{x}_b)^T \mathbf{B}^{-1}(\mathbf{x} - \mathbf{x}_b) + \frac{1}{2}(H(\mathbf{x}) - \mathbf{y})^T \mathbf{R}^{-1}(H(\mathbf{x}) - \mathbf{y}), \tag{11.7}$$

where $\mathbf{x}$ is the control vector, $\mathbf{x}_b$ is the background state vector; $\boldsymbol{B}$ is the background error covariance matrix; $\mathbf{y}$ includes brightness temperature observations from MSU or MSU-like AMSU-A channels; $H(\mathbf{x})$ represents the forward operator such as CRTM, which calculates the radiance at the top of the atmosphere for a given set of input parameters including the atmospheric state variables $\mathbf{x}$; $\boldsymbol{R}$ is the sum of observation error covariance matrix (O) and CRTM error covariance matrix (F). The state variable ($\mathbf{x}$) in Eq. (11.7) includes the atmospheric temperature profile, water vapor profile, and surface parameters (e.g., sea surface temperature (SST), surface emissivity). Here, a climatological atmospheric profile is taken as the background field ($\mathbf{x}_b$).

The minimum solution, $\mathbf{x}$, in Eq. (11.7) is obtained through an iterative procedure [320] in two sequences: one assuming a clear-sky condition for CRTM simulation and the other one cloudy conditions. The largest iteration numbers in the two sequences are set to 2 and 7, respectively. The minimization procedure is stopped if

$$\varepsilon = \frac{1}{4}(H(\mathbf{x}) - \mathbf{y})^T \mathbf{R}^{-1}(H(\mathbf{x}) - \mathbf{y}) \leq 1. \tag{11.8}$$

For AMSU and MSU instrument, which has a noise of 0.25 K, Eq. (11.8) mathematically means that the convergence is declared to have reached if the residuals between the measurements and the simulations at any given iteration are less than or equal to one standard deviation of the noise that is assumed in the radiances.

11.5.2 MSU and AMSU-A Cross-Calibration

The MSU and AMSU-A data set available from June 1979 to December 2009 were assimilated for obtaining microwave temperature retrieval. From NOAA-6 to NOAA-15, the data during the overlapping period of two satellites are intercalibrated, and the retrievals from two MSUs are almost identical. Thus, in our analysis, only the retrieval from the newer instrument is used in the time series. It is worth mentioning that NOAA-8 has only 2 years of data and NOAA-12 missed 8 months of data within a 5-year period. Note that MSU data at FOV 6 and AMSU-A data at FOV 15, which are all close to the nadir, were extracted for this study. Since the variational retrieval requires some ancillary data such as SST, the monthly mean SST analysis from the climate prediction center is used. This SST climatology has been only available for the months after 1981. The retrieval from MSU between June 1979 and October 1981 are based on SST and SSW climatology after 1981. As mentioned before, 1DVar was developed for AMSU-A applications, and thus, MSU data are also linearly remapped to AMSU-A channels through regression relationships.

The consistency in brightness temperature observations between MSU and AMSU-A can be illustrated using the SNO data between NOAA-14 MSU and NOAA-15 AMSU-A during their overlapping period. For SNO collocation, a spatial separation between two observations is set to less than 100 km, with a temporal separation being less than 100 s. From SNO data in 2002, Figure 11.21 shows that SNO data points are distributed along the diagonal with the MSU and AMSU-A brightness temperatures as coordinates, confirming a reliable bias correction for satellite observations from NOAA-14 and NOAA-15 by Zou and Wang [317].

11.5.3 Cloud Detection Algorithm for MSU Applications

Since the WF of AMSU-A channel 3 (or MSU channel 1) has a WF peak near the surface, the clouds in the atmosphere can affect the brightness temperature

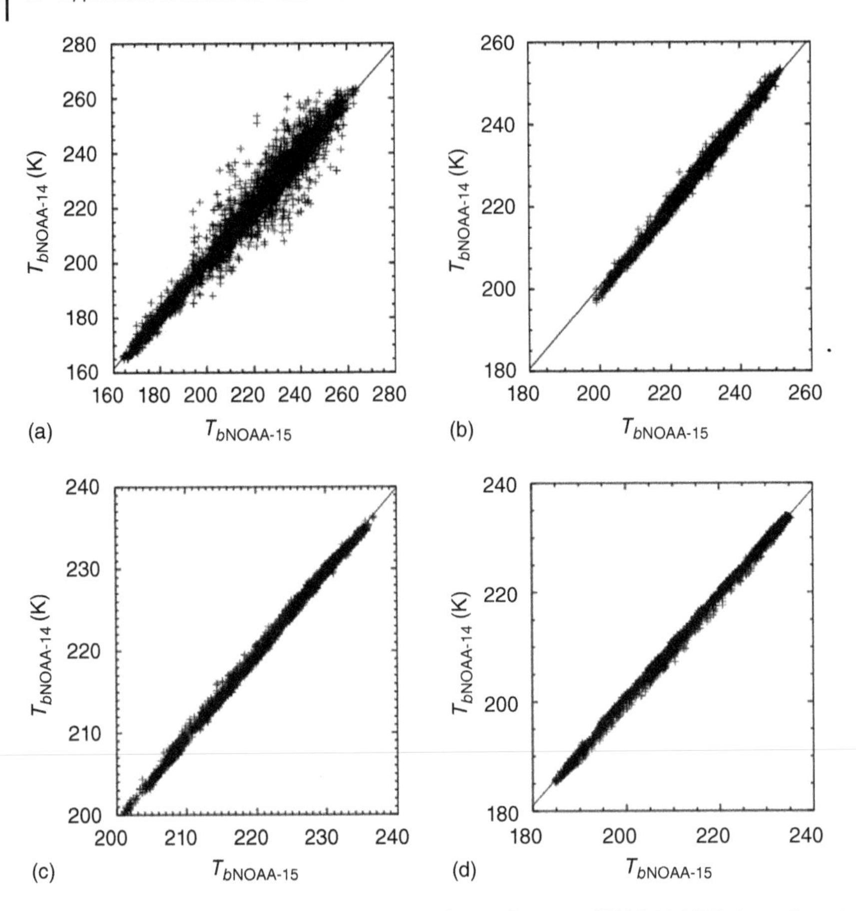

Figure 11.21 The brightness temperature correlations between NOAA-14 MSU channels and the corresponding NOAA-15 AMSUA subset channels for SNO data in 2002. The total number of data count is 5166. Collocation criteria: spatial separation <100 km and temporal separation <100 s. (Weng and Zou 2014 [243]. Reproduced with permission of Springer.)

through their emission and scattering. Thus, this channel is a dirty window channel in a clear atmosphere and is sensitive to both lower tropospheric temperature and surface temperature. It becomes opaque when clouds and precipitation occur in the atmosphere. Brightness temperatures at AMSU-A channel 5 (or MSU channel 2) are much less sensitive to the presence of clouds since most clouds occur below its weighting peak and have smaller impacts. Therefore, these two channels could be used for cloud detection due to their differential sensitivity. A cloud detection algorithm similar to that in [149] is developed. Firstly, cloud LWP is estimated from brightness temperatures at MSU channels 1 and 2 (or AMSU-A channels 3 and 5) using the following formula:

$$\mathrm{LWP}_{index} = c_0 + c_1 \log(290 - T_{b,Ch\,1}) + c_2 \log(290 - T_{b,Ch\,2}), \tag{11.9}$$

where $c_0 = 4.4313$, $c_1 = -1.3801$, and $c_2 = 0.4138$. The coefficients may also depend on the range of cloud liquid water as well as the scan angle.

A logarithmic form is selected in Eq. (11.9) due to an exponential relationship between the brightness temperature at channel 3 and cloud LWP. The coefficients c_1, c_2, and c_3 in the algorithm are derived from AMSU-A data simulated with a set of 1900 radiosonde profiles distributed over all the geographical regions. For each profile, a cloud layer below the freezing level with a randomly selected value of liquid water content within a range of $0-0.3\,\mathrm{g/m^3}$ is added to the profile. Since the brightness temperature over ocean is also sensitive to surface roughness, wind speed is also varied within $0-10\,\mathrm{m/s}$ for producing the simulated data set.

Cloud LWP can be estimated from MSU-like AMSU-A channels 3 and 5. Figure 11.22a compares LWP index derived from MSU-like AMSU-A channels

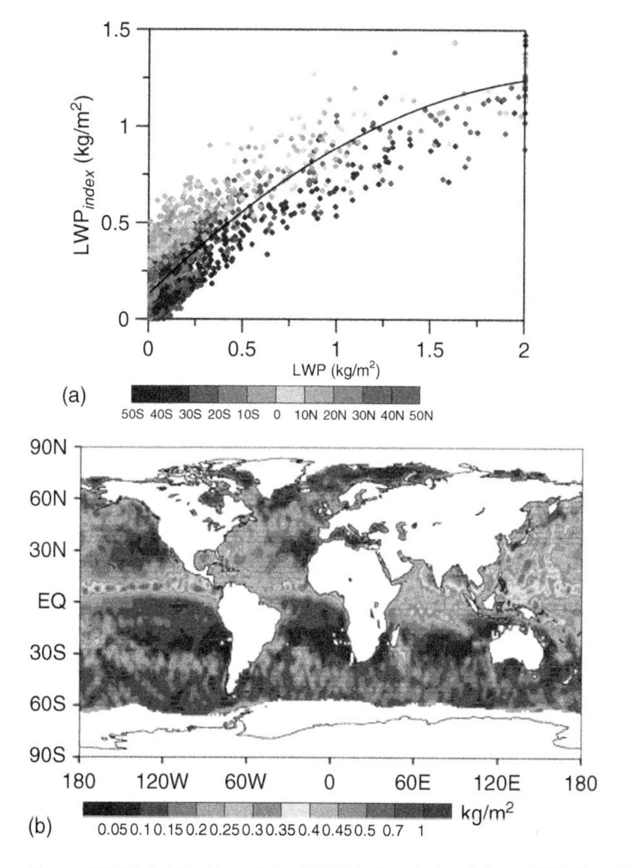

Figure 11.22 (a) Scatter plot of LWP index derived from MSU-like AMSU-A channels 3 and 5 using Eq. (11.4) (y-axis) and LWP derived from AMSU-A channels 1 and 2 at the nadir only over ocean on August 1, 2011. The black line represents a parabolic fitting: $LWP_{index} = -0.16 \times LWP^2 + 0.87 \times LWP + 0.15$. (b) Global distribution of monthly mean cloud LWP_{index} (unit: $\mathrm{kg/m^2}$) within $1° \times 1°$ grid box over ocean in August 2011. (Weng and Zou 2014 [243]. Reproduced with permission of Springer.)

3 and 5 using Eq. (11.9) with the LWP derived from AMSU-A channels 1 and 2 [211] at the nadir only over ocean on August 1, 2011. It is seen that LWP_{index} varies quadratically with LWP: $LWP_{index} = 0.87 \times LWP + 0.15 - 0.16 \times LWP^2$. The global distribution of monthly mean cloud LWP_{index} is presented in Figure 11.22b for August 2011. If LWP_{index} is greater than 0.5 kg/m^2, clouds are identified as a precipitating type within the satellite field of view. Any microwave observational data with an LWP_{index} greater than this threshold was not used for the 1DVar temperature profile retrieval.

Another technique for dealing with cloud effects in the temperature retrieval is to include the cloud liquid water in the state control variable. However, a limited amount of information from four-channel MSU data on clouds makes it difficult to simultaneously resolve all the profiles of temperature, water vapor, and cloud liquid water content.

11.5.4 Temperature Trend from 1DVar

Figure 11.23 provides the mean and standard deviation of the differences between observations and model simulations before and after the 1DVar data assimilation. Both mean and standard deviations are reduced by more than an order of differences for channels 3, 5, 7, and 9. Before examining the 30-year variations of atmospheric temperature deduced from satellite observations, a verification of the 1DVar results with GPS RO data is carried out first.

Figure 11.24 provides the 30-year variations of the global temperature anomaly at 10 pressure levels from June 1979 to December 2009 by a 1DVar approach. It is

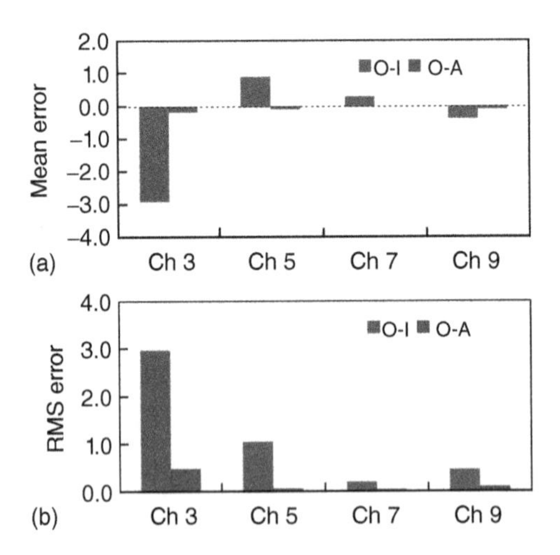

Figure 11.23 (a) Mean and (b) RMS differences of brightness temperatures between observations and model simulations from initial guess (O-I, unit: 10 K, left bars) and 1DVar analysis (O-A, unit: K, right bars) on August 28, 2011. (Weng and Zou 2014 [243]. Reproduced with permission of Springer.)

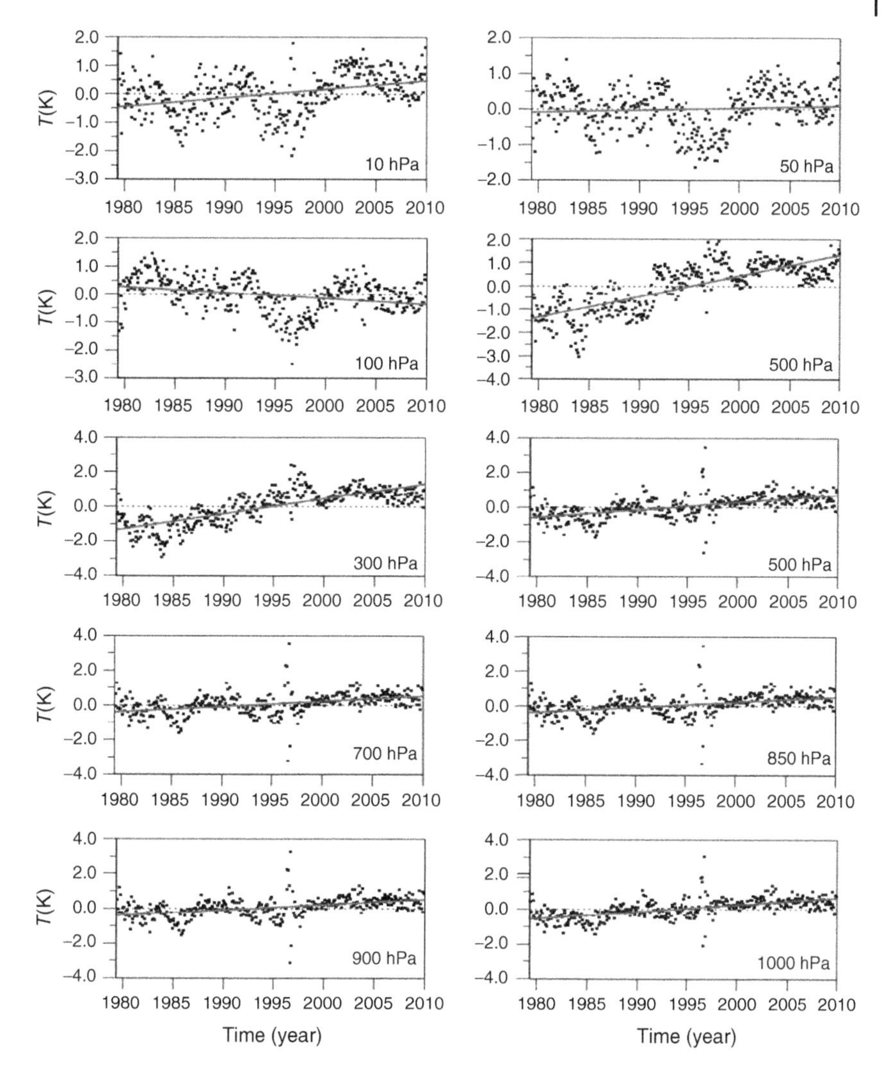

Figure 11.24 Monthly mean temperature anomaly (dark black dots) at 10 pressure levels over ocean and the linear trend (moderate gray line). (Weng and Zou 2014 [243]. Reproduced with permission of Springer.)

seen that the temperature anomaly is consistently negative in the earlier years and positive in the later years below 200 hPa, as a result of global warming. A reversed sign of temperature anomaly is seen between 50 and 100 hPa,

The vertical and latitudinal distribution of the global mean temperature linear trend calculated from the temperature retrieval from 1980 to 2009 is presented in Figure 11.25a. The decadal warming trend is about 0.5 K at 1000 hPa in the low and middle latitudes, decreases to 0.3–0.4 K/decade in the low troposphere (e.g., around 800 hPa), and increases to about 0.9 K between 200 and 300 hPa. A larger warming trend from 0.3 to 0.5 K/decade in the lower troposphere from

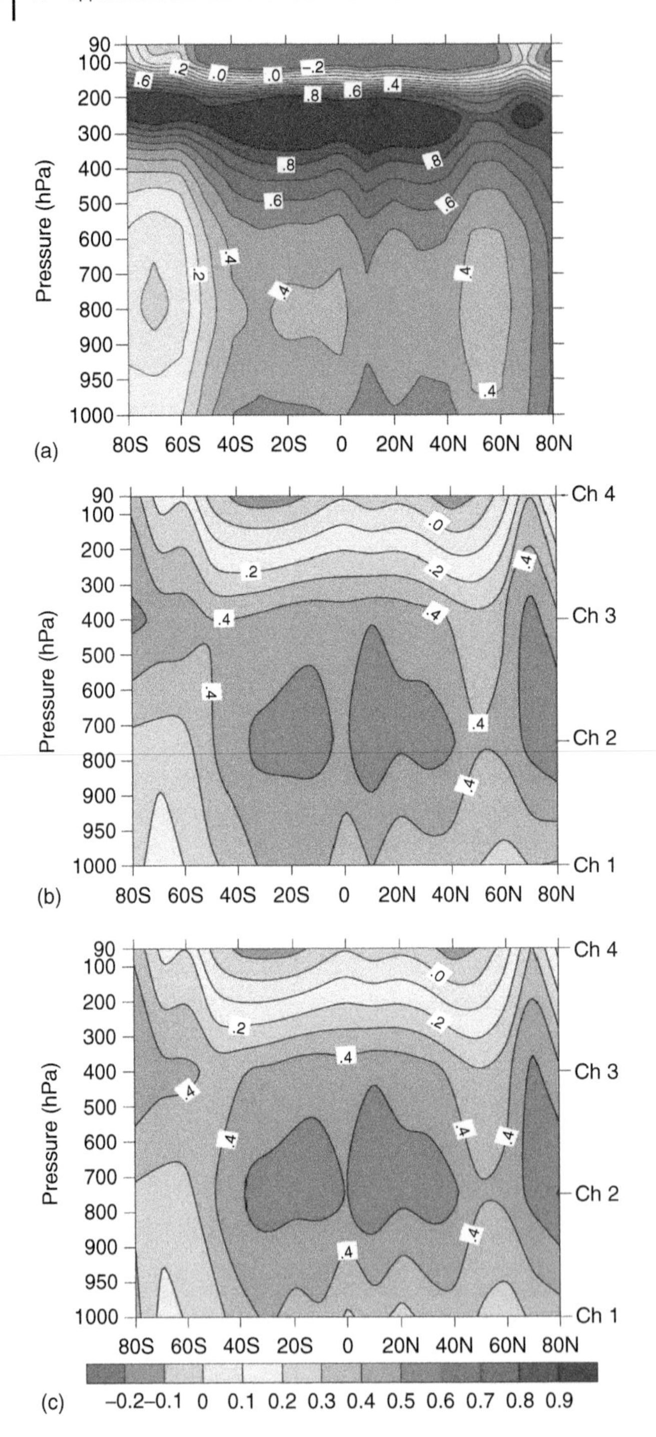

Figure 11.25 Decadal linear trend (°/decade) of (a) temperature over ocean from 1DVar retrieval, (b) brightness temperature observations, and (c) brightness temperature simulations using the 1DVar retrieval as input to CRTM. (Weng and Zou 2014 [243]. Reproduced with permission of Springer.)

1000 to 800 hPa is consistent with the climate-model-predicted trend [321, 322]. In Santer's studies [321], it was found that there is no longer a serious discrepancy between modeled and observed trends in tropical lapse rates. The constructed T2LT for MRI model ensemble mean shows about 0.371 C/decade in the lower tropospheric temperature warming over the tropical regions. Note from Figure 11.25 that a weak cooling trend (~0.2 K/decade) is seen at 100 hPa. Trends above 90 hPa are not reliable due to the lack of upper-level channels and are therefore not shown in this study.

As a self-validation, the retrieved profiles are used as input to CRTM to calculate synthetic MSU brightness temperatures. The long-term trends in the synthetic temperatures (Figure 11.25c) are then compared to those in the MSU/AMSU-A data (Figure 11.25b). The decadal linear trends in brightness temperature observations compared favorably with the trends derived from synthetic brightness temperature simulated using the 1DVar retrieval as input to CRTM, suggesting that the retrieval process does not introduce any spurious trends. The retrieved tropospheric warming trends (Figure 11.25a) in the upper troposphere are considerably larger than those directly inferred from MSU/AMSU brightness temperature data (Figure 11.25b). Since MSU channel brightness temperatures always represent deep-layer atmospheric temperatures, trends in MSU/AMSU-A brightness temperature data would be different from those of the retrieved temperatures. It is seen that the warming trends of MSU channel 3 are significantly smaller than the temperature trends in the upper troposphere where the peak WF of channel 3 is located; the warming trends of MSU channel 2 are slightly larger than the temperature trends in the low troposphere, and the cooling trends of MSU channel 4 are slightly smaller than the cooling trends of the retrieved atmospheric temperature trend.

11.6 Summary and Conclusions

The uncertainty in the climate trend from the observation is derived as a function of data record length and the noise of the measurements and the natural variability. To reduce the trend uncertainty, a longer data record is required. With a small measurement noise, and a small natural variability, the trend can be derived with a great accuracy within a shorter time.

SSM/I SDR intersensor calibration scheme is developed for constructing a long data record. This calibration scheme is built on the SCO technique, which collects observations as well as location from two SSM/I sensors that have an overlapping time period. The SCO intersensor calibration scheme requires an SSM/I sensor to be the reference radiometer so that other sensors can be calibrated against this reference sensor. F13 is selected as a reference sensor because of its most stable local equatorial crossing time and has the most interceptions with other satellites.

Results demonstrate that the newly developed SSM/I SDR intersensor calibration scheme substantially improves the consistency of the monthly mean oceanic rain-free T_b time series for all sensors, in terms of 58% reduced maximum intersensor bias, 30% declined mean absolute bias, and 21% decreased standard deviation. In addition, as we expected, the intersensor calibration has little impact on the mean rain-free oceanic monthly T_b; however, it exerts a dramatic influence on the climate trend of the monthly T_b series. Due to the consistency of the $\mathrm{T_b}$ series for all SSM/I sensors, the SDR's climate trend becomes more reliable. The SDR's climate trend over ocean varies from −0.70 to 0.36 K/decade before the intersensor calibration and from −0.44 to 0.09 K/decade after the calibration, resulting in a percentage change of 37% at 19v GHz channel and more than 70% at other channels.

The SSM/I SDR intersensor calibration has large impacts on the water vapor trend. The maximum TPW intersenor bias before and after the calibration changes from −1.5 to ±0.1 mm for the global ocean and from −3 to ±0.5 mm for the tropical ocean, resulting in a 75% and 20% deduction, respectively, of the mean absolute intersensor bias. The corresponding TPW standard deviation is also decreased by 40% and 21%. The TPW climate trend after the intersensor calibration is 0.34 mm/decade (or 1.59% per decade) for the global ocean and 0.63 mm/decade (or 1.39% per decade) for the tropical ocean, showing a trend decrease of 38% and 54%, respectively, before the calibration. The TPW trend with the intersensor-calibrated SSM/I measurements is in good agreement with previously published results.

Cloudy and cloud-free atmospheric temperature trends are derived from satellite microwave temperature sounding observations. The global daily mean brightness temperatures observed by the NOAA polar-orbiting satellite NOAA-15 AMSU-A instrument over the 13-year time period from August 1, 1999 to June 30, 2012 over ocean are used in our analysis. The traditional linear regression method is applied to the 13-year global brightness temperatures of AMSU-A channels 3, 5, 7, and 9 to obtain global and hemispheric averaged warming and cooling trends. By binning all AMSU-A channel brightness temperature measurements within 5° latitudinal bands, the latitudinal dependence of the cloud impact on global warming trend is calculated, and its relationship with global average warming/cooling is determined. It is shown that the atmospheric warming trends in the middle latitudes are significantly larger when cloud effects are removed from the 13-year microwave sounding data of AMSU-A channels 3 and 5 in both hemispheres. The scattering and emission effect of clouds and precipitation significantly reduces the values of the warming trends in the low and middle troposphere derived from the microwave data. The cooling trends are found in the high latitudes in the Southern Hemisphere and in all latitudes in the Northern Hemisphere. The high-altitude clouds tend to reduce the cooling trends of AMSU-A channel 9 in the Northern Hemisphere, especially in the middle and high latitudes in the Northern Hemisphere. The cloud impacts on the cooling trends in the Southern Hemisphere are negligible. However, their impacts on the temperature trends could be much larger in the latitudinal zones than on the global warming trend.

The 1DVar method is used for deriving atmospheric temperatures at each pressure level in the troposphere and low stratosphere from MSU and AMSU-A on board the NOAA polar-orbiting satellites over the time period from June 1978 to August 2010. The 30-year atmospheric temperature profiles are then derived from the satellite observations, detecting the global warming/cooling trends. The high accuracy of the CRTM for the forward simulation of satellite-observed microwave radiances makes the 1DVar approach extremely appropriate for deriving the TCDRs. As the first step, only temperatures over oceanic surfaces are derived and analyzed. The retrieved tropospheric warming trends in the upper troposphere are considerably larger than those directly inferred from MSU/AMSU brightness temperature data, an expected result of MSU channels representing a deep-layer average atmospheric temperature. The warming trends of MSU channel 2 are slightly larger than the temperature trends in the low troposphere where the peak WF of channel 2 is located. The cooling trends of the low stratospheric MSU channel 4 are slightly smaller than the cooling trends of the retrieved atmospheric temperature trend in the low stratosphere.

References

1. Dicke, R.H., Beringer, R., Kyhl, R.L., and Vane, A. (1946) Atmospheric absorption measurements with a microwave radiometer. *Phys. Rev.*, **70** (5–6), 340–348.
2. Weng, F. and Zou, X. (2013) Errors from Rayleigh–Jeans approximation in satellite microwave radiometer calibration systems. *Appl. Opt.*, **52** (3), 505–508.
3. Evans, K. and Stephens, G. (1991) A new polarized atmospheric radiative transfer model. *J. Quant. Spectrosc. Radiat. Transf.*, **46** (5), 413–423.
4. Chandrasekhar, S. (1949) *Radiative Transfer*, Dover Publications, Inc., New York, p. 35.
5. Weng, F., Zou, X., Sun, N., Yang, H., Tian, M., Blackwell, W.J., Wang, X., Lin, L., and Anderson, K. (2013) Calibration of Suomi national polar-orbiting partnership advanced technology microwave sounder. *J. Geophys. Res.: Atmos.*, **118** (19), 11187–11200.
6. Zou, X., Zhao, J., Weng, F., and Qin, Z. (2012) Detection of radio-frequency interference signal over land from FY-3B Microwave Radiation Imager (MWRI). *IEEE Trans. Geosci. Electron.*, **50** (12), 4994–5003.
7. Zhao, J., Zou, X., and Weng, F. (2013) WindSat radio frequency interference signature and its identification over Greenland and Antarctic. *IEEE Trans. Geosci. Electron.*, **51** (9), 4830–4839.
8. Zou, X., Lin, L., and Weng, F. (2014) Absolute calibration of ATMS upper level temperature sounding channels using GPS RO observations. *IEEE Trans. Geosci. Electron.*, **52** (2), 1397–1406.
9. Weng, F., Yang, H., and Zou, X. (2013) On convertibility from antenna to sensor brightness temperature for ATMS. *IEEE Trans. Geosci. Electron.*, **10** (4), 771–775.
10. Rothman, L.S., Gordon, I.E., Barbe, A., Benner, D.C., Bernath, P.F., Birk, M., Boudon, V., Brown, L.R., Campargue, A., and Champion, J.-P. (2009) The HITRAN 2008 molecular spectroscopic database. *J. Quant. Spectrosc. Radiat. Transf.*, **110** (9), 533–572.
11. Clough, S., Shephard, M., Mlawer, E., Delamere, J., Iacono, M., Cady-Pereira, K., Boukabara, S., and Brown, P. (2005) Atmospheric radiative transfer modeling: a summary of the AER codes. *J. Quant. Spectrosc. Radiat. Transf.*, **91** (2), 233–244.
12. McMillin, L.M. and Fleming, H.E. (1976) Atmospheric transmittance of an absorbing gas: a computationally fast and accurate transmittance model for absorbing gases with constant mixing ratios in inhomogeneous atmospheres. *Appl. Opt.*, **15** (2), 358–363.
13. Saunders, R., Matricardi, M., and Brunel, P. (1999) An improved fast radiative transfer model for assimilation of satellite radiance observations. *Q. J. R. Meteorol. Soc.*, **125** (556), 1407–1425.
14. Saunders, R., Rayer, P., Brunel, P., Von Engeln, A., Bormann, N., Strow, L., Hannon, S., Heilliette, S., Liu, X., and Miskolczi, F. (2007) A comparison of radiative transfer models for simulating Atmospheric Infrared Sounder (AIRS)

Passive Microwave Remote Sensing of the Earth: For Meteorological Applications, First Edition. Fuzhong Weng.

radiances. *J. Geophys. Res.: Atmos.*, **112** (D01S90), 1–17.

15. Ding, S., Xie, Y., Yang, P., Weng, F., Liu, Q., Baum, B., and Hu, Y. (2009) Estimates of radiation over clouds and dust aerosols: optimized number of terms in phase function expansion. *J. Quant. Spectrosc. Radiat. Transf.*, **110** (13), 1190–1198.
16. Liu, G. (2004) Approximation of single scattering properties of ice and snow particles for high microwave frequencies. *J. Atmos. Sci.*, **61** (20), 2441–2456.
17. Riley, W.J. (2007) *Handbook of Frequency Stability Analysis*, Hamilton Technical Services, Beaufort, SC, USA, p. 158.
18. Voigt, W. (1887) Ueber das Doppler'sche Princip. *Nachr. K. Gesel. Wiss. George-August-Universität*, **2**, 41–52.
19. Townes, C.H. and Schawlow, A.L. (2013) *Microwave Spectroscopy*, vol. **720**, Courier Corporation.
20. Van Vleck, J. (1947) The absorption of microwaves by oxygen. *Phys. Rev.*, **71** (7), 413.
21. Lenoir, W.B. (1968) Microwave spectrum of molecular oxygen in the mesosphere. *J. Geophys. Res.*, **73** (1), 361–376.
22. Rosenkranz, P.W. and Staelin, D.H. (1988) Polarized thermal microwave emission from oxygen in the mesosphere. *Radio Sci.*, **23** (5), 721–729.
23. Van Vleck, J.H. and Weisskopf, V.F. (1945) On the shape of collision-broadened lines. *Rev. Mod. Phys.*, **17** (2–3), 227–234.
24. Liebe, H.J. (1984) The atmospheric water vapor continuum below 300 GHz. *Int. J. Infrared Millimeter Waves*, **5** (2), 207–227.
25. Liebe, H.J. (1985) An updated model for millimeter wave propagation in moist air. *Radio Sci.*, **20** (5), 1069–1089.
26. Liebe, H.J. (1987) A contribution to modeling atmospheric millimeter-wave properties. *Frequenz*, **41** (1–2), 31–36.
27. Rosenkranz, P.W. (1993) Absorption of microwaves by atmospheric gases, in *Atmospheric Remote Sensing by Microwave Radiometry* (ed. M.A. Janssen), John Wiley, New York, pp. 37–79.
28. Liebe, H., Rosenkranz, P., and Hufford, G. (1992) Atmospheric 60-GHz oxygen spectrum: new laboratory measurements and line parameters. *J. Quant. Spectrosc. Radiat. Transf.*, **48** (5), 629–643.
29. Han, Y., Weng, F., Liu, Q., and van Delst, P. (2007) A fast radiative transfer model for SSMIS upper atmosphere sounding channels. *J. Geophys. Res.: Atmos.*, **112** (D11121), 1–12.
30. Mie, G. (1908) Beiträge zur Optik trüber Medien, speziell kolloidaler Metallösungen. *Ann. Phys.*, **330** (3), 377–445.
31. Evans, K.F. and Stephens, G.L. (1995) Microwave radiative transfer through clouds composed of realistically shaped ice crystals. Part I. Single scattering properties. *J. Atmos. Sci.*, **52** (11), 2041–2057.
32. Evans, K.F., Walter, S.J., Heymsfield, A.J., and Deeter, M.N. (1998) Modeling of submillimeter passive remote sensing of cirrus clouds. *J. Appl. Meteorol.*, **37** (2), 184–205.
33. Liu, G. (2008) A database of microwave single-scattering properties for nonspherical ice particles. *Bull. Am. Meteorol. Soc.*, **89** (10), 1563–1570.
34. Hong, G., Yang, P., Baum, B.A., Heymsfield, A.J., Weng, F., Liu, Q., Heygster, G., and Buehler, S.A. (2009) Scattering database in the millimeter and submillimeter wave range of 100–1000 GHz for nonspherical ice particles. *J. Geophys. Res.: Atmos.*, **114** (D6).
35. King, M.D., Platnick, S., Yang, P., Arnold, G.T., Gray, M.A., Riedi, J.C., Ackerman, S.A., and Liou, K.-N. (2004) Remote sensing of liquid water and ice cloud optical thickness and effective radius in the Arctic: application of airborne multispectral MAS data. *J. Atmos. Ocean. Technol.*, **21** (6), 857–875.
36. Yang, P., Wei, H., Huang, H.-L., Baum, B.A., Hu, Y.X., Kattawar, G.W., Mishchenko, M.I., and Fu, Q. (2005) Scattering and absorption property

database for nonspherical ice particles in the near-through far-infrared spectral region. *Appl. Opt.*, **44** (26), 5512–5523.

37. Baum, B.A., Heymsfield, A.J., Yang, P., and Bedka, S.T. (2005) Bulk scattering properties for the remote sensing of ice clouds. Part I: microphysical data and models. *J. Appl. Meteorol.*, **44** (12), 1885–1895.
38. Liu, G. and Curry, J.A. (2000) Determination of ice water path and mass median particle size using multichannel microwave measurements. *J. Appl. Meteorol.*, **39** (8), 1318–1329.
39. Zhao, L. and Weng, F. (2002) Retrieval of ice cloud parameters using the Advanced Microwave Sounding Unit. *J. Appl. Meteorol.*, **41** (4), 384–395.
40. Liu, Q. and Weng, F. (2006) Combined Henyey-Greenstein and Rayleigh phase function. *Appl. Opt.*, **45** (28), 7475–7479.
41. Wilheit, T.T., Chang, A., Rao, M.V., Rodgers, E., and Theon, J.S. (1977) A satellite technique for quantitatively mapping rainfall rates over the oceans. *J. Appl. Meteorol.*, **16** (5), 551–560.
42. Wu, R. and Weinman, J. (1984) Microwave radiances from precipitating clouds containing aspherical ice, combined phase, and liquid hydrometeors. *J. Geophys. Res.: Atmos.*, **89** (D5), 7170–7178.
43. Kummerow, C. and Weinman, J. (1988) Determining microwave brightness temperatures from precipitating horizontally finite and vertically structured clouds. *J. Geophys. Res.: Atmos.*, **93** (D4), 3720–3728.
44. Liu, Q. and Weng, F. (2006) Advanced doubling-adding method for radiative transfer in planetary atmospheres. *J. Atmos. Sci.*, **63** (12), 3459–3465.
45. Stamnes, K., Tsay, S.-C., Wiscombe, W., and Jayaweera, K. (1988) Numerically stable algorithm for discrete-ordinate-method radiative transfer in multiple scattering and emitting layered media. *Appl. Opt.*, **27** (12), 2502–2509.
46. Weng, F. (1992) A multi-layer discrete-ordinate method for vector radiative transfer in a vertically-inhomogeneous, emitting and scattering atmosphere—II. Application. *J. Quant. Spectrosc. Radiat. Transfer*, **47** (1), 35–42.
47. Weng, F. and Liu, Q. (2003) Satellite data assimilation in numerical weather prediction models. Part I: forward radiative transfer and Jacobian modeling in cloudy atmospheres. *J. Atmos. Sci.*, **60** (21), 2633–2646.
48. Schulz, F., Stamnes, K., and Weng, F. (1999) VDISORT: an improved and generalized discrete ordinate method for polarized (vector) radiative transfer. *J. Quant. Spectrosc. Radiat. Transf.*, **61** (1), 105–122.
49. Dave, J. (1970) Intensity and polarization of the radiation emerging from a plane-parallel atmosphere containing monodispersed aerosols. *Appl. Opt.*, **9** (12), 2673–2684.
50. Liu, Q. and Ruprecht, E. (1996) Radiative transfer model: matrix operator method. *Appl. Opt.*, **35** (21), 4229–4237.
51. Liu, Q. and Weng, F. (2002) A microwave polarimetric two-stream radiative transfer model. *J. Atmos. Sci.*, **59** (15), 2396–2402.
52. Wentz, F.J. and Spencer, R.W. (1998) SSM/I rain retrievals within a unified all-weather ocean algorithm. *J. Atmos. Sci.*, **55** (9), 1613–1627.
53. Weng, F. and Grody, N.C. (2000) Retrieval of ice cloud parameters using a microwave imaging radiometer. *J. Atmos. Sci.*, **57** (8), 1069–1081.
54. Stogryn, A. (1967) The apparent temperature of the sea at microwave frequencies. *IEEE Trans. Antennas Propag.*, **15** (2), 278–286.
55. Rosenkranz, P. and Staelin, D. (1972) Microwave emissivity of ocean foam and its effect on nadiral radiometric measurements. *J. Geophys. Res.*, **77** (33), 6528–6538.
56. Cox, C. and Munk, W. (1954) Measurement of the roughness of the sea surface from photographs of the sun's glitter. *J. Opt. Soc. Am.*, **44** (11), 838–850.
57. Wilheit, T.T., (1979) A model for the microwave emissivity of the ocean's surface as a function of wind speed. *IEEE Trans. Geosci. Electron.*, **17** (4), 244–249.

58. English, S.J. and T.J. Hewison. (1998) Fast Generic Millimeter-Wave Emissivity Model. Asia-Pacific Symposium on Remote Sensing of the Atmosphere, Environment, and Space. International Society for Optics and Photonics.
59. Wentz, F.J., Christensen, E.J., and Richardson, K.A. (1981) Dependence of sea-surface microwave emissivity on friction velocity as derived from SMMR/SASS, in *Oceanography from Space*, Springer, pp. 741–749.
60. Ellison, W., English, S., Lamkaouchi, K., Balana, A., Obligis, E., Deblonde, G., Hewison, T., Bauer, P., Kelly, G., and Eymard, L. (2003) A comparison of ocean emissivity models using the Advanced Microwave Sounding Unit, the Special Sensor Microwave Imager, the TRMM Microwave Imager, and airborne radiometer observations. *J. Geophys. Res.: Atmos.*, **108** (D21), 4663, ACL 1-5.
61. Debye, P. (1929) *Polar Molecules*, The Chemical Catalog Co., New York.
62. Klein, L.A. and Swift, C.T. (1977) An improved model for the dielectric constant of sea water at microwave frequencies. *IEEE Trans. Antennas Propag.*, **25** (1), 104–111.
63. Liebe, H.J., Hufford, G.A., and Manabe, T. (1991) A model for the complex permittivity of water at frequencies below 1 THz. *Int. J. Infrared Millimeter Waves*, **12** (7), 659–675.
64. Stogryn, A., Bull, H., Rubayi, K., and Iravanchy, S. (1995) *The Microwave Permittivity of Sea and Fresh Water*, GenCorp Aerojet, Azusa, CA.
65. Durden, S.L. and Vesecky, J.F. (1985) A physical radar cross-section model for a wind-driven sea with swell. *IEEE J. Oceanic Eng.*, **10** (4), 445–451.
66. Liu, Q., Weng, F., and English, S.J. (2011) An improved fast microwave water emissivity model. *IEEE Trans. Geosci. Electron.*, **49** (4), 1238–1250.
67. Somaraju, R. and Trumpf, J. (2006) Frequency, temperature and salinity variation of the permittivity of seawater. *IEEE Trans. Antennas Propag.*, **54** (11), 3441–3448.
68. Guillou, C., Ellison, W., Eymard, L., Lamkaouchi, K., Prigent, C., Delbos, G., Balana, G., and Boukabara, S. (1998) Impact of new permittivity measurements on sea surface emissivity modeling in microwaves. *Radio Sci.*, **33**, 649–668.
69. Ellison, W., Balana, A., Delbos, G., Lamkaouchi, K., Eymard, L., Guillou, C., and Prigent, C. (1998) New permittivity measurements of seawater. *Radio Sci.*, **33** (3), 639–648.
70. Barthel, J., Bachhuber, K., Buchner, R., Hetzenauer, H., and Kleebauer, M. (1991) A computer-controlled system of transmission lines for the determination of the complex permittivity of lossy liquids between 8.5 and 90 GHz. *Ber. Bunsenges. Phys. Chem.*, **95** (8), 853–859.
71. Stogryn, A. (1971) Equations for calculating the dielectric constant of saline water (correspondence). *IEEE Trans. Microwave Theory Tech.*, **19** (8), 733–736.
72. Meissner, T. and Wentz, F.J. (2004) The complex dielectric constant of pure and sea water from microwave satellite observations. *IEEE Trans. Geosci. Electron.*, **42** (9), 1836–1849.
73. Berger, M., Camps, A., Font, J., Kerr, Y., Miller, J., Johannessen, J.A., Boutin, J., Drinkwater, M.R., Skou, N., Floury, N. (2002) Measuring Ocean Salinity with ESA's SMOS Mission–Advancing the Science, *ESA Bulletin*, **111**, 113–121.
74. Ho, W. and Hall, W. (1973) Measurements of the dielectric properties of seawater and NaCl solutions at 2.65 GHz. *J. Geophys. Res.*, **78** (27), 6301–6315.
75. Ho, W., Love, A., VanMelle, M. (1974) Measurements of the dielectric properties of sea water at 1.43 GHz, NASA, Washington DC, p. 35.
76. Goodberlet, M., Swift, C., and Wilkerson, J. (1989) Remote sensing of ocean surface winds with the Special Sensor Microwave/Imager. *J. Geophys. Res.: Oceans*, **94** (C10), 14547–14555.
77. Wu, S.-T. and Fung, A.K. (1972) A noncoherent model for microwave emissions and backscattering from

the sea surface. *J. Geophys. Res.*, **77** (30), 5917–5929.

78. Bjerkaas, A. and Riedel, F. (1979) *Proposed Model for the Elevation Spectrum of a Wind-Roughened Sea Surface,* DTIC Document.
79. Donelan, M.A. and Pierson, W.J. (1987) Radar scattering and equilibrium ranges in wind-generated waves with application to scatterometry. *J. Geophys. Res.: Oceans,* **92** (C5), 4971–5029.
80. Apel, J.R. (1994) An improved model of the ocean surface wave vector spectrum and its effects on radar backscatter. *J. Geophys. Res.: Oceans,* **99** (C8), 16269–16291.
81. Elfouhaily, T., Chapron, B., Katsaros, K., and Vandemark, D. (1997) A unified directional spectrum for long and short wind-driven waves. *J. Geophys. Res.: Oceans,* **102** (C7), 15781–15796.
82. Dinnat, E.P., Boutin, J., Caudal, G., and Etcheto, J. (2003) Issues concerning the sea emissivity modeling at L band for retrieving surface salinity. *Radio Sci.*, **38** (4), 25-1–25-4.
83. Liu, Q., Simmer, C., and Ruprecht, E. (1998) Monte Carlo simulations of the microwave emissivity of the sea surface. *J. Geophys. Res.: Oceans,* **103** (C11), 24983–24989.
84. Yueh, S.H., W.J. Wilson, S.V. Nghiem, F.K. Li, W.B. Ricketts. (1994) Polarimetric Passive Remote Sensing of Ocean Wind Vectors. Geoscience and Remote Sensing Symposium, 1994. IGARSS'94. Surface and Atmospheric Remote Sensing: Technologies, Data Analysis and Interpretation., International. IEEE.
85. Pierson, W.J. and Moskowitz, L. (1964) A proposed spectral form for fully developed wind seas based on the similarity theory of SA Kitaigorodskii. *J. Geophys. Res.*, **69** (24), 5181–5190.
86. Yueh, S.H. (1997) Modeling of wind direction signals in polarimetric sea surface brightness temperatures. *IEEE Trans. Geosci. Electron.*, **35** (6), 1400–1418.
87. Wentz, F.J. (1975) A two-scale scattering model for foam-free sea microwave brightness temperatures. *J. Geophys. Res.*, **80** (24), 3441–3446.
88. Ulaby, F.T., Moore, R.K., Fung, A.K., and House, A. (1981) *Microwave Remote Sensing: Active and Passive,* Addison-Wesley, Reading, MA.
89. Guissard, A. and Sobieski, P. (1987) An approximate model for the microwave brightness temperature of the sea. *Int. J. Remote Sens.*, **8** (11), 1607–1627.
90. Tang, C. (1974) The effect of droplets in the air-sea transition zone on the sea brightness temperature. *J. Phys. Oceanogr.*, **4** (4), 579–593.
91. Schrader, M. (1995) *Ein dreiskalenmodell zur Berechnung der Reflektivitaet der Ozeanoberflaeche im Mikrowellenfrequenzbereich,* Berichte-Institut fur Meereskunde an der Christian Albrechts Universitat KIEL.
92. Stogryn, A. (1972) The emissivity of sea foam at microwave frequencies. *J. Geophys. Res.*, **77** (9), 1658–1666.
93. Rose, L.A., Asher, W.E., Reising, S.C., Gaiser, P.W., Germain, K.M.S., Dowgiallo, D.J., Horgan, K.A., Farquharson, G., and Knapp, E. (2002) Radiometric measurements of the microwave emissivity of foam. *IEEE Trans. Geosci. Electron.*, **40** (12), 2619–2625.
94. Kazumori, M., Liu, Q., Treadon, R., and Derber, J.C. (2008) Impact study of AMSR-E radiances in the NCEP global data assimilation system. *Mon. Weather Rev.*, **136** (2), 541–559.
95. Yueh, S.H. (2000) Estimates of Faraday rotation with passive microwave polarimetry for microwave remote sensing of Earth surfaces. *IEEE Trans. Geosci. Electron.*, **38** (5), 2434–2438.
96. Germain, K.S., G. Poe, P. Gaiser, Polarimetric Emission Model of the Sea at Microwave Frequencies and Comparison with Measurements. 2002, DTIC Document.
97. Choudhury, B. and Chang, A.T. (1979) Two stream theory of reflectance of snow. *IEEE Trans. Geosci. Electron.*, **17** (3), 63–68.
98. Wang, J.R. and Schmugge, T.J. (1980) An empirical model for the complex dielectric permittivity of soils as a function of water content. *IEEE Trans. Geosci. Electron.*, **4**, 288–295.

99. Mo, T., Choudhury, B., Schmugge, T., Wang, J., and Jackson, T. (1982) A model for microwave emission from vegetation-covered fields. *J. Geophys. Res.: Oceans*, **87** (C13), 11229–11237.
100. Kerr, Y.H. and Njoku, E.G. (1990) A semiempirical model for interpreting microwave emission from semiarid land surfaces as seen from space. *IEEE Trans. Geosci. Electron.*, **28** (3), 384–393.
101. Fung, A.K. (1994) *Microwave Scattering and Emission Models and Applications*, Artech House Publishers, Norwood, MA, p. 592.
102. Choudhury, B.J. and Pampaloni, P. (1995) *Passive Microwave Remote Sensing of Land–Atmosphere Interactions*, VSP.
103. Tsang, L., Kong, J.A., Shin, R.T. (1985) Theory of microwave remote sensing, Wiley-Interscience, New York, p. 632.
104. Jones, A.S. and Vonder Haar, T.H. (1997) Retrieval of microwave surface emittance over land using coincident microwave and infrared satellite measurements. *J. Geophys. Res.: Atmos.*, **102** (D12), 13609–13626.
105. Prigent, C., Rossow, W.B., and Matthews, E. (1997) Microwave land surface emissivities estimated from SSM/I observations. *J. Geophys. Res.: Atmos.*, **102** (D18), 21867–21890.
106. Weng, F., Yan, B., and Grody, N.C. (2001) A microwave land emissivity model. *J. Geophys. Res.: Atmos.*, **106** (D17), 20115–20123.
107. Prigent, C., Aires, F., and Rossow, W.B. (2006) Land surface microwave emissivities over the globe for a decade. *Bull. Am. Meteorol. Soc.*, **87** (11), 1573.
108. Karbou, F., Gérard, É., and Rabier, F. (2006) Microwave land emissivity and skin temperature for AMSU-A and -B assimilation over land. *Q. J. R. Meteorol. Soc.*, **132** (620), 2333–2355.
109. Karam, M.A., Fung, A.K., Lang, R.H., and Chauhan, N.S. (1992) A microwave scattering model for layered vegetation. *IEEE Trans. Geosci. Electron.*, **30** (4), 767–784.
110. Ulaby, F.T., Sarabandi, K., Mcdonald, K., Whitt, M., and Dobson, M.C. (1990) Michigan microwave canopy scattering model. *Int. J. Remote Sens.*, **11** (7), 1223–1253.
111. Fung, A., Chen, M., and Lee, K. (1987) Fresnel field interaction applied to scattering from a vegetation layer. *Remote Sens. Environ.*, **23** (1), 35–50.
112. Şeker, Ş. (1986) Microwave Backscattering from a Layer of Randomly Oriented Discs with Application to Scattering from Vegetation. IEE Proceedings H (Microwaves, Antennas and Propagation). IET.
113. Tsang, L., Kong, J., and Shin, R. (1984) Radiative transfer theory for active remote sensing of a layer of non-spherical particles. *Radio Sci.*, **19** (2), 629–642.
114. Lang, R.H. and Sighu, J.S. (1983) Electromagnetic backscattering from a layer of vegetation: a discrete approach. *IEEE Trans. Geosci. Electron.*, **1**, 62–71.
115. Karam, M. and Fung, A. (1982) Vector forward scattering theorem. *Radio Sci.*, **17** (4), 752–756.
116. Durden, S.L., Van Zyl, J.J., and Zebker, H.A. (1989) Modeling and observation of the radar polarization signature of forested areas. *IEEE Trans. Geosci. Electron.*, **27** (3), 290–301.
117. Ulaby, F.T. and El-rayes, M.A. (1987) Microwave dielectric spectrum of vegetation. Part II: dual-dispersion model. *IEEE Trans. Geosci. Electron.*, **5**, 550–557.
118. Mätzler, C. (1994) Microwave (1–100 GHz) dielectric model of leaves. *IEEE Trans. Geosci. Electron.*, **32** (4), 947–949.
119. Grody, N.C. and Weng, F. (2008) Microwave emission and scattering from deserts: theory compared with satellite measurements. *IEEE Trans. Geosci. Electron.*, **46** (2), 361–375.
120. Liang, D., Xu, X., Tsang, L., Andreadis, K.M., and Josberger, E.G. (2008) The effects of layers in dry snow on its passive microwave emissions using dense media radiative transfer theory based on the quasicrystalline approximation (QCA/DMRT). *IEEE Trans. Geosci. Electron.*, **46** (11), 3663–3671.
121. Sancer, M.I. (1969) Shadow-corrected electromagnetic scattering from a

randomly rough surface. *IEEE Trans. Antennas Propag.*, **17** (5), 577–585.
122. Leader, J.C. (1978) Incoherent backscatter from rough surfaces: the two-scale model reexamined. *Radio Sci.*, **13** (3), 441–457.
123. Brown, G.S. (1978) Backscattering from a Gaussian-distributed perfectly conducting rough surface. *IEEE Trans. Antennas Propag.*, **26** (3), 472–482.
124. Dobson, M.C., Ulaby, F.T., Hallikainen, M.T., and El-Rayes, M.A. (1985) Microwave dielectric behavior of wet soil—Part II: dielectric mixing models. *IEEE Trans. Geosci. Remote Sens.*, **23** (1), 35–46.
125. Ulaby, F.T., Moore, R.K., and Fung, A.K. (1986) *Microwave Remote Sensing: Active and Passive*, Artech House Inc., Norwood, MA, pp. 1065–2162.
126. Wang, J. and Choudhury, B. (1981) Remote sensing of soil moisture content, over bare field at 1.4 GHz frequency. *J. Geophys. Res.: Oceans*, **86** (C6), 5277–5282.
127. Coppo, P., Luzi, G., Paloscia, S., Pampaloni, P. (1991) Effect of Soil Roughness on Microwave Emission: Comparison Between Experimental Data and Models. Geoscience and Remote Sensing Symposium, 1991. IGARSS'91. Remote Sensing: Global Monitoring for Earth Management., International. IEEE.
128. Wegmuller, U. and Matzler, C. (1999) Rough bare soil reflectivity model. *IEEE Trans. Geosci. Electron.*, **37** (3), 1391–1395.
129. Marshall, J.S. and Palmer, W.M.K. (1948) The distribution of raindrops with size. *J. Meteorol.*, **5** (4), 165–166.
130. Spencer, R.W. (1986) A satellite passive 37-GHz scattering-based method for measuring oceanic rain rates. *J. Clim. Appl. Meteorol.*, **25** (6), 754–766.
131. Mugnai, A. and Smith, E.A. (1988) Radiative transfer to space through a precipitating cloud at multiple microwave frequencies. Part I: model description. *J. Appl. Meteorol.*, **27** (9), 1055–1073.
132. Smith, E.A. and Mugnai, A. (1988) Radiative transfer to space through a precipitating cloud at multiple microwave frequencies. Part II: results and analysis. *J. Appl. Meteorol.*, **27** (9), 1074–1091.
133. Marzano, F., Di Michele, S., Tassa, A., Mugnai, A. (2000) Bayesian Techniques for Precipitation Profiles Retrieval from Spaceborne Microwave Radiometers. Proceedings of PORSEC-2000, pp. 221–228.
134. Kummerow, C. and Giglio, L. (1994) A passive microwave technique for estimating rainfall and vertical structure information from space. Part I: algorithm description. *J. Appl. Meteorol.*, **33** (1), 3–18.
135. Olson, W.S., Bauer, P., Kummerow, C.D., Hong, Y., and Tao, W.-K. (2001) A melting-layer model for passive/active microwave remote sensing applications. Part II: simulation of TRMM observations. *J. Appl. Meteorol.*, **40** (7), 1164–1179.
136. Han, Y., van Delst, P., Liu, Q., Weng, F., Yan, B., Treadon, R., Derber, J. JCSDA Community Radiative Transfer Model-Version 1 (CRTM-V1). NESDIS: NOAA Tech. Rep. p. 40.
137. Mishchenko, M.I. and Travis, L.D. (1994) Light scattering by polydispersions of randomly oriented spheroids with sizes comparable to wavelengths of observation. *Appl. Opt.*, **33** (30), 7206–7225.
138. Yang, P., Feng, Q., Hong, G., Kattawar, G.W., Wiscombe, W.J., Mishchenko, M.I., Dubovik, O., Laszlo, I., and Sokolik, I.N. (2007) Modeling of the scattering and radiative properties of nonspherical dust-like aerosols. *J. Aerosol Sci.*, **38** (10), 995–1014.
139. King, M.D., Platnick, S., Hubanks, P.A., Arnold, G.T., Moody, E.G., Wind, G., and Wind, B. (2006) Collection 005 change summary for the MODIS cloud optical property (06_OD) algorithm. *MODIS Atmos.*, NASA report, p. 8.
140. Yang, P., Liou, K., Wyser, K., and Mitchell, D. (2000) Parameterization of the scattering and absorption properties of individual ice crystals. *J. Geophys. Res.: Atmos.*, **105** (D4), 4699–4718.
141. Kazumori, M. and English, S.J. (2015) Use of the ocean surface wind direction

signal in microwave radiance assimilation. *Q. J. R. Meteorol. Soc.*, **141** (689), 1354–1375.

142. Chen, M. and Weng, F. (2016) Modeling land surface roughness effect on soil microwave emission in community surface emissivity model. *IEEE Trans. Geosci. Remote Sens.*, **54** (3), 1716–1726 (URL: http://ieeexplore.ieee.org/stamp/stamp.jsp?arnumber=7312969).
143. Van Delst, P. (2003) JCSDA Infrared Sea Surface Emissivity Model. Proceedings of the 13th International TOVS study conference. Citeseer.
144. Wu, X. and Smith, W.L. (1997) Emissivity of rough sea surface for 8–13 μm: modeling and verification. *Appl. Opt.*, **36** (12), 2609–2619.
145. Carter, C., Liu, Q., Yang, W., Hommel, D., Emery, W. (2002) Net Heat Flux, Visible/Infrared Imager/Radiometer Suite Algorithm Theoretical Basis Document, http://npoesslib.ipo.noaa.gov/ulistcategoryv3.php35.
146. Yan, B., Weng, F., Okamoto, K. (2004) A Microwave Snow Emissivity Model. 8th Specialist Meeting on microwave radiometry and remote sensing applications, Rome, Italy.
147. Weng, F., Zou, X., Wang, X., Yang, S., and Goldberg, M. (2012) Introduction to Suomi national polar-orbiting partnership advanced technology microwave sounder for numerical weather prediction and tropical cyclone applications. *J. Geophys. Res.: Atmos.*, **117** (D112), 1–14.
148. Greenwald, T.J., Stephens, G.L., Vonder Haar, T.H., and Jackson, D.L. (1993) A physical retrieval of cloud liquid water over the global oceans using Special Sensor Microwave/Imager (SSM/I) observations. *J. Geophys. Res.: Atmos.*, **98** (D10), 18471–18488.
149. Weng, F. and Grody, N.C. (1994) Retrieval of cloud liquid water using the special sensor microwave imager (SSM/I). *J. Geophys. Res.: Atmos.*, **99** (D12), 25535–25551.
150. Weng, F., Grody, N.C., Ferraro, R., Basist, A., and Forsyth, D. (1997) Cloud liquid water climatology from the Special Sensor Microwave/Imager. *J. Clim.*, **10** (5), 1086–1098.
151. Wentz, F.J. (1997) A well-calibrated ocean algorithm for special sensor microwave/imager. *J. Geophys. Res.: Oceans*, **102** (C4), 8703–8718.
152. Harris, B. and Kelly, G. (2001) A satellite radiance-bias correction scheme for data assimilation. *Q. J. R. Meteorol. Soc.*, **127** (574), 1453–1468.
153. Healy, S. and Eyre, J.R. (2000) Retrieving temperature, water vapour and surface pressure information from refractive-index profiles derived by radio occultation: a simulation study. *Q. J. R. Meteorol. Soc.*, **126** (566), 1661–1683.
154. Palmer, P.I., Barnett, J., Eyre, J., Healy, S., (2000) A non-linear optimal estimation inverse method for radio occultation measurements of temperature, humidity and surface pressure. *arXiv preprint physics/0003010.*
155. Kishore, P., Namboothiri, S., Jiang, J., Sivakumar, V., and Igarashi, K. (2009) Global temperature estimates in the troposphere and stratosphere: a validation study of COSMIC/FORMOSAT-3 measurements. *Atmos. Chem. Phys.*, **9** (3), 897–908.
156. Anthes, R.A., Ector, D., Hunt, D., Kuo, Y., Rocken, C., Schreiner, W., Sokolovskiy, S., Syndergaard, S., Wee, T., and Zeng, Z. (2008) The COSMIC/FORMOSAT-3 mission: early results. *Bull. Am. Meteorol. Soc.*, **89** (3), 313–333.
157. Weng, F. (2002) Microwave polarimetric signal from hurricane environment. *J. Electromagn. Waves Appl.*, **16** (4), 467–480.
158. Chen, Y., Weng, F., Han, Y., and Liu, Q. (2008) Validation of the community radiative transfer model by using CloudSat data. *J. Geophys. Res.: Atmos.*, **113** (D00a03), 1–15.
159. Vinnikov, K.Y., Grody, N.C., Robock, A., Stouffer, R.J., Jones, P.D., and Goldberg, M.D. (2006) Temperature trends at the surface and in the troposphere. *J. Geophys. Res.: Atmos.*, **111** (D03106), 1–14.

160. Christy, J.R., Spencer, R.W., Norris, W.B., Braswell, W.D., and Parker, D.E. (2003) Error estimates of version 5.0 of MSU-AMSU bulk atmospheric temperatures. *J. Atmos. Ocean. Technol.*, **20** (5), 613–629.
161. Mears, C.A. and Wentz, F.J. (2005) The effect of diurnal correction on satellite-derived lower tropospheric temperature. *Science*, **309** (5740), 1548–1551.
162. Zou, C.Z., Goldberg, M.D., Cheng, Z., Grody, N.C., Sullivan, J.T., Cao, C., and Tarpley, D. (2006) Recalibration of microwave sounding unit for climate studies using simultaneous nadir overpasses. *J. Geophys. Res.: Atmos.*, **111** (D114), 1–24.
163. Grody, N.C., Vinnikov, K.Y., Goldberg, M.D., Sullivan, J.T., and Tarpley, J.D. (2004) Calibration of multisatellite observations for climatic studies: Microwave Sounding Unit (MSU). *J. Geophys. Res.: Atmos.*, **109** (D104), 1–12.
164. ATBD, J.A. (2011) *Joint Polar Satellite System (JPSS) Advanced Technology Microwave Sounder (ATMS) SDR Radiometric Calibration Algorithm Theoretical Basic Document (ATBD).*
165. Wiltshire, W.M., Nilsson, M. D. and Campbell, F. B. (2005) Application for Authorization to Launch and Operate DirectTV 11, a Partial Replacement Ka-Band Satellite, at 99 deg. Federal Communications Commission, WL, FCC form 312.
166. Allan, D.W. (1987) Should the classical variance be used as a basic measure in standards metrology? *IEEE Trans. Instrum. Meas.*, **1001** (2), 646–654.
167. Tian, M., Zou, X., and Weng, F. (2015) Use of Allan deviation for characterizing satellite microwave sounder noise equivalent differential temperature (NEDT). *IEEE Geosci. Remote Sens. Lett.*, **12** (12), 2477–2480.
168. Wu, Y. and Weng, F. (2011) Detection and correction of AMSR-E radio-frequency interference. *Acta Meteorol. Sin.*, **25**, 669–681.
169. Njoku, E.G., Ashcroft, P., Chan, T.K., and Li, L. (2005) Global survey and statistics of radio-frequency interference in AMSR-E land observations. *IEEE Trans. Geosci. Electron.*, **43** (5), 938–947.
170. Kawanishi, T., Sezai, T., Ito, Y., Imaoka, K., Takeshima, T., Ishido, Y., Shibata, A., Miura, M., Inahata, H., and Spencer, R.W. (2003) The Advanced Microwave Scanning Radiometer for the Earth Observing System (AMSR-E), NASDA's contribution to the EOS for global energy and water cycle studies. *IEEE Trans. Geosci. Electron.*, **41** (2), 184–194.
171. Li, L., Njoku, E.G., Im, E., Chang, P.S., and Germain, K.S. (2004) A preliminary survey of radio-frequency interference over the US in Aqua AMSR-E data. *IEEE Trans. Geosci. Electron.*, **42** (2), 380–390.
172. Li, L., Gaiser, P.W., Bettenhausen, M.H., and Johnston, W. (2006) WindSat radio-frequency interference signature and its identification over land and ocean. *IEEE Trans. Geosci. Electron.*, **44** (3), 530–539.
173. Zou, X., Tian, X., and Weng, F. (2014) Detection of television frequency interference with satellite microwave imager observations over oceans. *J. Atmos. Ocean. Technol.*, **31** (12), 2759–2776.
174. Adams, I.S., Bettenhausen, M.H., Gaiser, P.W., and Johnston, W. (2010) Identification of ocean-reflected radio-frequency interference using WindSat retrieval chi-square probability. *IEEE Trans. Geosci. Electron.*, 7 (2), 406–410.
175. Wiltshire, W.M., Nilsson, M.D. Campbell, F.B., Application for Authorization to Launch and Operate DirectTV 10, a Partial Replacement Ka-Band Satellite, at 103 deg. Federal Communications Commission, WL, FCC form 312.
176. Palmen, E. (1948) On the formation and structure of tropical hurricanes *Geophysica*, **3** (1), 26–38.
177. Gray, W.M. (1979) Hurricanes: their formation, structure and likely role in the tropical circulation. *Meteorol. Over Trop. Ocean*, 77, 155–218.
178. Emanuel, K.A. (1986) An air-sea interaction theory for tropical cyclones. Part

I: steady-state maintenance. *J. Atmos. Sci.*, **43** (6), 585–605.
179. Brown, O.B., Brown, J.W., and Evans, R.H. (1985) Calibration of advanced very high resolution radiometer infrared observations. *J. Geophys. Res.: Oceans*, **90** (C6), 11667–11677.
180. Emery, W.J., Yu, Y., Wick, G.A., Schluessel, P., and Reynolds, R.W. (1994) Correcting infrared satellite estimates of sea surface temperature for atmospheric water vapor attenuation. *J. Geophys. Res.: Oceans*, **99** (C3), 5219–5236.
181. Paul McClain, E. (1989) Global sea surface temperatures and cloud clearing for aerosol optical depth estimates. *Int. J. Remote Sens.*, **10** (4–5), 763–769.
182. Petty, G. (1994) Physical retrievals of over-ocean rain rate from multichannel microwave imagery. Part I: theoretical characteristics of normalized polarization and scattering indices. *Meteorog. Atmos. Phys.*, **54** (1–4), 79–99.
183. Wentz, F.J., Gentemann, C., Smith, D., and Chelton, D. (2000) Satellite measurements of sea surface temperature through clouds. *Science*, **288** (5467), 847–850.
184. Wentz, F.J., Ashcroft, P., and Gentemann, C. (2001) Post-launch calibration of the TRMM microwave imager. *IEEE Trans. Geosci. Electron.*, **39** (2), 415–422.
185. Gentemann, C.L., Wentz, F.J., Mears, C.A., and Smith, D.K. (2004) In situ validation of tropical rainfall measuring mission microwave sea surface temperatures. *J. Geophys. Res.: Oceans*, **109** (C04021), 1–9.
186. Price, J.C. (1984) Land surface temperature measurements from the split window channels of the NOAA 7 Advanced Very High Resolution Radiometer. *J. Geophys. Res.: Atmos.*, **89** (D5), 7231–7237.
187. Prata, A.J. (1993) Land surface temperatures derived from the advanced very high resolution radiometer and the along-track scanning radiometer: 1 theory. *J. Geophys. Res.: Atmos.*, **98** (D9), 16689–16702.
188. Wang, J. (1995) Some features observed by the L-band push broom microwave radiometer over the Konza Prairie during 1985–1989. *J. Geophys. Res.: Atmos.*, **100** (D12), 25469–25479.
189. Jackson, T. and O'Neill, P. (1987) Temporal observations of surface soil moisture using a passive microwave sensor. *Remote Sens. Environ.*, **21** (3), 281–296.
190. Schmugge, T., Jackson, T., Kustas, W., and Wang, J. (1992) Passive microwave remote sensing of soil moisture: results from HAPEX, FIFE and MONSOON 90. *ISPRS J. Photogramm. Remote Sens.*, **47** (2), 127–143.
191. Choudhury, B.J. (1995) Synergism of optical and microwave observations for land surface studies, in *Passive Microwave Remote Sensing of Land–Atmosphere Interactions, Microwave Remote Sensing of Land–Atmosphere Interactions* (eds B.J. Choudhury, Y.H. Kerr, E.G. Njoku, and P. Pampaloni), VSP BV, Utrecht: the Netherlands, pp. 155–191.
192. Mcfarland, M.J., Miller, R.L., and Neale, C.M. (1990) Land surface temperature derived from the SSM/I passive microwave brightness temperatures. *IEEE Trans. Geosci. Electron.*, **28** (5), 839–845.
193. Njoku, E. (1993) Surface Temperature Estimation Over Land Using Satellite Microwave Radiometry, NASA Technical Reports, p. 28.
194. Weng, F. and Grody, N.C. (1998) Physical retrieval of land surface temperature using the special sensor microwave imager. *J. Geophys. Res.: Atmos.*, **103** (D8), 8839–8848.
195. Karbou, F., Aires, F., Prigent, C., and Eymard, L. (2005) Potential of Advanced Microwave Sounding Unit: a (AMSU-A) and AMSU-B measurements for atmospheric temperature and humidity profiling over land. *J. Geophys. Res.: Atmos.*, **110** (D07109), 1–16.
196. Liu, Q. and Weng, F. (2003) Retrieval of sea surface wind vectors from simulated satellite microwave polarimetric measurements. *Radio Sci.*, **38** (4), 8078, MAR 43-1-43-8.

197. Reynolds, R. and Roberts, L. (1987) A global sea surface temperature climatology from in situ, satellite and ice data. *Trop. Ocean-Atmos. Newslett*, **37**, 15–17.
198. Yan, B. and Weng, F. (2008) Applications of AMSR-E measurements for tropical cyclone predictions part I: retrieval of sea surface temperature and wind speed. *Adv. Atmos. Sci.*, **25** (2), 227–245.
199. Jones, A.S. and Vonder Haar, T.H. (1990) Passive microwave remote sensing of cloud liquid water over land regions. *J. Geophys. Res.: Atmos.*, **95** (D10), 16673–16683.
200. Betts, A.K. and Ball, J. (1995) The FIFE surface diurnal cycle climate. *J. Geophys. Res.: Atmos.*, **100** (D12), 25679–25693.
201. Grody, N.C. (1991) Classification of snow cover and precipitation using the Special Sensor Microwave Imager. *J. Geophys. Res.: Atmos.*, **96** (D4), 7423–7435.
202. Neale, C.M., Mcfarland, M.J., and Chang, K. (1990) Land-surface-type classification using microwave brightness temperatures from the Special Sensor Microwave/Imager. *IEEE Trans. Geosci. Electron.*, **28** (5), 829–838.
203. Stephens, G.L., Randall, D.A., Wittmeyer, I.L. (1991) Observations of the Earth's Radiation Budget in Relation to Atmospheric Hydrology, II Cloud Effects and Cloud Feedback. Spaceborne Observations of Columnar Water Vapor: SSMI Observations. Citeseer. *J. Geophys. Res.*, **96**, 15325–15340.
204. Grody, N.C. (1976) Remote sensing of atmospheric water content from satellites using microwave radiometry. *IEEE Trans. Antennas Propag.*, **24** (2), 155–162.
205. Grody, N., Gruber, A., and Shen, W. (1980) Atmospheric water content over the tropical Pacific derived from the Nimbus-6 scanning microwave spectrometer. *J. Appl. Meteorol.*, **19** (8), 986–996.
206. Prabhakara, C., Wang, I., Chang, A., Gloersen, P. (1982) A Statistical Examination of Nimbus 7 SMMR Data and Remote Sensing of Sea Surface Temperature, Liquid Water Content in the Atmosphere and Surfaces Wind Speed, NASA technical memorandum 84927, p. 53.
207. Takeda, T. and Guosheng, L. (1987) Estimation of atmospheric liquid-water amount by NIMBUS 7 SMMR data: a new method and its application to the western North-Pacific region. *J. Meteorol. Soc. Jpn.*, **65** (6), 931–947.
208. Alishouse, J.C., Snider, J.B., Westwater, E.R., Swift, C.T., Ruf, C.S., Snyder, S.A., Vongsathorn, J., and Ferraro, R.R. (1990) Determination of cloud liquid water content using the SSM/I. *IEEE Trans. Geosci. Electron.*, **28** (5), 817–822.
209. Hargens, U., Simmer, C., and Ruprecht, E. (1992) Remote sensing of cloud liquid water during ICE'89. Proceedings of the specialist meeting on microwave radiometry and remote sensing applications, Boulder, pp. 27-31.
210. Liu, G. and Curry, J.A. (1993) Determination of characteristic features of cloud liquid water from satellite microwave measurements. *J. Geophys. Res.: Atmos.*, **98** (D3), 5069–5092.
211. Weng, F., Zhao, L., Ferraro, R.R., Poe, G., Li, X., and Grody, N.C. (2003) Advanced microwave sounding unit cloud and precipitation algorithms. *Radio Sci.*, **38** (4), 8068, MAR 33-1-33-13.
212. Stephens, G.L. and Webster, P.J. (1981) Clouds and climate: sensitivity of simple systems. *J. Atmos. Sci.*, **38** (2), 235–247.
213. Liou, K.-N. (1986) Influence of cirrus clouds on weather and climate processes: a global perspective. *Mon. Weather Rev.*, **114** (6), 1167–1199.
214. Ebert, E.E. and Curry, J.A. (1992) A parameterization of ice cloud optical properties for climate models. *J. Geophys. Res.: Atmos.*, **97** (D4), 3831–3836.
215. Fu, Q. and Liou, K.N. (1993) Parameterization of the radiative properties

of cirrus clouds. *J. Atmos. Sci.*, **50** (13), 2008–2025.

216. Heymsfield, A.J. and Donner, L.J. (1990) A scheme for parameterizing ice-cloud water content in general circulation models. *J. Atmos. Sci.*, **47** (15), 1865–1877.
217. Vivekanandan, J., Turk, J., and Bringi, V. (1991) Ice water path estimation and characterization using passive microwave radiometry. *J. Appl. Meteorol.*, **30** (10), 1407–1421.
218. Racette, P., Adler, R., Wang, J., Gasiewski, A., and Zacharias, D. (1996) An airborne millimeter-wave imaging radiometer for cloud, precipitation, and atmospheric water vapor studies. *J. Atmos. Ocean. Technol.*, **13** (3), 610–619.
219. Wang, J.R., Zhan, J., and Racette, P. (1997) Storm-associated microwave radiometric signatures in the frequency range of 90–220 GHz. *J. Atmos. Ocean. Technol.*, **14** (1), 13–31.
220. Arkin, P.A. and Meisner, B.N. (1987) The relationship between large-scale convective rainfall and cold cloud over the western hemisphere during 1982–84. *Mon. Weather Rev.*, **115** (1), 51–74.
221. Liu, G. and Curry, J.A. (1998) Remote sensing of ice water characteristics in tropical clouds using aircraft microwave measurements. *J. Appl. Meteorol.*, **37** (4), 337–355.
222. Deeter, M.N. and Evans, K.F. (2000) A novel ice-cloud retrieval algorithm based on the Millimeter-Wave Imaging Radiometer (MIR) 150- and 220-GHz channels. *J. Appl. Meteorol.*, **39** (5), 623–633.
223. Ulbrich, C.W. (1983) Natural variations in the analytical form of the raindrop size distribution. *J. Clim. Appl. Meteorol.*, **22** (10), 1764–1775.
224. Cumming, W. (1952) The dielectric properties of ice and snow at 3.2 centimeters. *J. Appl. Phys.*, **23** (7), 768–773.
225. Stowe, L.L., Davis, P.A., and McClain, E.P. (1999) Scientific basis and initial evaluation of the CLAVR-1 global clear/cloud classification algorithm for the Advanced Very High Resolution Radiometer. *J. Atmos. Ocean. Technol.*, **16** (6), 656–681.
226. Heymsfield, A.J. and Platt, C. (1984) A parameterization of the particle size spectrum of ice clouds in terms of the ambient temperature and the ice water content. *J. Atmos. Sci.*, **41** (5), 846–855.
227. Blackwell, W.J., Barrett, J.W., Chen, F.W., Leslie, R.V., Rosenkranz, P.W., Schwartz, M.J., and Staelin, D.H. (2001) NPOESS Aircraft Sounder Testbed-Microwave (NAST-M): instrument description and initial flight results. *IEEE Trans. Geosci. Electron.*, **39** (11), 2444–2453.
228. Bauer, P. and Mugnai, A. (2003) Precipitation profile retrievals using temperature-sounding microwave observations. *J. Geophys. Res.: Atmos.*, **108** (D23), 4730, 1–7.
229. Han, Y., Zou, X., and Weng, F. (2015) Cloud and precipitation features of Super Typhoon Neoguri revealed from dual oxygen absorption band sounding instruments on board FengYun-3C satellite. *Geophys. Res. Lett.*, **42** (3), 916–924.
230. Gasiewski, A. and Staelin, D. (1990) Numerical modeling of passive microwave O2 observations over precipitation. *Radio Sci.*, **25** (3), 217–235.
231. Zhu, T., Zhang, D.-L., and Weng, F. (2002) Impact of the Advanced Microwave Sounding Unit measurements on hurricane prediction. *Mon. Weather Rev.*, **130** (10), 2416–2432.
232. Zhu, T. and Weng, F. (2013) Hurricane Sandy warm-core structure observed from advanced Technology Microwave Sounder. *Geophys. Res. Lett.*, **40** (12), 3325–3330.
233. Zou, X., Weng, F., Zhang, B., Lin, L., Qin, Z., and Tallapragada, V. (2013) Impacts of assimilation of ATMS data in HWRF on track and intensity forecasts of 2012 four landfall hurricanes. *J. Geophys. Res.: Atmos.*, **118** (20), 11558–11576.
234. Zhang, P., Yang, J., Dong, C., Lu, N., Yang, Z., and Shi, J. (2009) General introduction on payloads, ground segment and data application of Fengyun

3A. *Front. Earth Sci. China*, **3** (3), 367–373.

235. Dong, C., Yang, J., Yang, Z., Lu, N., Shi, J., Zhang, P., Liu, Y., Cai, B., and Zhang, W. (2009) An overview of a new Chinese weather satellite FY-3A. *Bull. Am. Meteorol. Soc.*, **90** (10), 1531–1544.
236. Goodrum, G., Kidwell, K., Winston, W. (2009) NOAA KLM User's Guide with NOAA-N,-N-Prime supplement. *NOAA*. http://www2. ncdc. noaa. gov/docs/klm/-cover. htm (accessed 10 Feburary 2017).
237. Mo, T. and Liu, Q. (2008) A study of AMSU-A measurement of brightness temperatures over the ocean. *J. Geophys. Res.: Atmos.*, **113** (D120), 1–15.
238. Weng, F. (2009) *Advances in radiative transfer modeling in support of satellite data assimilation*, Hyperspectral Imaging and Sensing of the Environment, Optical Society of America, Technical Digest, paper HWD1.
239. Payne, V.H., Mlawer, E.J., Cady-Pereira, K.E., and Moncet, J.-L. (2011) Water vapor continuum absorption in the microwave. *IEEE Trans. Geosci. Electron.*, **49** (6), 2194–2208.
240. Gu, S., Guo, Y., Wang, Z., and Lu, N. (2012) Calibration analyses for sounding channels of MWHS onboard FY-3A. *IEEE Trans. Geosci. Electron.*, **50** (12), 4885–4891.
241. You, R., Gu, S., Guo, Y., Wu, X., Yang, H., and Chen, W. (2012) Long-term calibration and accuracy assessment of the FengYun-3 Microwave Temperature Sounder radiance measurements. *IEEE Trans. Geosci. Electron.*, **50** (12), 4854–4859.
242. Chevallier, F., Di Michele, S., McNally, A.P. (2006) Diverse Profile Datasets from the ECMWF 91-Level Short-Range Forecasts. European Centre for Medium-Range Weather Forecasts.
243. Weng, F. and Zou, X. (2014) 30-Year atmospheric temperature record derived by one-dimensional variational data assimilation of MSU/AMSU-A observations. *Clim. Dyn.*, **43** (7–8), 1857–1870.
244. Kidder, S.Q., Goldberg, M.D., Zehr, R.M., and DeMaria, M. (2000) Satellite analysis of tropical cyclones using the Advanced Microwave Sounding Unit (AMSU). *Bull. Am. Meteorol. Soc.*, **81** (6), 1241.
245. Knaff, J.A., Zehr, R.M., Goldberg, M.D., and Kidder, S.Q. (2000) An example of temperature structure differences in two cyclone systems derived from the Advanced Microwave Sounder Unit. *Weather Forecast.*, **15** (4), 476–483.
246. Spencer, R.W. and Braswell, W.D. (2001) Atlantic tropical cyclone monitoring with AMSU-A: estimation of maximum sustained wind speeds. *Mon. Weather Rev.*, **129** (6), 1518–1532.
247. Demuth, J.L., DeMaria, M., Knaff, J.A., and Vonder Haar, T.H. (2004) Evaluation of Advanced Microwave Sounding Unit tropical-cyclone intensity and size estimation algorithms. *J. Appl. Meteorol.*, **43** (2), 282–296.
248. Demuth, J.L., DeMaria, M., and Knaff, J.A. (2006) Improvement of advanced microwave sounding unit tropical cyclone intensity and size estimation algorithms. *J. Appl. Meteorol. Climatol.*, **45** (11), 1573–1581.
249. Zhu, Y. and Gelaro, R. (2008) Observation sensitivity calculations using the adjoint of the Gridpoint Statistical Interpolation (GSI) analysis system. *Mon. Weather Rev.*, **136** (1), 335–351.
250. Rodgers, C.D. (1976) Retrieval of atmospheric temperature and composition from remote measurements of thermal radiation. *Rev. Geophys.*, **14** (4), 609–624.
251. Weng, F., Han, Y., van Delst, P., Liu, Q., Yan, B. (2005) JCSDA community radiative transfer model (CRTM). Proc. 14th Int. ATOVS Study Conf.
252. McMillin, L., Crone, L., and Kleespies, T. (1995) Atmospheric transmittance of an absorbing gas. 5. Improvements to the OPTRAN approach. *Appl. Opt.*, **34** (36), 8396–8399.
253. Chevallier, F., Bauer, P., Mahfouf, J.F., and Morcrette, J.J. (2002) Variational retrieval of cloud profile from ATOVS observations. *Q. J. R. Meteorol. Soc.*, **128** (585), 2511–2525.
254. Moreau, E., Bauer, P., and Chevallier, F. (2003) Variational retrieval of rain profiles from spaceborne passive

microwave radiance observations. *J. Geophys. Res.: Atmos.*, **108** (D16), 4521, 1–11.

255. Boukabara, S.-A., Weng, F., and Liu, Q. (2007) Passive microwave remote sensing of extreme weather events using NOAA-18 AMSUA and MHS. *IEEE Trans. Geosci. Electron.*, **45** (7), 2228–2246.
256. Liu, Q. and Weng, F. (2005) One-dimensional variational retrieval algorithm of temperature, water vapor, and cloud water profiles from advanced microwave sounding unit (AMSU). *IEEE Trans. Geosci. Electron.*, **43** (5), 1087–1095.
257. Li, J., Wolf, W.W., Menzel, W.P., Zhang, W., Huang, H.-L., and Achtor, T.H. (2000) Global soundings of the atmosphere from ATOVS measurements: the algorithm and validation. *J. Appl. Meteorol.*, **39** (8), 1248–1268.
258. Ferraro, R.R., Weng, F., Grody, N.C., Zhao, L., Meng, H., Kongoli, C., Pellegrino, P., Qiu, S., and Dean, C. (2005) NOAA operational hydrological products derived from the Advanced Microwave Sounding Unit. *IEEE Trans. Geosci. Electron.*, **43** (5), 1036–1049.
259. Hawkins, H.F. and Rubsam, D.T. (1968) Hurricane Hilda, 1964: II. Structure and budgets of the hurricane on October 1, 1964. *Mon. Weather Rev.*, **96** (9), 617–636.
260. Lorenc, A.C. (1997) Development of an operational variational assimilation scheme. *J. Meteorol. Soc. Jpn.*, **2** (75), 229–236.
261. Zhou, D.K., Smith, W.L., Liu, X., Li, J., Larar, A.M., and Mango, S.A. (2005) Tropospheric CO observed with the NAST-I retrieval methodology, analyses, and first results. *Appl. Opt.*, **44** (15), 3032–3044.
262. Smith, W. and Woolf, H. (1976) The use of eigenvectors of statistical covariance matrices for interpreting satellite sounding radiometer observations. *J. Atmos. Sci.*, **33** (7), 1127–1140.
263. Moncet, J.L., Boukabara, S., D'Entremont, R., Snell, D., Rieu-Isaacs, H., Lipton, A., Evans, F., and Mitchell, D. (2001) *Algorithm Theoretical Basis Document (ATBD) for the Conical-Scanning Microwave Imager/Sounder (CMS) Environmental Data Records (EDRS)*, AER Internal Document.
264. Phalippou, L. (1996) Variational retrieval of humidity profile, wind speed and cloud liquid-water path with the SSM/I: potential for numerical weather prediction. *Q. J. R. Meteorol. Soc.*, **122** (530), 327–355.
265. Deblonde, G. and English, S. (2003) One-dimensional variational retrievals from SSMIS-simulated observations. *J. Appl. Meteorol.*, **42** (10), 1406–1420.
266. Boukabara, S.-A., Garrett, K., Chen, W., Iturbide-Sanchez, F., Grassotti, C., Kongoli, C., Chen, R., Liu, Q., Yan, B., and Weng, F. (2011) MiRS: an all-weather 1DVAR satellite data assimilation and retrieval system. *IEEE Trans. Geosci. Electron.*, **49** (9), 3249–3272.
267. Yan, B. and Weng, F. (2012) Assimilation of F-16 Special Sensor Microwave Imager/Sounder data in the NCEP global forecast system. *Weather Forecast.*, **27** (3), 700–714.
268. Jung, J. (2014) Addtion to Baseline Observing System Experiments. 12th JCSDA Workshop on Satellite Data Assimilation. College Park, Maryland, USA.
269. Wu, W.-S., Purser, R.J., and Parrish, D.F. (2002) Three-dimensional variational analysis with spatially inhomogeneous covariances. *Mon. Weather Rev.*, **130** (12), 2905–2916.
270. Derber, J.C. and Wu, W.-S. (1998) The use of TOVS cloud-cleared radiances in the NCEP SSI analysis system. *Mon. Weather Rev.*, **126** (8), 2287–2299.
271. Purser, R.J., Wu, W.-S., Parrish, D.F., and Roberts, N.M. (2003) Numerical aspects of the application of recursive filters to variational statistical analysis. Part I: spatially homogeneous and isotropic Gaussian covariances. *Mon. Weather Rev.*, **131** (8), 1524–1535.
272. Purser, R.J., Wu, W.-S., Parrish, D.F., and Roberts, N.M. (2003) Numerical aspects of the application of recursive filters to variational statistical analysis. Part II: spatially inhomogeneous and

anisotropic general covariances. *Mon. Weather Rev.*, **131** (8), 1536–1548.

273. Janjic, Z.I., Gerrity, J. Jr., and Nickovic, S. (2001) An alternative approach to nonhydrostatic modeling. *Mon. Weather Rev.*, **129** (5), 1164–1178.

274. Janjic, Z., Black, T., Rogers, E., Chuang, H., Dimego, G. (2003) The NCEP Nonhydrostatic Meso Model and First Experiences with its Applications. EGS-AGU-EUG Joint Assembly.

275. Gopalakrishnan, S.G., Marks, F. Jr.,, Zhang, X., Bao, J.-W., Yeh, K.-S., and Atlas, R. (2011) The experimental HWRF system: a study on the influence of horizontal resolution on the structure and intensity changes in tropical cyclones using an idealized framework. *Mon. Weather Rev.*, **139** (6), 1762–1784.

276. Zhang, X., Quirino, T., Yeh, K.-S., Gopalakrishnan, S., Marks, F., Goldenberg, S., and Aberson, S. (2011) HWRFx: improving hurricane forecasts with high-resolution modeling. *Comput. Sci. Eng.*, **13** (1), 13–21.

277. Yeh, K.-S., Zhang, X., Gopalakrishnan, S., Aberson, S., Rogers, R., Marks, F., and Atlas, R. (2012) Performance of the experimental HWRF in the 2008 hurricane season. *Nat. Hazards*, **63** (3), 1439–1449.

278. Bozeman, M.L., Niyogi, D., Gopalakrishnan, S., Marks, F.D. Jr., Zhang, X., and Tallapragada, V. (2012) An HWRF-based ensemble assessment of the land surface feedback on the post-landfall intensification of Tropical Storm Fay (2008). *Nat. Hazards*, **63** (3), 1543–1571.

279. Pattanayak, S., Mohanty, U., and Gopalakrishnan, S. (2012) Simulation of very severe cyclone Mala over Bay of Bengal with HWRF modeling system. *Nat. Hazards*, **63** (3), 1413–1437.

280. Gopalakrishnan, S.G., Goldenberg, S., Quirino, T., Zhang, X., Marks, F. Jr., Yeh, K.-S., Atlas, R., and Tallapragada, V. (2012) Toward improving high-resolution numerical hurricane forecasting: Influence of model horizontal grid resolution, initialization, and physics. *Weather Forecast.*, **27** (3), 647–666.

281. Grody, N., Zhao, J., Ferraro, R., Weng, F., and Boers, R. (2001) Determination of precipitable water and cloud liquid water over oceans from the NOAA 15 advanced microwave sounding unit. *J. Geophys. Res.*, **106** (D3), 2943–2953.

282. Eyre, J. (1992) A Bias Correction Scheme for Simulated TOVS Brightness Temperatures. European Centre for Medium-Range Weather Forecasts.

283. Auligné, T., McNally, A., and Dee, D. (2007) Adaptive bias correction for satellite data in a numerical weather prediction system. *Q. J. R. Meteorol. Soc.*, **133** (624), 631–642.

284. Swadley, S., Poe, G., Baker, N., Blankenship, C., Campbell, W., Ruston, B., Kunkee, D., Boucher, D., Bell, W. (2006) Calibration anomalies and radiance assimilation correction strategies for the Defense Meteorological Satellite Program (DMSP) Special Sensor Microwave Imager Sounder (SSMIS). Proceedings of the 15th International TOVS Studies Conference.

285. Yan, B. and Weng, F. (2009) Assessments of F16 Special Sensor Microwave Imager and sounder antenna temperatures at lower atmospheric sounding channels. *Adv. Meteorol.*, **2009**, 420985, 1–18.

286. Bell, W. and Swadley, S. (2005) SSMIS Calibration Issues. First SSMIS Working Group Meeting. Monterey, CA, USA.

287. Yan, B. and Weng, F. (2008) Intercalibration between special sensor microwave imager/sounder and special sensor microwave imager. *IEEE Trans. Geosci. Electron.*, **46** (4), 984–995.

288. Rosenkranz, P., Hutchison, K., Hardy, K., and Davis, M. (1997) An assessment of the impact of satellite microwave sounder incidence angle and scan geometry on the accuracy of atmospheric temperature profile retrievals. *J. Atmos. Ocean. Technol.*, **14** (3), 488–494.

289. Boucher, D. and Poe, G. (2005) *F16 SSMIS Calibration/Validation Executive Summary*, Polamax, Silver Spring, MA, USA.

290. Zou, C.-Z., Gao, M., and Goldberg, M.D. (2009) Error structure and atmospheric temperature trends in observations from the microwave sounding unit. *J. Clim.*, **22** (7), 1661–1681.
291. Wang, L., Zou, C.-Z., and Qian, H. (2012) Construction of stratospheric temperature data records from stratospheric sounding units. *J. Clim.*, **25** (8), 2931–2946.
292. Spencer, R.W. and Christy, J.R. (1992) Precision and radiosonde validation of satellite gridpoint temperature anomalies. Part I; MSU Channel 2. . Pt. 1; MSU Channel 2, *J. Climate*, **5** (8), 847–857.
293. Spencer, R.W. and Christy, J.R. (1992) Precision and radiosonde validation of satellite gridpoint temperature anomalies. Part II: a tropospheric retrieval and trends during 1979–90. *J. Clim.*, **5** (8), 858–866.
294. Christy, J.R., Spencer, R.W., and Lobl, E.S. (1998) Analysis of the merging procedure for the MSU daily temperature time series. *J. Clim.*, **11** (8), 2016–2041.
295. Christy, J.R., Spencer, R.W., and Braswell, W.D. (2000) MSU tropospheric temperatures: dataset construction and radiosonde comparisons. *J. Atmos. Ocean. Technol.*, **17** (9), 1153–1170.
296. Mears, C.A., Schabel, M.C., and Wentz, F.J. (2003) A reanalysis of the MSU channel 2 tropospheric temperature record. *J. Clim.*, **16** (22), 3650–3664.
297. Mears, C.A. and Wentz, F.J. (2009) Construction of the Remote Sensing Systems V3. 2 atmospheric temperature records from the MSU and AMSU microwave sounders. *J. Atmos. Ocean. Technol.*, **26** (6), 1040–1056.
298. Qin, Z., Zou, X., and Weng, F. (2012) Comparison between linear and nonlinear trends in NOAA-15 AMSU-A brightness temperatures during 1998–2010. *Clim. Dyn.*, **39** (7–8), 1763–1779.
299. Spencer, R.W., Christy, J.R., and Grody, N.C. (1990) Global atmospheric temperature monitoring with satellite microwave measurements: method and results 1979–84. *J. Clim.*, **3** (10), 1111–1128.
300. Spencer, R.W. (1993) Global oceanic precipitation from the MSU during 1979–91 and comparisons to other climatologies. *J. Clim.*, **6** (7), 1301–1326.
301. Prabhakara, C., Nucciarone, J., and Yoo, J.-M. (1995) Examination of 'global atmospheric temperature monitoring with satellite microwave measurements': 1) theoretical considerations. *Clim. Chang.*, **30** (3), 349–366.
302. Prabhakara, C., Yoo, J.-M., Maloney, J., Nucciarone, J., Arking, A., Cadeddu, M., and Dalu, G. (1996) Examination of 'global atmospheric temperature monitoring with satellite microwave measurements': 2) analysis of satellite data. *Clim. Chang.*, **33**, 459–476.
303. Spencer, R.W., Christy, J.R., and Grody, N.C. (1996) Global atmospheric temperature monitoring with satellite microwave measurements. *Climate Change*, **33**, 477–489.
304. Wentz, F.J., Ricciardulli, L., Hilburn, K., and Mears, C. (2007) How much more rain will global warming bring? *Science*, **317** (5835), 233–235.
305. Yang, S., Weng, F., Yan, B., Sun, N., and Goldberg, M. (2011) Special Sensor Microwave Imager (SSM/I) intersensor calibration using a simultaneous conical overpass technique. *J. Appl. Meteorol. Climatol.*, **50** (1), 77–95.
306. Colton, M.C. and Poe, G.A. (1999) Intersensor calibration of DMSP SSM/I's: F-8 to F-14, 1987–1997. *IEEE Trans. Geosci. Electron.*, **37** (1), 418–439.
307. Hilburn, K. and Wentz, F. (2008) Intercalibrated passive microwave rain products from the unified microwave ocean retrieval algorithm (UMORA). *J. Appl. Meteorol. Climatol.*, **47** (3), 778–794.
308. Kunkee, D.B., Swadley, S.D., Poe, G.A., Hong, Y., and Werner, M.F. (2008) Special sensor microwave imager sounder (SSMIS) radiometric calibration anomalies—Part I: identification and characterization. *IEEE Trans. Geosci. Electron.*, **46** (4), 1017–1033.
309. Cao, C., Weinreb, M., and Xu, H. (2004) Predicting simultaneous nadir

overpasses among polar-orbiting meteorological satellites for the inter-satellite calibration of radiometers. *J. Atmos. Ocean. Technol.*, **21** (4), 537–542.

310. Poe, G., Uliana, E., Gardiner, B., vonRentzell, T. (2006) Mitigation of DMSP F-15 RADCAL Interference with SSM. I, Technical report, NRL, Monterey, CA. Code 7541.

311. Yan, B. and Weng, F. (2006) Recalibration of DMPS SSM/I for Weather and Climate Applications. Proceedings of the 15th International TOVS Studies Conference.

312. Sun, N. and Weng, F. (2008) Evaluation of special sensor microwave imager/sounder (SSMIS) environmental data records. *IEEE Trans. Geosci. Electron.*, **46** (4), 1006–1016.

313. Sohn, B.-J. and Smith, E.A. (2003) Explaining sources of discrepancy in SSM/I water vapor algorithms. *J. Clim.*, **16** (20), 3229–3255.

314. Ebert, E.E. and Manton, M.J. (1998) Performance of satellite rainfall estimation algorithms during TOGA COARE. *J. Atmos. Sci.*, **55** (9), 1537–1557.

315. Scott, C. (2005) Climate Change & Tropospheric Temperature Trends Part I - What Do We Know Today and Where Is It Taking Us? http://www.scottchurchdirect.com/docs/MSU-Troposphere-Review01.pdf (accessed 10 Feburary 2017).

316. Nash, J. and Forrester, G. (1986) Long-term monitoring of stratospheric temperature trends using radiance measurements obtained by the TIROS-N series of NOAA spacecraft. *Adv. Space Res.*, **6** (10), 37–44.

317. Zou, C.Z. and Wang, W. (2011) Intersatellite calibration of AMSU-A observations for weather and climate applications. *J. Geophys. Res.: Atmos.*, **116** (D113), 1–20.

318. Weng, F., Zou, X., and Qin, Z. (2014) Uncertainty of AMSU-A derived temperature trends in relationship with clouds and precipitation over ocean. *Clim. Dyn.*, **43** (5–6), 1439–1448.

319. Bohren, C.F. and Huffman, D.R. (2008) *Absorption and Scattering of Light by Small Particles*, John Wiley & Sons.

320. Zou, X., Navon, I., and Sela, J. (1993) Variational data assimilation with moist threshold processes using the NMC spectral model. *Tellus A*, **45** (5), 370–387.

321. Santer, B.D., Thorne, P., Haimberger, L., Taylor, K.E., Wigley, T., Lanzante, J., Solomon, S., Free, M., Gleckler, P.J., and Jones, P. (2008) Consistency of modelled and observed temperature trends in the tropical troposphere. *Int. J. Climatol.*, **28** (13), 1703–1722.

322. Santer, B.D., Wigley, T.M., Mears, C., Wentz, F.J., Klein, S.A., Seidel, D.J., Taylor, K.E., Thorne, P.W., Wehner, M.F., and Gleckler, P.J. (2005) Amplification of surface temperature trends and variability in the tropical atmosphere. *Science*, **309** (5740), 1551–1556.

Index